Phase Resetting
in Medicine and Biology

Springer
Berlin
Heidelberg
New York
Barcelona
Hong Kong
London
Milan
Paris
Singapore
Tokyo

Springer Series in Synergetics

An ever increasing number of scientific disciplines deal with complex systems. These are systems that are composed of many parts which interact with one another in a more or less complicated manner. One of the most striking features of many such systems is their ability to spontaneously form spatial or temporal structures. A great variety of these structures are found, in both the inanimate and the living world. In the inanimate world of physics and chemistry, examples include the growth of crystals, coherent oscillations of laser light, and the spiral structures formed in fluids and chemical reactions. In biology we encounter the growth of plants and animals (morphogenesis) and the evolution of species. In medicine we observe, for instance, the electromagnetic activity of the brain with its pronounced spatio-temporal structures. Psychology deals with characteristic features of human behavior ranging from simple pattern recognition tasks to complex patterns of social behavior. Examples from sociology include the formation of public opinion and cooperation or competition between social groups.

In recent decades, it has become increasingly evident that all these seemingly quite different kinds of structure formation have a number of important features in common. The task of studying analogies as well as differences between structure formation in these different fields has proved to be an ambitious but highly rewarding endeavor. The Springer Series in Synergetics provides a forum for interdisciplinary research and discussions on this fascinating new scientific challenge. It deals with both experimental and theoretical aspects. The scientific community and the interested layman are becoming ever more conscious of concepts such as self-organization, instabilities, deterministic chaos, nonlinearity, dynamical systems, stochastic processes, and complexity. All of these concepts are facets of a field that tackles complex systems, namely synergetics. Students, research workers, university teachers, and interested laymen can find the details and latest developments in the Springer Series in Synergetics, which publishes textbooks, monographs and, occasionally, proceedings. As witnessed by the previously published volumes, this series has always been at the forefront of modern research in the above mentioned fields. It includes textbooks on all aspects of this rapidly growing field, books which provide a sound basis for the study of complex systems.

Peter A. Tass

Phase Resetting in Medicine and Biology

Stochastic Modelling and Data Analysis

With 129 Figures, 7 in Color

 Springer

Dr.Dr. Peter A. Tass

Neurologische Klinik
Heinrich-Heine-Universität
Moorenstrasse 5
D-40225 Düsseldorf
Germany
E-Mail: tass@neurologie.uni-duesseldorf.de

ISSN 0172-7389

ISBN 3-540-65697-9 Springer-Verlag Berlin Heidelberg New York

Library of Congress Cataloging-in-Publication Data

Tass, Peter A., 1963-
Phase resetting in medicine and biology: stochastic modelling and data analysis / Peter A. Tass.
p. cm. – (Springer series in synergetics, ISSN 0172-7389)
Includes bibliographical references and index.
ISBN 3-540-65697-9 (hardcover : alk. paper)
1. Computational neuroscience. 2. Stochastic analysis. 3. Oscillations. 4. Synchronization. I. Title. II. Series.
QP357.5.T37 1999 612.8'13–dc21 99-25203

© Springer-Verlag Berlin Heidelberg 1999
Printed in Germany

The use of general descriptive names, registered names, trademarks, etc. in this publication does not imply, even in the absence of a specific statement, that such names are exempt from the relevant protective laws and regulations and therefore free for general use.

Typesetting: Camera ready by the author using a Springer TEX macro package
Cover design: *design & production*, Heidelberg
Computer to film: Mercedesdruck, Berlin

SPIN: 10708993   55/3144/di - 5 4 3 2 1 0 – Printed on acid-free paper

To my mother,
Gertraud Tass,

to my sister,
Ute Tass,

and to the memory of my father,
Alexander Tass

Preface

Synchronization processes are of great interest and importance in biology, medicine and physics. In particular, for the comprehension of brain functioning it appears inevitable that one should analyze neuronal synchronization processes. This book presents a new understanding of how a stimulus influences synchronization patterns of a population of oscillators. On the one hand, a variety of stimulation-induced dynamical phenomena will be presented; on the other hand, new data analysis tools will be developed which will serve as a link between theory and experiment. In this way it will be possible to use the theory presented here as a basis for the design and evaluation of stimulation experiments and stimulation techniques in medicine and biology. We shall focus particularly on applications concerning the analysis of magnetoencephalography (MEG) and electroencephalography (EEG) data as well as deep brain stimulation techniques used in Parkinsonian patients.

This book addresses graduate students, professors and scientists in various fields including biology, mathematics, medicine, neuroscience, physiology and physics. Besides mathematically involved parts, the book also provides the reader with numerous illustrations and explications of the deep dynamical principles governing stimulation-induced desynchronization and synchronization processes. Therefore this book will be of interest to a general readership, and those who are not familiar with mathematics should not be deterred by the formulas. Indeed, some parts of the book are written particularly for neurologists, neuroscientists, neurosurgeons, and physiologists who may profit from this new approach, e.g., by applying it to MEG and EEG analysis or to the improvement of stimulation techniques in neurology and neurosurgery.

I hope that this book will bear fruit in medicine and that it will contribute to a physiology which appropriately takes into account the importance of regulatory and self-organizing processes. In my opinion Hermann Haken's synergetics is a perfectly suitable theoretical basis for the study of such physiological processes. Since modern computer facilities make it possible to apply these theoretical tools to biological data very effectively, we have good prospects of revealing the tremendous beauty and significance of holistic regulatory dynamics in physiology.

It is my desire to thank my teachers, friends and colleagues: First of all my thanks go to Prof. Hermann Haken for being an outstanding and inspiring teacher, for his kind and continuous support, for much friendly advice, for our stimulating discussions, and for numerous fruitful comments on the manuscript. I consider it an honour and a pleasure to publish this book in the Springer Series of Synergetics.

When I came back to medicine after studying physics and mathematics, Prof. H.-J. Freund made it possible for me to do research in a neurological department with excellent scientific activities and equipment. I am very grateful for his visionary confidence and his superb support which enabled me to perform the studies presented in this book. Moreover, I would like to thank him for his marvellous teaching and for our inspiring discussions.

I express my special thanks to all colleagues in Düsseldorf, particularly, to PD Dr.Dr. H. Hefter for his deep and stimulating interest in investigating oscillatory processes in neurology, for his very kind support, and for our wonderful collaboration. My thanks go to Dr. A. Schnitzler and Dr. J. Volkmann for numerous fruitful discussions and a marvellous collaboration in the field of magnetoencephalography. Additionally, I am very grateful for the very friendly and fruitful collaboration with Dr. G. Fink, Dr. K. Müller, J. Salomon, F. Schmitz, Dr. P. Weiß, and Dr. J. Weule. I gratefully acknowledge the financial support of my studies by the German Science Foundation (SFB 194, A5).

My warmest thanks go to Prof. A. Wunderlin, for his great and sophisticated teaching, for carefully reading the manuscript, and for our enriching discussions.

I am very grateful to Prof. J. Kurths, Dr. M. Rosenblum, Prof. A. Pikovsky, G. Guasti, C. Raab, and Dr. H. Voss for our successful and stimulating common studies. The intensive collaboration with Prof. J. Kurths and Dr. M. Rosenblum is a pure joy.

Prof. K. Kirchgässner accompanied my work in an inspiring, tireless and very friendly way for which I would like to express my special thanks.

I am very indebted to Prof. H. Schmid-Schönbein for his encouraging and important support and his enormous efforts to establish an integrative and synergetic physiology.

Many thanks go to Prof. D. Epstein for our collaboration, for the many stimulating discussions, and for his kind and wise advice.

I am very grateful to Dr. D. Ebert for the many interesting discussions and for his permanent and friendly interest in my studies.

I thank the staff of Springer-Verlag for the excellent cooperation, in particular, Prof. W. Beiglböck, Ms. G. Dimler, Mr. F. Holzwarth, Dr. A. Lahee, Ms. E. Pfendbach, and Ms. B. Reichel-Mayer.

Last but not least I thank G. Burghardt, H. Hefter, I. Kupisch, L. Raiber, P. Weiß, and S. Wosch for their humour, support and sympathy.

Düsseldorf, January 1999   *Peter A. Tass*

Contents

1. Introduction

1.1 Goal

Oscillators abound in physics (Haken 1970, 1983a) and many other branches of the natural sciences such as chemistry (Kuramoto 1984), biology and medicine (Freeman 1975, Freund 1983, Steriade, Jones, Llinás 1990). These days the theory of oscillators is a well-established branch of nonlinear physics and mathematics (cf. Haken 1977, 1983a, Winfree 1980, Arnold 1983, Kuramoto 1984, Guckenheimer and Holmes 1990). It revealed many important insights into the interactions of oscillators and the reactions of an oscillator to external perturbations.

For instance, the impact of a pulsatile stimulus on a single oscillator is understood in detail (cf. Winfree 1980). This knowledge is important for studying the reactions of a single neuron to a pulsatile electrical stimulus since both a firing and a bursting neuron can be modelled as an oscillator (cf. Murray 1989). Best (1979) theoretically analyzed phase resetting of a single neuron: He investigated the stimulus' impact on the neuron's phase dynamics characterized by the relationship between the phase before and after application of a pulsatile stimulus. Among other effects Best predicted that the repetitive firing of an axon is stopped by a well-timed electrical stimulus of the right intensity and duration. These predictions were experimentally verified by Guttman, Lewis, Rinzel (1980).

In the meantime the behavior of populations of neurons became the focus of attention. This is due to a large number of animal experiments which clearly indicate that the synchronization of oscillatory neuronal activity is a fundamental mechanism for combining related information within a brain area and between different brain areas (Eckhorn et al. 1988, Gray and Singer 1987, 1989, for a review see Singer and Gray 1995). Moreover, neuronal synchronization plays a significant role under pathological conditions, for example, during epochs of tremor (Freund 1983, Elble and Koller 1990) or epileptic seizures (Engel and Pedley 1997).

It is of great importance to understand how a stimulus influences synchronized neuronal activity: On the one hand stimulation is a major experimental tool for investigating neuronal synchronization processes. For instance, natural sensory stimulation is used for probing neuronal information processing (cf. Hari and Salmelin 1997), whereas experimental electrical or magnetic stimulation of the nervous system serves for analyzing the dynamical interactions of different brain areas (cf. Freund 1987, Steriade, Jones, Llinás 1990). On the other hand, these days deep brain stimulation with chronically implanted electrodes is a promising therapy, e.g., for patients suffering from advanced Parkinson's disease that do not respond any more to drug therapy (Benabid et al. 1991, Blond et al. 1992, for a review see Volkmann, Sturm, Freund 1998). The knowledge concerning stimulation of synchronized neuronal activity is mainly based on experimental results and clinical observations (cf. Sears and Stagg 1976, Freund 1987, Steriade, Jones, Llinás 1990, Benabid 1996).

Accordingly, one may expect that a deeper theoretical understanding of the relevant dynamical phenomena may improve both the study of brain functioning as well as therapeutic stimulation techniques. For instance, understanding characteristic transient responses of a population of neurons to external perturbations may provide us with invaluable clues for investigating how the central nervous system reacts and adapts to rapidly changing external conditions. For the therapeutic applications, however, it is, e.g., a great challenge to design desynchronizing stimulation techniques as effective and gentle as possible.

The theoretical investigations of the spontaneously emerging dynamics of populations of interacting oscillators have revealed numerous significant results (cf. Kuramoto 1984). Nevertheless, there is still an enormous need for studies addressing the impact of stimulation on a cluster of oscillators. Accordingly, this book has three principal goals:

1. It wants to provide the reader with new insights into stimulation induced desynchronization and synchronization processes which are relevant for biology, medicine and, in particular, neuroscience. For this reason this book presents a stochastic approach to phase resetting of a population of interacting oscillators in the presence of noise. The results of this study can be applied to different branches of the natural sciences. However, the motivation for this study comes from neurophysiological considerations, and throughout the whole book all results will be interpreted in terms of neuroscience.

2. This book wants to contribute to a theoretical framework for the interpretation of stimulation induced transient dynamics of neuronal populations as measured, for instance, by means of magnetoencephalography (MEG) and electroencephalography (EEG). Therefore data analysis methods will be elaborated which make it possible to use the theoretical results for the interpretation and evaluation of neurophysiological processes. In particular, new tools will be presented that make it possible to detect stimulus induced short-term epochs of synchronized cerebral activity from MEG and EEG data.

3. Moreover this book is written to provide neurologists and neurosurgeons with a theoretical basis for the improvement and design of stimulation

techniques used, for example, for the therapy of patients suffering from movement disorders. Correspondingly, a new stimulation mode, namely a single pulse deep brain stimulation with feedback control, will be suggested for the treatment of parkinsonian resting tremor in patients with chronically implanted electrodes.

The next section is dedicated to the therapeutic significance of phase resetting in neurology and neurosurgery. Finally, it will be outlined how this new approach to phase resetting originated from Winfree's (1980) pioneering studies and how it was developed using methods and strategies from the interdisciplinary field of synergetics founded by Haken (1977, 1983a).

1.2 Physiological Motivation

In the human body there is a multitude of physiological rhythms acting on time scales ranging from milliseconds to months, for instance, neuronal activity, heart beat, respiration, blood circulation, energy metabolism in cells, peristalsis, repetitive motor behavior (walking, running, flying, swimming, chewing), smooth muscle tone, sleep cycles, autonomous equilibrium, reproduction, seasonal growth and involution (cf. von Holst 1935, 1939, Winfree 1980, Hildebrandt 1982, 1987, Freund 1983, Glass and Mackey 1988, Steriade, Jones, Llinás 1990, Haken and Koepchen 1991, Schmid-Schönbein and Ziege 1991, Schmid-Schönbein et al. 1992, Perlitz et al. 1995, Hari and Salmelin 1997, Basar 1998a, 1998b). In this context the term *rhythm* stands for the coordinated, often synchronized collective action of populations of oscillatory subsystems, for example, oscillatory neurons.

Under pathological conditions these rhythms may vanish, they may be modified, or qualitatively different rhythms may occur (cf. Freund 1983, Glass and Mackey 1988, Engel and Pedley 1997). Moreover, pathologically altered rhythms may adapt to environmental changes in characteristically different ways. To study these rhythms and their interactions one uses different stimulation techniques (cf. Niedermeyer and Lopes da Silva 1987, Glass and Mackey 1988, Hari and Salmelin 1997). The analysis of the rhythms' responses to stimulation may certainly be improved by probing suitable theoretical models in which the oscillators' interactions and the influence of noise are appropriately taken into account. In several cases one might benefit from the results presented in this book.

1.2.1 Resetting Cerebral Rhythms

The performance of complex tasks in humans requires a well-coordinated interaction of complex networks of neuronal clusters. While the anatomical aspects of these networks, e.g., the localization and the connections of different neuronal populations are already known in detail (cf. Creutzfeldt 1983,

Nieuwenhuys, Voogd, van Huijzen 1991), the complex dynamics of the neuronal interactions within and between clusters is not yet sufficiently understood. Accordingly, the dynamical aspects of neuronal processing have to be studied in order to cope with the admirable regulatory and control functions of the central nervous system. This branch of science was strongly promoted by Haken (1983b, 1996) who suggested that cerebral regulation essentially relies on self-organizing processes. More precisely, he proposed that the brain operates close to instabilities where typically only a few modes of activity dominate the system's behavior.

In this context oscillatory processes turned out to be particularly relevant. One encounters a plethora of neuronal oscillatory activity in the nervous system (cf. Cohen, Rossignol, Grillner 1988, Steriade, Jones, Llinás 1990). Various types of rhythmic activity are observed in different areas of the brain. These rhythms play an important role in physiological as well as pathological processes such as movement control and tremor (cf. Freund 1983, 1987a, Elble and Koller 1990, Rothwell 1994). Particularly during the last decade a large number of animal experiments addressed the interactions of neurons within a cluster and between clusters located in different brain areas (Gray and Singer 1987, 1989, Eckhorn et al. 1988). In both cases synchronization in terms of coincident firing turned out to be a basic mechanism for combining neuronal processing which may be widely distributed in the brain (for a review see Singer and Gray 1995).

Concerning regulatory and synchronization processes stimulation is of great importance:

1. *Natural sensory stimulation:* The brain is permanently exposed to floods of incoming information. Continuously innumerable sensory inputs have to be quickly evaluated so that the behavior can be adjusted to environmental changes. To probe the brain's processing of sensory information, stimuli of different modalities such as tones or flashlights are administered and the evoked brain activity is registered by means of EEG or MEG. In this way the role of different brain areas in sensory information processing was studied in detail (cf. Hari and Salmelin 1997). On the other hand characteristic changes of brain responses to sensory stimuli are valuable diagnostic hints (cf. Niedermeyer and Lopes da Silva 1987).

2. *Experimental electrical and magnetic stimulation:* Today one cannot imagine to study interacting neuronal oscillators without using stimulation. With this aim in view one administers a stimulus to a small volume of brain tissue and observes the resulting changes of the brain activity and the corresponding behavior. Examinations of this kind are performed in animal experiments (Fritsch and Hitzig 1870, for a review see Creutzfeldt 1983, Steriade, Jones, Llinás 1990) and in patients during neurosurgery (Foerster 1936, Penfield, Boldrey 1937, for a review see Freund 1987b). Moreover, in neurology stimulation is applied for diagnostic

(cf. Barker, Jalinous, Freeston 1986, Meyer 1992, Schnitzler and Benecke 1994, Classen et al. 1997) as well as therapeutic purposes.

1.2.2 Deep Brain Stimulation

The brain consists of an enormous multitude of neuronal clusters densely interconnected in a complex way. By means of bidirectional or unidirectional anatomical connections these clusters form groups or loops (cf. Creutzfeldt 1983, Steriade, Jones, Llinás 1990). In certain movement disorders, for instance, in Parkinson's disease within particular loops neuronal clusters are rhythmically active in an abnormal way (cf. Llinás and Jahnsen 1982, Pare, Curro'Dossi, Steriade 1990, Volkmann et al. 1996). For the treatment of such diseases electrodes may be chronically implanted within the brain with millimeter precision within a particular neuronal cluster. These electrodes are used to suppress pathological neuronal activity by means of high-frequency stimulation (Benabid et al. 1991, Blond et al. 1992). Such treatment is now employed for patients suffering from advanced Parkinson's disease that do not benefit any more from drug therapy (for a review see Volkmann and Sturm 1998).

High-frequency deep brain stimulation of Prakinson's disease, a new and encouraging treatment (Benabid et al. 1991, Blond et al. 1992), was developed along the lines of an empirical approach motivated by observations during stereotactic surgery (cf. Benabid et al. 1996). Up to now it remains unclear how this type of stimulation suppresses the rhythmic neuronal activity which generates the parkinsonian tremor (cf. Andy 1983, Benabid et al. 1991, 1993, Blond et al. 1992, Caparros-Lefebvre et al. 1994, Strafella et al. 1997). Accordingly, a sound theoretical comprehension of the stimulation induced dynamics of synchronized neuronal clusters will probably contribute to an improvement of such stimulation techniques.

Chapter 10 will be devoted to the disease mechanism giving rise to parkinsonian resting tremor and the therapeutic stimulation techniques. Based on the results presented in this book a single pulse deep brain stimulation with feedback control will be suggested for the treatment of parkinsonian resting tremor in patients with chronically implanted electrodes. This new stimulation mode may have important advantages for both the patient and the stimulating device as explained below.

As yet, there are no experimental studies which are based on a suitable theory aiming at improving stimluation techniques used in neurology and neurosurgery. It is one of the goals of this book to initiate a combination of mutually stimulating theoretical and experimental studies addressing this very issue. To illustrate how fruitful combined efforts of this kind may be let us briefly consider a cardiological problem which is of great current interest: the sudden cardiac death. The latter, one of the leading causes of death in the industrialized world, is primarily due to ventricular fibrillation (for a review see Winfree 1994). Many theoretical as well as experimental studies addressed this issue since a better understanding of this pathological type of spatiotemporal dynamics is necessary, for example, to improve implanted defibrillators.

To study the propagation of excitation in the heart and its responses to stimulation the heart muscle is modeled as excitable but not oscillatory medium (cf. Winfree 1987). Numerous theoretical studies were dedicated to rotors, that means sources of rotating spiral waves, which were suggested to be the dynamical mechanism giving rise to the ventricular tachycardias that immediately precede fibrillation (Moe and Abildskov 1959, Moe, Rheinboldt, Abildskov 1964, Krinskii 1966, 1968, 1978 Krinskii and Kholopov 1967a, 1967b, Smith and Cohen 1984, Adam et al. 1984, Allessie et al. 1985, Winfree 1987, Pertsov et al. 1993, Panfilov and Holden 1997). In animal experiments transient eruptions of rotors were finally verified (Witkowski et al. 1998). Up to now, the transition from the rotors' regime to fibrillation is not sufficiently understood (Winfree 1987, 1994). Accordingly, this problem requires further studies until it might be possible, for instance, to control the motion of the rotor's spiral wave by means of an effective low-voltage cardiac defibrillation which elegantly forces the wave tip out of the heart (Holden 1997).

This brief survey of studies addressing ventricular fibrillation should illustrate both the tremendous efforts as well as the brilliant results obtained in this way. For the improvement of stimulation techniques in neurology a similar combination of theoretical and experimental studies is necessary.

1.3 Stochastic Approach

In the 1960s Winfree (1970) started his famous theoretical investigation of phase resetting with his studies on circadian rhythms. Winfree (1970, 1980) showed that an oscillation can be annihilated by a stimulus of a critical intensity and duration administered at a critical initial phase. In particular, using a deep and elegant topological reasoning he performed a detailed analysis of phase resetting in populations of non-interacting phase oscillators as well as non-interacting attractor cycle oscillators both without considering random forces (for a review see Winfree 1980).

Noise is inevitable at a level of description more macroscopic than that of quantum mechanics. For this reason I developed a stochastic approach to phase resetting (Tass 1996a): Taking into account noise, phase resetting was studied in an *ensemble* of phase oscillators, that is a large population of non-interacting phase oscillators. The ensemble model will be presented in Chap. 2 and will serve as an introduction to the stochastic analysis of stimulation induced synchronization and desynchronization processes. Moreover, the investigation of the ensemble's dynamics will reveal dynamical phenomena like burst splitting which are important for the evaluation of neurophysiological data (Tass 1996b).

A very important feature, however, is missing in the ensemble model since the oscillators in that model do not interact via deterministic couplings. Obviously, the latter have to be taken into account, for example, in order to study the interplay between stimulus and couplings during the stimulation on the one hand and the resynchronization after the stimulation on the other hand. In the model presented in this book we shall consider a *cluster* of phase oscillators, i.e. a population of interacting phase oscillators. The interactions will be modeled by smooth global couplings. Certainly, in physiological networks of oscillators, e.g., within a cluster of oscillatory neurons coupling coefficients scatter around a mean. Nevertheless, global couplings may be considered as a reasonable first approximation because the synchronization behavior in such a model typically hardly changes by introducing variations of the couplings provided these variations are moderate compared to the couplings' mean (Tass and Haken 1996).

Despite the homogeneous type of interaction various and qualitatively different synchronized states may emerge in globally coupled clusters of oscillators as will be explained in Sect. 3.3. Thus, before we can start to analyze effects of stimulation we have to clarify which synchronized states are important in neurophysiological perspective. Guided by the fact that neurons may act as coincidence detectors (von der Malsburg and Schneider 1986, König, Engel, Singer 1996) we shall focus on cluster states. The latter are synchronized states displaying distinct clusters of phase locked oscillators (Sakaguchi, Shinomoto, Kuramoto 1987, 1988, Strogatz and Mirollo 1988a, 1988b, Golomb, Hansel, Shraiman, Sompolinsky 1992, Hakim and Rappel 1992, Hansel, Mato, Meunier 1993, Nakagawa, Kuramoto 1993, Okuda 1993, Tass and Haken 1996, Tass 1997).

One may expect that as a result of a stimulation, for example, one perfectly synchronized cluster may break in two or more distinct clusters each containing phase locked oscillators. In other words, the stimulus may change the configuration of the cluster state which is given by the number of oscillators within each cluster. For this reason it was necessary to investigate how the configuration influences relevant dynamical features such as the synchronization frequency (Tass 1997): With this aim in view clustering was studied in a population of globally coupled phase oscillators with randomly distributed eigenfrequencies without fluctuations. It turned out that depending on the configuration the synchronization frequency may decisively differ from the mean of the eigenfrequencies. For certain configurations one may even encounter so-called frozen states which are cluster states with vanishing synchronization frequency. This investigation will be presented in Chap. 3 where the possible role of cluster states concerning neural coding will be discussed, too.

Starting from these preparatory studies Chap. 4 will serve for deriving a stochastic model equation for a large population of globally coupled phase oscillators subjected to a stimulus and to noise. Chapter 5 is devoted to the

dynamics emerging in the absence of a stimulus. Put otherwise, we shall analyze how the clustering is influenced by random forces, in this way obtaining
a collection of noisy cluster states. In Chaps. 6 and 7 the impact of different
types of stimuli on these particular cluster states will be studied. Chapter
6 will be dedicated to single pulse stimulation and, correspondingly, it will
present a detailed phase resetting analysis where we shall encounter a variety
of stimulation induced desynchronization and synchronization processes. On
the contrary, Chap. 7 provides an investigation of the effects of periodic stimuli which may be either smooth or pulstile. In that chapter we shall especially
focus on phase locking between the stimulus and the cluster's dynamics.

The model presented in this study consists of phase oscillators. This means
that we assume that the stimulus affects the oscillators in a way that we are
allowed to neglect effects due to the amplitude dynamics. Accordingly, if
one wants to consider very strong stimuli one has to take into account the
amplitude dynamics, too. As discussed in Chap. 9 strong stimuli will cause
additional dynamical phenomena such as amplitude death, that means due
to the stimulus the single oscillator is stopped and remains within a stable
state with vanishing amplitude. However, if one wants to study stimulation of
neuronal activity with low or moderate intensity the phase oscillator approach
is in fact appropriate as will be explained in detail below. The stimulation
with lower intensities is interesting from the neurophysiological point of view
for two main reasons:

1. An important functional mode of cortical neurons is coincidence detection (von der Malsburg and Schneider 1986, König, Engel, Singer 1996).
   In this mode neurons preferentially respond to synchronized input. Correspondingly, it is to be expected that the physiological or pathological
   function of the activity of a synchronized neuronal cluster is severely perturbed by a stimulation induced desynchronization. In general, a stimulus
   does not need to provoke the amplitude death of each neuronal oscillator
   within a cluster. Rather it may be sufficient to change the state of synchronization appropriately in order to qualitatively modify the cluster's
   functional impact on other neurons.
2. On the other hand therapeutic stimulation of neuronal activity should
   be as gentle as possible in order to avoid side effects. Thus, also from a
   therapeutic standpoint one may favour low stimulation intensities.

Finally, what would all the results obtained with our stochastic approach
be worth provided they could not contribute to a better understanding of
physiological mechanisms? For this reason in Chap. 8 appropriate data analysis tools will be presented which are designed for investigating stimulation
induced processes of collective oscillatory activity as, e.g., registered with
MEG and EEG. Such methods are necessary, for instance, to study the course
of short-term epochs of synchronized cerebral activity evoked by sensory stimulation.

1.4 Synergetics

As a consequence of the discovery of deep dynamical principles underlying self-organization processes in physics, chemistry, biology and the social sciences Haken (1977, 1983a) brought the interdisciplinary field of synergetics into being. Inspired and guided by synergetics huge numbers of studies addressed the dynamics of self-organizing systems which typically consist of a large number of subsystems. Irrespective of the nature of a particular system under consideration one encounters unifying dynamical principles. In particular, in the neighbourhood of bifurcations the system's dynamics is often governed by only few variables, the so-called order parameters (for a review see Haken 1977, 1983a, 1996).

By means of the slaving principle (Haken 1975, Wunderlin and Haken 1975) one can understand the action of the order parameters: A bifurcation, that means a qualitative change of the system's dynamics occurs when a control parameter, also called bifurcation parameter, is, e.g., increased and exceeds a critical value. Close to the bifurcation the system's variables split into two groups, the slowly varying order parameters and the quickly relaxing stable modes, the so-called enslaved modes. In that situation the order parameters obey the order parameter equation which is typically low-dimensional compared to the entire system. The stable modes, on the other hand, are directly given by the order parameters.

A rigorous derivation and proof of the order parameter equation requires that the order parameters which emerge due to the bifurcation are within a small but finite neighbourhood around zero (Haken 1975, Wunderlin and Haken 1975, cf. Pliss 1964, Kelley 1967). In this context it is important that several characteristic stimulation induced dynamical phenomena presented in this book are due to the fact that the slaving principle is actually valid also far outside this small neighbourhood. This point will be addressed especially in Chap. 6. We shall there investigate how single pulse stimulation affects a cluster of synchronized phase oscillators. It will be shown that the single pulses may remove the system far away from its stationary state which in mathematics is called the center manifold (Pliss 1964, Kelley 1967, cf. Wunderlin and Haken 1981), and, accordingly, as a result of the stimulation the state of synchronization will change considerably. Due to the oscillators' synchronizing interactions the initial synchronized state may reappear. Remarkably, the process of resynchronization is typically governed by the slaving principle: Acccording to our numerical simulations, the enslaved modes quickly relax to values which are given by the slowly recovering order parameters. Put otherwise, the system's relaxation to the center manifold is governed by the slaving principle.

2. Resetting an Ensemble of Oscillators

2.1 Introductory Remarks

This chapter will introduce to the issue of phase resetting of populations of synchronized oscillators. The studies of Winfree (1980) will serve as a starting point for discussing advantages and drawbacks of different types of deterministic models. This will prepare us for the stochastic approach to phase resetting (Tass 1996a, 1996b). Along the lines of this approach a population of biological oscillators will be modeled by an ensemble of phase oscillators. The latter consists of identical oscillators subjected to fluctuations and the stimulus' impact. No deterministic interactions between the oscillators are taken into account. This rather simple model will allow us to assess the consequences of investigating phase resetting in terms of statistical physics. In particular, in comparison with the deterministic models one encounters qualitatively different phenomena which are important for the understanding of stimulation experiments. The influence of deterministic couplings on the stimulation induced dynamics in a noisy setting will be postponed and analyzed in detail in subsequent chapters.

2.2 Deterministic Models

Many biological oscillators consist of a large number of oscillating subsystems. In particular, in the nervous system one typically encounters groups of neurons exhibiting synchronized oscillatory activity, in this way behaving like single giant oscillators (see, for instance, Winfree 1980, Cohen, Rossignol and Grillner 1988, Elble and Koller 1990, Steriade, Jones and Llinás 1990, Singer and Gray 1995). As yet, in many biological systems the microscopic details of the oscillators' dynamics and their mutual interactions are not completely understood and, thus, no microscopic equations are available.

However, along the lines of a *top-down approach* the system under consideration can often be described by means of only a few appropriate macroscopic variables. This approach is justified by the *slaving principle* of synergetics. According to the latter, in particular, in the neighborhood of an instability the dynamics of complex and high-dimensional systems may be determined by a low dimensional equation, the so-called *order parameter equation*

(Haken 1977). The idea behind the top-down approach is that starting from a macroscopic level of description based on the interaction of modelling and experiments one reaches a microscopic level of understanding. In physics this approach turned out to be fruitful several times. A detailed foundation of top-down modelling in neuroscience was presented by Haken (1996).

There are many examples from different branches of the natural sciences where the dynamics of a cluster of limit cycle oscillators could appropriately be described by only two variables, namely cluster amplitude and cluster phase (see, for instance, Winfree 1980, Haken 1983, and further references in Sect. 3.3). In some cases even a one-dimensional dynamics of the cluster phase turned out to reflect the main dynamical features. Quite in the spirit of the top-down reasoning for the time being let us consider a cluster of limit cycle oscillators displaying a collective phase dynamics which can appropriately be described by a single variable.

2.2.1 Macroscopic Level

At the macroscopic level of description the cluster of oscillators is considered as a single giant phase oscillator. In doing this, in particular, it is assumed that the dynamics of the cluster amplitude can be neglected in a first approximation. For the sake of shortness in this section the cluster of oscillators will simply be denoted as oscillator. A particular value of the cluster phase ψ may be assigned to a marker event within a period of the collective oscillatory activity. Thus, the marker event is observed whenever ψ equals, e.g., $k \cdot 2\pi$, where k is an integer. A marker event might, for instance, be an upstrike of the cardiac action potential or the onset of synchronously bursting neuronal activity. A rather simple example of a phase dynamics is given by

$$\dot{\psi} = \Omega \ , \tag{2.1}$$

where the dot denotes differentiation with respect to time t, i.e. $\dot{\psi} = d\psi/dt$. (2.1) can be integrated in a straightforward way yielding $\psi(t) = \Omega t + \psi_0$, where ψ_0 is a constant. So, (2.1) governs the dynamics of a rigid rotator periodically generating its marker event. Although (2.1) is definitely not a complicated differential equation, in the 1960's it was, however, the starting point of Winfree's (1970) phase resetting analysis (for a review see Winfree 1980). According to a top-down approach he modeled the resetting stimulus by adding an additional term on the right hand side of (2.1) obtaining

$$\dot{\psi} = \Omega + S(\psi, I) \ . \tag{2.2}$$

S takes into account that the vulnerability to stimulation of a large number of different biological oscillators depends on the oscillator's phase (cf. Winfree 1980). Therefore the stimulation S depends on the phase ψ. Moreover the effect of stimulation obviously depends on the stimulus' intensity I. The stimulation starts at time t_B and ends at time t_E, and the stimulation duration will be denoted by $T = t_E - t_B$. (2.1) and (2.2) govern the whole stimulation scenario: During stimulation (2.2) is the evolution equation of the oscillator's phase, whereas (2.1) is valid before and after stimulation.

Winfree (1980) suggested to model the stimulation by setting

$$S(\psi, I) = I \cos \psi \; . \tag{2.3}$$

By the way, (2.2) and (2.3) were also used in a different context. For example, in laser physics they are well-known phase-locking equations (Gardner 1966, Haken et al. 1967).

Let us introduce a generalization of (2.3), where S depends on the phase ψ and on a set of stimulation parameters denoted by $\{X_m\}$ and $\{Y_m\}$. Identifying 0 and 2π of the phase ψ, in general, one can expand S in terms of Fourier modes:

$$S(\psi, \{X_m\}, \{Y_m\}) = \sum_{m=1}^{\infty} [X_m \sin(m\psi) + Y_m \cos(m\psi)] \; . \tag{2.4}$$

To motivate this generalization we should briefly dwell on the question as to where the harmonic terms, that means the higher order terms like $\sin(2\psi)$, $\cos(2\psi)$, $\sin(3\psi)$, $\cos(3\psi)$, etc in (2.4) may come from. With this aim in view it is important to take into account that the stimulus primarily acts on the oscillator as a whole. In other words, the stimulus influences both the oscillator's phase as well as amplitude dynamics. In the phase model (2.2) we obviously neglect the amplitude dynamics. The idea behind this approximation is well-known from the concept of interacting phase oscillators (Kuramoto 1984). While the latter will be explained in detail in Chap. 3, for the time being it is sufficient to realize the main principle: By only taking into account the phase dynamics we assume that the oscillator moves along a stable limit cycle and, in particular, that the stimulus removes the oscillator from its limit cycle in a way that can be neglected in a first approximation.

To illustrate this principle let us consider a simple model of a limit cycle oscillator subjected to a stimulus which exclusively acts on the oscillator's phase. In this way we shall additionally understand the origin of the harmonic terms of $S(\psi, \{X_m\}, \{Y_m\})$. Without stimulation let the oscillator obey the simple evolution equation

$$\dot{z} = (\alpha + i\Omega)z - \beta z^2 z^* \; , \tag{2.5}$$

where z is a complex variable, and z^* denotes the complex conjugate of z. Ω is the eigenfrequency, α is a real and β a nonnegative real parameter. Polar coordinates r and ψ are introduced by inserting the hypothesis

$$z(t) = r(t) \exp[i\psi(t)] \; , \tag{2.6}$$

where r and ψ are real and r is nonnegative. In this way we can immediately derive the equations governing the dynamics of the oscillator's amplitude r and the phase ψ:

$$\dot{r} = \alpha r - \beta r^3 , \tag{2.7}$$

$$\dot{\psi} = \Omega . \tag{2.8}$$

For negative α the amplitude vanishes, i.e. $r = 0$ is a stable fixed point. When α becomes positive the dynamics given by (2.6) and (2.7) undergoes a Hopf bifurcation giving rise to a stable limit cycle oscillation with frequency Ω and amplitude $\sqrt{\alpha/\beta}$ (cf. Haken 1983). From (2.8) we can easily determine the oscillator's phase $\psi(t) = \Omega t + \theta$, where θ is a constant. We shall analyze this bifurcation in a similar context in detail in Chap. 5.

(2.5) is a normal form , which means that the limit cycle dynamics of many and even more complicated oscillators can be transformed onto or can be approximated by the dynamics given by (2.5) (Elphik et al. 1987, Iooss and Adelmeyer 1992). For instance, applying the rotating wave approximation and the slowly varying wave approximation to the Van der Pol oscillator or to the neurophysiological HKB oscillator we end up with (2.7) and (2.8) (Haken, Kelso, Bunz 1985). Hence, (2.5) provides us with a suitable minimal model of a limit cycle oscillator.

Next, to additionally model the stimulus' impact on the oscillator we recall that the vulnerability to stimulation of numerous biological oscillators depends on the oscillator's dynamical state, in particular, on its phase (cf. Winfree 1980). Accordingly, we assume that the stimulus $\tilde{S}$ depends on z and z^*. Thus, in general, the model may be written in the form

$$\dot{z} = (\alpha + i\Omega)z - \beta z^2 z^* + \tilde{S}(z, z^*) , \tag{2.9}$$

where $\alpha > 0$. To illustrate the action of a stimulus which exclusively acts on the phase let us choose a special form of $\tilde{S}$ given by

$$\tilde{S}(z, z^*) = z \sum_{m=1}^{\infty} \left[\eta_m z^{m+1} z^* - \eta_m^* (z^*)^{m+1} z \right] \tag{2.10}$$

with complex coefficients

$$\eta_m = w_m + i v_m \quad (w_m, v_m \in \mathbb{R}) . \tag{2.11}$$

In the sense of the normal form theorem $\tilde{S}(z, z^*)$ from (2.10) represents a class of stimulation mechanisms which merely influence the phase. Inserting (2.6) and (2.11) into (2.10) one obtains the evolution equations of the amplitude r and the phase ψ:

$$\dot{r} = \alpha r - \beta r^3 , \tag{2.12}$$

$$\dot{\psi} = \Omega + \sum_{m=1}^{\infty} \left[2r^{m+2} w_m \sin(m\psi) + 2r^{m+2} v_m \cos(m\psi) \right] . \tag{2.13}$$

From (2.12) and (2.13) we read off that the stimulus $\tilde{S}(z, z^*)$ only acts on the oscillator's phase whereas the amplitude remains unaffected. For this reason the amplitude r relaxes towards the stable value $\sqrt{\alpha/\beta}$ irrespective of the initial value $r(0)$ and irrespective of the stimulus' action. To determine how the stimulus influences the oscillator's motion along its limit cycle we simply have to insert the limit cycle amplitude $r = \sqrt{\alpha/\beta}$ into (2.13). Putting

$$X_m = 2w_m \left(\frac{\alpha}{\beta}\right)^{1+m/2} \quad, \quad Y_m = 2v_m \left(\frac{\alpha}{\beta}\right)^{1+m/2} \tag{2.14}$$

we then immediately obtain the phase dynamics (2.2) with the stimulation mechanism (2.4). These considerations show that the harmonics $\sin(m\psi)$, $\cos(m\psi)$ with $m = 2, 3, \ldots$ in (2.4) originate from higher order terms $z^{m+2}z^*$, $z^2(z^*)^{m+1}$ of the stimulation mechanism modeled by (2.10).

Now, let us come back to the analysis of the phase model given by (2.2) and (2.4). By introducing

$$I_m = \sqrt{X_m^2 + Y_m^2} \,, \quad \cos\gamma_m = \frac{Y_m}{I_m} \,, \quad \sin\gamma_m = -\frac{X_m}{I_m} \tag{2.15}$$

(2.4) can be written as

$$S(\psi, \{I_m\}, \{\gamma_m\}) = \sum_{m=1}^{\infty} I_m \cos(m\psi + \gamma_m) \,, \tag{2.16}$$

where in this general form S depends on infinitely many parameters $(I_1, I_2, \ldots$ and $\gamma_1, \gamma_2, \ldots)$. For the sake of brevity I will mostly omit the parameters below, thus, briefly wirting $S(\psi)$ instead of $S(\psi, \{I_m\}, \{\gamma_m\})$. Additionally $K(\psi)$, an abbreviation of the vector field of (2.2) is introduced, which will be used below. With this notation the model takes the form

$$\dot{\psi} = K(\psi) = \Omega + S(\psi) \,. \tag{2.17}$$

Obviously (2.17) has a potential $V(\psi)$ given by

$$V(\psi) = -\int_0^\psi K(\xi) \, \mathrm{d}\xi \,, \quad \text{where } \dot{\psi} = -\frac{\mathrm{d}V(\psi)}{\mathrm{d}\psi} \,. \tag{2.18}$$

To point out some important features of macroscopic phase oscillator models in this section it is sufficient to restrict our consideration to the simple stimulation model (2.3), i.e. in terms of (2.16) we set $I_1 = I$, $I_j = 0$ for $j > 1$ and $\gamma_1, \gamma_2, \gamma_3 \ldots = 0$. So, we focuss on

$$\dot{\psi} = \Omega + I \cos\psi \,. \tag{2.19}$$

Moreover it is sufficient to analyze (2.19) for $I \geq 0$. For $I < 0$ one has to insert the shifted phase $\phi = \psi + \pi$ into (2.19), in this way obtaining $\dot{\phi} = \Omega + |I| \cos\phi$. For (2.19) the potential reads

$$V(\psi) = -\Omega\psi - I\sin\psi \ . \tag{2.20}$$

The dynamics of (2.19) crucially depends on the stimulation intensity I.

If the *stimulation* is *subcritical*, i.e. if $0 < I < \Omega$ holds, (2.19) has no fixed points. In this case the phase increases with increasing time. The bifurcation diagram in Fig. 2.1 shows the saddle-node bifurcation occurring if the stimulation intensity is increased. At the *critical* intensity $I = \Omega$ a fixed point occurs. If the stimulation becomes *supercritical* $(I > \Omega)$ this fixed point splits into a phase attractor (ψ_a) and a phase repellor (ψ_r).

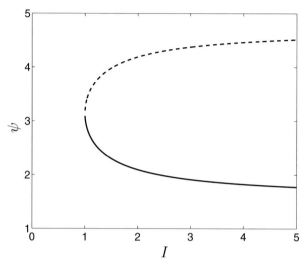

Fig. 2.1. *Bifurcation diagram:* The phases of the fixed points of (2.19) are plotted versus intensity I for $\Omega = 1$. A stable fixed point (ψ_a, *solid line*) and an unstable fixed point (ψ_r, *dashed line*) occur for supercritical stimulation intensity $I > \Omega = 1$.

To understand the stimulation process modelled by (2.19) it is important to focus on how the outcome of the stimulation depends on the phase at the beginning of the stimulation $\psi(t_B)$, the stimulation intensity I, and the stimulation duration T. For the sake of shortness the phase at the beginning and at the end of the stimulation will be denoted by ψ_B and ψ_E respectively:

$$\psi_B = \psi(t_B) \ , \quad \psi_E = \psi(t_E) \ . \tag{2.21}$$

ψ_B will be called *old phase*, and ψ_E *new phase*. Figure 2.2 illustrates the effect of the stimulation parameters I and T and the initial phase ψ_B on the outcome of the stimulation. 2000 numerical integrations of (2.19) were performed for $\Omega = 1$ and for four different stimulation intensities I using a 4th order Runge-Kutta algorithm. In each series the initial values of ψ_B were equally spaced in $[0, 2\pi]$.

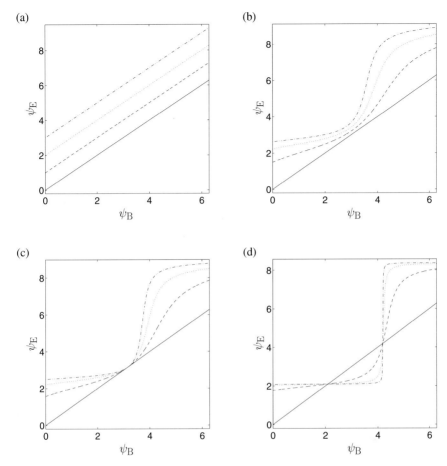

Fig. 2.2a–d. *Phase resetting curves:* (2.19) was integrated numerically 2000 times with initial values of ψ_B (*solid line*) equally spaced in $[0, 2\pi]$ for $\Omega = 1$. ψ_E is plotted for different stimulation durations: $T = 1$ (*dashed line*), $T = 2$ (*dotted line*), $T = 3$ (*dot-dashed line*). The series of simulations were performed for four different stimulation intensities I. (**a**) No stimulation ($I = 0$): Phase increases homogeneously. (**b**) Subcritical stimulation ($I = 0.8 < \Omega = 1$): Increase of the phase depends on ψ_B. However, there is no fixed point of the phase. (**c**) Critical stimulation ($I = 1 = \Omega$): A fixed point occurs at π. (**d**) Supercritical stimulation ($I = 2 > \Omega = 1$): There are two fixed points, attractor ψ_a (*left one*) and repellor ψ_r (*right one*).

For vanishing stimulation intensity ($I = 0$) the phase increases homogeneously with increasing duration T (Fig. 2.2a). During subcritical stimulation ($0 < I < \Omega$) no fixed point occurs. Nevertheless the increase of the phase depends on ψ_B (Fig. 2.2b). For critical stimulation ($I = \Omega$) the phase increases except for a fixed point at π (Fig. 2.2c). As a consequence of a supercritical stimulation ($I > \Omega$) the phase may increase or decrease depending on the initial value ψ_B. Correspondingly two fixed points, attractor ψ_a and repellor ψ_r, show up (Fig. 2.2d). Comparing Fig. 2.2a with Figs. 2.2b–d clearly shows that stimulation causes a phase shift. In other words, it resets the phase. For this reason the plots in Fig. 2.2 are called *phase resetting curves*.

All curves in Fig. 2.2 share one common topological feature: In all these curves $d\psi_\mathrm{E}/d\psi_\mathrm{B}$ is positive for all I, T and ψ_B. Moreover the average gradient over a cycle ranging from 0 to 2π is 1. For this reason this kind of resetting behavior is called *type 1 resetting* (Winfree 1980). A single phase oscillator governed by (2.17) always exhibits this kind of resetting behavior (Winfree 1980). Note that the so-called *winding number* used in topology is the same as the average value of the slope $d\psi_\mathrm{E}/d\psi_\mathrm{B}$ (cf. Milnor 1965, Winfree 1980).

A large number of stimulation experiments clearly revealed that biological oscillators typically show two different types of resetting behavior. For an extensive review I refer to Glass and Mackey (1988) and, in particular, to Winfree (1980). Depending on the stimulation intensity both types of resetting occur: type 1 resetting and the so-called *type 0 resetting*, where in the phase resetting curve the average gradient over a cycle equals 0 (Winfree 1980). A simple example for type 0 resetting is provided by a stimulation which is strong enough to simply restart the oscillator. In this case ψ_E, the phase at the end of the stimulation, is always the same no matter at what initial phase ψ_B the stimulus is presented.

The question as to whether type 0 or type 1 resetting is observed is not just of topological interest. Actually, annihilation of synchronized oscillatory activity is intimately related with type 0 and type 1 resetting and transitions between them (Winfree 1980). Therefore the model given by (2.17) has to be improved in order to account for both types of resetting. Remaining on a macroscopic level of description model (2.17) can be extended by additionally taking into account the dynamics of the cluster amplitude. In this way the giant oscillator which consists of a large number of single oscillators is modeled by two variables: phase and amplitude. Such models have been extensively studied (cf. Winfree 1980, Glass and Mackey 1988, Murray 1989). A main result of these studies is that in two-dimensional models one encounters both, type 0 as well as type I resetting behavior. Moreover certain phase singularities, the so-called *black holes*, show up when there is a transition between the two types of resetting (Winfree 1980, Glass Mackey 1988). In the context of phase resetting the term black hole has two related meanings depending on the level of description:

1. *Single oscillator:* Let us consider an oscillator which has both a stable limit cycle as well as a stable fixed point connected with a vanishing amplitude. Assume that a well-timed stimulus of appropriate strength knocks the oscillator out of the basin of attraction of the limit cycle into the basin of attraction of the stable fixed point. In this way the stimulus induces an annihilation of the oscillation or, in other words, an amplitude

death. A stable state with vanishing amplitude was called black hole by Winfree (1980).

2. *Populations of oscillators:* A population of synchronized oscillators may undergo a stimulation induced desynchronization. In this case a stable desynchronized state is denoted as a black hole. However, according to the slaving principle (Haken 1977) synchronization processes in populations of oscillators may be described by means of low-dimensional order parameter equations (Haken 1977, Kuramoto 1984). Such an order parameter may behave like a simple single oscillators. Since the amplitude of the order parameter typically reflects the strength of the oscillators' synchronization, a black hole, that means a desynchronized state corresponds to an amplitude death of the order parameter. For this reason the two meanings, namely amplitude death of a single oscillator on the one hand and desynchronization of a population of oscillators on the other hand, are deeply connected. This will be explained in detail below, especially in Sect. 6.6. Of course, the term black hole as used in the context of phase resetting must not be confused with the well-known black holes of cosmology.

Although two-dimensional models are certainly more appropriate than one-dimensional models, in this book, however, a different approach is chosen as outlined in the next section. At the end of this chapter a detailed comparison between this approach and the results from two-dimensional deterministic models will motivate this choice.

2.2.2 Cluster of Oscillators

This book is dedicated to the analysis of how stimulation affects synchronized oscillatory activity. Thus, the idea suggests itself to take into account that the biological oscillator under consideration consists of a large number of single oscillators. In the former section such a cluster of oscillators was modeled as one giant oscillator. Obviously this approach may be improved by choosing a more microscopic level of description. With this aim in view Winfree considered a cluster of N limit cycle oscillators (for a review see Winfree 1980). For the sake of simplicity all oscillators are assumed to have the same eigenfrequency. Winfree neglected both mutual interactions of the oscillators and noise. Hence, without stimulation his model reads

$$\dot{\psi}_j = \Omega \quad (j = 1, \ldots, N) , \tag{2.22}$$

where ψ_j denotes the phase of the jth oscillator. So, (2.22) models N non-interacting rigid rotators which are not subjected to fluctuations. On this level of description every single oscillator generates a marker event whenever its phase equals, e.g., $k \cdot 2\pi$ (with integer k). A marker event might, thus, be the spike of a single neuron.

To model the impact of the stimulation the stimulus is assumed to affect every single oscillator instantaneously with equal strength starting at time t_B and ending at time t_E. Apart from the stimulation intensity the effect of the stimulation on the jth oscillator depends on the phase of this particular oscillator. For this reason, according to (2.19) during the stimulation, i.e. for $t_B \leq t \leq t_E$, the phase dynamics obeys the evolution equation

$$\dot{\psi}_j = \Omega + S(\psi_j, I) = \Omega + I \cos \psi_j \quad (j = 1, \ldots, N) \qquad (2.23)$$

(cf. Winfree 1980). To analyze the cluster's collective dynamics and, in particular, to describe synchronization processes, it is convenient to introduce the complex *cluster variable* Z

$$Z(t) = R(t) \, \exp[i\varphi(t)] = \frac{1}{N} \sum_{k=1}^{N} \exp[i\psi_k(t)] \, , \qquad (2.24)$$

where ψ_k is the phase of the kth oscillator, and N is the number of oscillators (see, for instance, Aizawa 1976, Winfree 1980, Kuramoto 1984, Yamaguchi and Shimizu 1984, Shiino and Frankowicz 1989, Matthews and Strogatz 1990, Tass and Haken 1996, Tass 1997). The real quantities R and φ will be denoted as *cluster amplitude* and *cluster phase*, respectively. The more the cluster is synchronized the larger is R, where $0 \leq R(t) \leq 1$ holds for all times t. The cluster amplitude serves for investigating the extent of the cluster's in-phase synchronization. According to (2.23) the stimulus acts on a microscopic level by affecting every single oscillator, in this way inducing macroscopic changes of the cluster's synchronization. The latter can appropriately be described by means of the macrovariables R and φ.

As already mentioned in Sect. 2.2.1 a single phase oscillator only exhibits type 1 resetting. In contrast to that a cluster consisting of at least three phase oscillators also shows type 0 resetting for suitable stimulation intensity and stimulation duration (Winfree 1980). Comparing the diversity of the dynamics of the cluster model (2.23) with that one of the macroscopic model (2.19) it, thus, turns out to be worth choosing the microscopic level of description. Anyhow, one may still be bothered about two severe drawbacks of model (2.23): Noise as well as the oscillators' mutual interactions are neglected.

To be aware of the consequences of taking into account fluctuations, it is important to recall that Winfree (1980) mainly used topological arguments in his phase resetting analysis. In particular, he used a topological theorem which says that the only continuous maps from the disk to the circle have winding number 0 around the border of the disk (cf., e.g., Milnor 1965). This topological approach is both powerful and elegant because it does not depend on an explicit model for the dynamics of the stimulated oscillators. Rather it is sufficient that some qualitative assumptions are fulfilled, for instance, concerning the smoothness of the oscillators' dynamics. However, this rather general way of reasoning does not work if one takes into account fluctuating forces. That is why this book is dedicated to an approach based on statistical physics.

2.3 Stochastic Model

The modification of the model given by (2.22) and (2.23) will be performed step by step so that it becomes apparent which resetting phenomena are noise induced and which are due to the oscillators' mutual (deterministic) interactions. First of all, in this chapter fluctuations are additionally taken into account along the lines of the ensemble model which was presented in previous studies (Tass 1996a, 1996b). Accordingly the biological oscillator consisting of a large number of single oscillators is approximated by an ensemble of infinitely many identical phase oscillators. Consequently all oscillators of the ensemble have the same eigenfrequency. Moreover the stimulation affects every single oscillator in the same way, starting at time t_B and ending at time t_E. The random forces are modeled by additive noise denoted by $F(t)$. For the sake of simplicity Gaussian white noise is chosen. Thus,

$$\langle F(t)\rangle = 0 \quad \text{and} \quad \langle F(t)F(t')\rangle = Q\delta(t - t') \tag{2.25}$$

hold, where Q is the constant *noise amplitude* . Based on the deterministic model (2.22) the dynamics of a single oscillator before and after the stimulation is governed by

$$\dot{\psi} = \Omega + F(t) . \tag{2.26}$$

As the ensemble consists of identical oscillators it is convenient to drop the index, in this way writing ψ instead of ψ_j. Note that (deterministic) interactions are neglected in this model.

Analogously the impact of the stimulation is modelled by adding noise on the right hand side of (2.23). Hence, during the stimulation, i.e. for $t_\mathrm{B} \le t \le t_\mathrm{E}$, the phase dynamics of the single oscillator is determined by

$$\dot{\psi} = \Omega + S(\psi, \{I_m\}, \{\gamma_m\}) + F(t) \tag{2.27}$$

with

$$S(\psi, \{I_m\}, \{\gamma_m\}) = \sum_{m=1}^{\infty} I_m \cos(m\psi + \gamma_m) , \tag{2.28}$$

where $\{I_m\}$ and $\{\gamma_m\}$ are stimulation parameters (cf. Tass 1996a, 1996b). From the model equations (2.26) and (2.27) it follows that the noise amplitude is assumed not to be affected by the stimulus.

Equations (2.26) and (2.27) are stochastic differential equations, so-called *Langevin equations* (see, for instance, Haken 1977, 1983, Gardiner 1985, Risken 1989). Both may be written in the form

$$\dot{\psi} = K(\psi) + F(t) , \tag{2.29}$$

where $K(\psi)$ denotes the deterministic part of the vector field of (2.26) and (2.27), respectively, whereas $F(t)$ denotes the stochastic forces fulfilling (2.25). In order to investigate the dynamics determined by Langevin equation (2.29) the corresponding *Fokker–Planck equation* has to be analyzed. The latter is a partial differential equation for the probability density $f(\psi, t)$, where the probability of finding the phase of a single oscillator at time t in the interval $\psi \ldots \psi + d\psi$ is given by $f(\psi, t)\, d\psi$. The Fokker–Planck equation which corresponds to (2.29) reads

$$\frac{\partial f(\psi, t)}{\partial t} + \frac{\partial}{\partial \psi}\left[K(\psi) f(\psi, t) - \frac{Q}{2}\frac{\partial f(\psi, t)}{\partial \psi}\right] = 0 \qquad (2.30)$$

with Q from (2.25) (cf. Haken 1977, 1983, Gardiner 1985, Risken 1989). $K(\psi)$ is known as *drift-coefficient*, while Q is called *diffusion coefficient*. By means of the *probability current*

$$j(\psi, t) = K(\psi) f(\psi, t) - \frac{Q}{2}\frac{\partial f(\psi, t)}{\partial \psi} \qquad (2.31)$$

(2.30) may be written in the form

$$\frac{\partial f(\psi, t)}{\partial t} + \frac{\partial}{\partial \psi} j(\psi, t) = 0 \,. \qquad (2.32)$$

The probability current will be useful for the interpretation of stimulation induced desynchronization processes.

Another important aspect concerning the interpretation of the ensemble's dynamics should be stressed from the very beginning: As the ensemble consists of infinitely many identical oscillators, the dynamics of every single oscillator is determined by the Langevin equation (2.29). Consequently the Fokker–Planck equation (2.30) holds for every single oscillator, and, thus, the probability $f(\psi, t)$ serves as a *distribution function* of the ensemble of oscillators. Therefore on the analogy of the cluster variables introduced in (2.24) it is possible to introduce an *ensemble variable* $Z(t)$, an *ensemble amplitude* $R(t)$ and an *ensemble phase* $\varphi(t)$ by setting

$$Z(t) = R(t)\exp[i\varphi(t)] = \int_0^{2\pi} f(\psi, t)\,\exp(i\psi)\,d\psi \qquad (2.33)$$

where R and φ are real quantities. The more the ensemble is synchronized the larger is $R(t)$, where $0 \le R(t) \le 1$ holds for all times t. By means of the ensemble variables several important features of synchronization processes can be described.

2.4 Fokker–Planck Equation

This section is devoted to a purely mathematical investigation of (2.30). The results will be discussed in terms of physiology in subsequent sections. In order to analyze (2.30) one has to take into account two boundary conditions:

1. As ψ is a phase one is allowed to identify 0 and 2π. Therefore the boundary condition

$$f(0,t) = f(2\pi,t) \qquad (2.34)$$

   has to be fulfilled for all times t.
2. f is a probability distribution. Consequently,

$$\int_0^{2\pi} f(\psi,t)\,\mathrm{d}\psi = 1 \qquad (2.35)$$

   holds for all times t.

Equation (2.30) will be investigated in two different ways:

1. The *stationary solution* of (2.30) will be derived analytically. Stationary solutions of analogous Fokker–Planck equations were already used in other branches of the natural sciences. For instance, a similar analysis was performed by Haken et al. (1967) to investigate phase locking in the context of laser physics.
2. To analyze transient phenomena occurring during and after stimulation a numerical analysis will be carried out based on a *Fourier transformation* of (2.30). The accuracy of the numerical analysis will be checked by comparing analytically and numerically derived stationary solutions.

2.4.1 Stationary Solution

The stationary solution of (2.30), denoted by f_{st}, satisfies the condition

$$\frac{\partial f_{st}}{\partial t} = 0 \ . \qquad (2.36)$$

Inserting (2.36) into (2.30) one obtains

$$K(\psi)f_{st}(\psi) - \frac{Q}{2}\frac{\mathrm{d}f_{st}(\psi)}{\mathrm{d}\psi} = c \ , \qquad (2.37)$$

where c denotes a real constant which will be determined below by taking into account boundary condition (2.34). Comparing (2.37) with (2.31) one immediately reads off that in the stationary state the probability current j_{st} is constant with respect to time and phase:

$$j_{st} = c \ . \qquad (2.38)$$

With a little calculation the solution of (2.37) is obtained as a sum of the solution of the homogenous equation (i.e. for $c = 0$) and a particular solution of the inhomogenous equation (2.37). This yields

$$f_{st}(\psi) = \mathcal{N}\exp\left[-\frac{2}{Q}V(\psi)\right]\left\{1 + \frac{B-1}{C}\int_0^\phi \exp\left[\frac{2}{Q}V(\psi')\right]\mathrm{d}\psi'\right\} , \qquad (2.39)$$

where

$$B = \exp\left[\frac{2}{Q}V(2\pi)\right] \quad \text{and} \quad C = \int_0^{2\pi} \exp\left[\frac{2}{Q}V(\psi')\right] d\psi' . \tag{2.40}$$

$\mathcal{N}$ is a normalization factor which guarantees that condition (2.35) is fulfilled.

$$V(\psi) = -\Omega\psi - \sum_{m=1}^{\infty} \frac{I_m}{m}[\sin(m\psi + \gamma_m) - \sin(\gamma_m)] \tag{2.41}$$

is the potential from (2.18), i.e. the potential governing the dynamics for vanishing noise amplitude Q (cf. (2.16), (2.17) and (2.29)). For this reason one obtains

$$B = \exp\left\{-\frac{4\pi\Omega}{Q}\right\} . \tag{2.42}$$

Consequently B equals 1 if and only if Ω vanishes. In this case the potential $V(\psi)$ is 2π-periodic, and the stationary solution simply reads

$$f_{\text{st}}(\psi) = N \exp\left\{-\frac{2}{Q}V(\psi)\right\} . \tag{2.43}$$

This is the solution of the homogenous equation (2.37) (i.e. for $c = 0$). Thus, according to (2.38) the constant probability current vanishes if and only if the ensemble's eigenfrequency Ω vanishes. In our model Ω is positive and therefore the probability current j_{st} is constant and does not vanish.

Concerning the uniqueness of the stationary solution f_{st} it is important to notice that according to (2.16), (2.17) and (2.25) the drift-coefficient $K(\psi)$ and the diffusion coefficient Q do not explicitly depend on time. Therefore every distribution $f(\psi, t)$ fulfilling boundary conditions (2.34) and (2.35) finally decays to the stationary solution given by (2.39). For a proof I refer to Risken (1989) (Sect. 6.1). So, the stationary solution f_{st} describes the ensemble's state which evolves provided the stimulation duration tends to infinity.

2.4.2 Fourier Transformation

In order to investigate the impact of short-term stimuli Fokker–Planck equation (2.30) has to be integrated numerically. It will turn out to be convenient to perform a numerical analysis of the Fourier transformed Fokker–Planck equation. With this aim in view both f and K are expanded in terms of Fourier modes:

$$f(\psi, t) = \sum_{k\in\mathbb{Z}} \hat{f}(k, t)\, e^{ik\psi} , \quad K(\psi) = \sum_{k\in\mathbb{Z}} \hat{K}(k)\, e^{ik\psi} , \tag{2.44}$$

where $\mathbb{Z}$ denotes the set of integer numbers $\{0, \pm1, \pm2, \pm3\dots\}$. f and K are real functions. For this reason the relations

$$\hat{K}(-k) = \hat{K}(k)^* \quad \text{and} \quad \hat{f}(-k,t) = \hat{f}(k,t)^* \quad \text{(for all } t\text{)} \tag{2.45}$$

are fulfilled for all wave numbers k, where y^* denotes the complex conjugate of y. Due to condition (2.35)

$$\hat{f}(0,t) = \frac{1}{2\pi} \tag{2.46}$$

holds for all times t. The evolution equation for the Fourier modes with non-vanishing wave numbers is derived by inserting (2.44) into (2.30). In this way one obtains the Fourier transformed Fokker–Planck equation

$$\frac{\partial \hat{f}(k,t)}{\partial t} = -\frac{Q}{2}k^2 \hat{f}(k,t) - ik \sum_{l \in \mathbb{Z}} \hat{f}(k-l,t)\hat{K}(l) . \tag{2.47}$$

This equation will be the starting point for the numerical analysis presented below.

2.5 Spontaneous Behavior

Before the effect of stimulation will be investigated it should first be considered what happens provided there is *no stimulation*. In this case K equals Ω, and one immediately obtains

$$f(\psi, t) = \sum_{k \in \mathbb{Z}} \underbrace{\hat{f}(k,0)}_{\text{I}} \underbrace{\exp\left(-\frac{Q}{2}k^2 t\right)}_{\text{II}} \underbrace{\exp\left(ik(\phi - \Omega t)\right)}_{\text{III}} . \tag{2.48}$$

The interpretation of this equation is straightforward: The initial distribution is determined by the Fourier modes for $t = 0$ (term I). As a consequence of term III $f(\psi, t)$ moves along the ψ-axis as travelling wave with velocity Ω. The latter is the ensemble's eigenfrequency. Term II gives rise to an exponential damping of the Fourier modes with non-vanishing wave numbers. For this reason with increasing time bumps of the distribution f decrease, while f decays to the uniform distribution given by (2.46):

$$f(\psi, t) \longrightarrow \frac{1}{2\pi} \quad (t \longrightarrow \infty) . \tag{2.49}$$

So, as a consequence of the fluctuations an initially synchronized ensemble becomes desynchronized with increasing time. This desynchronization process can likewise be described by means of the ensemble variables. With (2.33) one derives

$$Z(t) = R(t)e^{i\varphi(t)} = 2\pi \hat{f}(-1,t)$$

$$= \underbrace{2\pi \hat{f}(-1,0)}_{\text{I}} \underbrace{\exp\left(-\frac{Q}{2}t\right)}_{\text{II}} \underbrace{\exp(-i\Omega t)}_{\text{III}} . \qquad (2.50)$$

Terms I up to III in (2.49) and (2.50) correspond to each other respectively. In (2.50) the initial ensemble variable is given by term I, term II causes the damping of the ensemble amplitude, and term III induces a travelling wave.

From (2.50) it follows that once the ensemble amplitude $R(t)$ vanishes it will be equal to zero for all further times. It is important to note that the ensemble variables are determined by only one Fourier mode, no matter whether or not the other modes vanish. For this reason a vanishing ensemble amplitude does not necessarily correspond to the uniform distribution as illustrated in Fig. 2.3. Equations (2.48) and (2.50) are important, in particular, for understanding the ensemble's relaxation behavior after stimulation: Let us assume that due to a suitable stimulation the ensemble amplitude vanishes ($R = 0$). After turning off the stimulator the ensemble amplitude will remain zero. On the other hand peaks of the distribution which might occur due to the stimulation vanish during the relaxation process according to (2.49).

2.6 Black Holes Without Noise

This and the next section are devoted to the relationship between the annihilation of synchronized oscillatory activity and the occurence of phase singularities. Before we analyze singular behavior of the ensemble of oscillators subjected to fluctuations let us first dwell on stimulation induced desynchronization in a noise-free setting.

The topological investigation of the dynamics of the deterministic model given by (2.22) and (2.23) revealed that apart from type 0 and type 1 resetting behavior one additionally encounters an important phenomenon: Stimuli of suitable intensity and timing cause a singularity of the cluster phase (Winfree 1980). According to the topological level of description the occurence of a *phase singularity* means that the stimulation's outcome cannot be predicted. However, a multitude of experiments with clusters of synchronized oscillators showed that a phase singularity is typically related with an *annihilation of the synchronized periodic oscillation* (for a review see Winfree 1980).

Winfree (1980) pointed out the intimate relationship between annihilation of synchronization and phase singularity. To this end he analyzed the desynchronization process caused by the stimulation mechanism

$$S(\psi_j, I) = \Omega + I \cos \psi_j \qquad (j = 1, \dots, N) \qquad (2.51)$$

(cf. (2.23)) for supercritical stimulation intensity, i.e. for $I > 1$. Suppose the stimulus is suitably timed so that the oscillators' phases are centered around the repellor ψ_r when the stimulation starts (cf. Fig. 2.1). As illustrated in Fig.

(a)

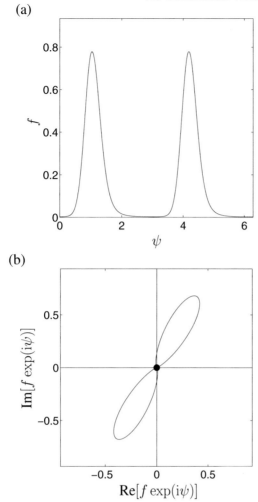

(b)

Fig. 2.3a,b. *Ensemble variable Z:* The position of the center of mass of the distribution f is given by the ensemble variable Z defined by (2.33). This relationship is illustrated by means of a distribution f which has a symmetric double peak. In (**a**) f is plotted over ψ, whereas (**b**) shows $f(\psi)\exp(i\psi)$ in the Gaussian plane. The black dot indicates that the center of mass is located in $(0,0)$.

2.4 in the course of the stimulation the cluster straddles ψ_r, in this way two groups of oscillators evolve. One group advances ($\psi_j > \psi_r$), whereas the other group delays ($\psi_j < \psi_r$) towards the attractor ψ_a. At a critical stimulation time T_{crit}, when the mean phase difference of both groups is $\approx \pi$, the center of mass of all oscillators is located in the origin of the Gaussian plane, i.e. the cluster is desynchronized. In this situation the cluster variable Z from (2.24) vanishes, and accordingly one cannot assign a phase to the cluster. The phase singularity corresponding to the annihilation of the synchronization was called *black hole* (Winfree 1980).

It should be emphasized that the phase stimulation of a cluster of oscillators according to (2.23) does not stop the activity of the single oscillators. Accordingly when the stimulation ends at the critical stimulation time T_{crit} every single oscillator is still active, and its dynamics is governed by (2.22). However, for suitable stimulation intensity and timing the initial pattern of synchrony has disappeared at time T_{crit}.

2.7 Ensemble Dynamics During Stimulation

In order to understand desynchronization in the presence of noise it is necessary to analyze how patterns of synchrony vanish in terms of statistical physics. With this aim in view this section concerns the impact of a simple type of stimulation on the ensemble dynamics. Investigating stimulation induced frequency shifts will additionally clarify the behavior of the single oscillators during stimulation. Finally the role of harmonics of the stimulation mechanism will be explained.

2.7.1 Black Holes in the Presence of Noise

In order to understand how the ensemble of the stochastic model described by (2.26) and (2.27) is forced into a black hole the motion of the ensemble variable $Z(t)$ and the motion of the distribution function $f(\psi, t)$ will be investigated. While section 2.7.3 will be concerned with the impact of higher harmonics of the stimulation mechanism, let us first dwell on the simple stimulation mechanism

$$S(\psi, I) = \Omega + I \cos \psi \qquad (2.52)$$

(cf. (2.27) and (2.28)). The ensemble's dynamics during stimulation will be considered for subcritical, critical and supercritical stimulation intensity I. The terms subcritical, critical and supercritical stimulation intensity refer to the oscillators' stimulation induced dynamics without noise as illustrated in Fig. 2.2.

With S given by (2.52) the Fourier transformed Fokker–Planck equation (2.47) takes the form

$$\frac{\partial \hat{f}(k, t)}{\partial t} = -\left(\frac{Q}{2} k^2 + \mathrm{i}\Omega k\right) \hat{f}(k, t)$$
$$- \mathrm{i}\frac{kI}{2} \hat{f}(k - 1, t) - \mathrm{i}\frac{kI}{2} \hat{f}(k + 1, t) . \qquad (2.53)$$

For (2.52) the potential $V(\psi)$ from (2.41), i.e. the potential governing the dynamics for zero noise amplitude Q (cf. (2.16), (2.17), (2.18) and (2.29)) reads

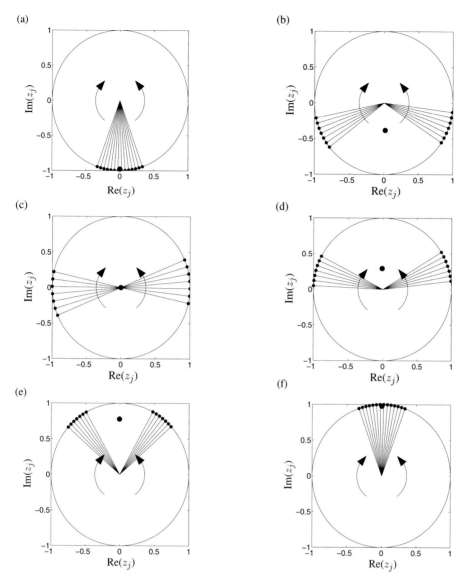

Fig. 2.4a–f. *Desynchronization and phase singularity:* The plots schematically illustrate how desynchronization gives rise to a phase singularity of the cluster phase φ from (2.24). A small dot (connected with the origin) on the unit circle in the complex plane represents a single oscillator according to $z_j = \exp(i\psi_j)$. The big dot corresponds to the cluster variable $Z = N^{-1}\sum_{j=1}^{N}\exp(i\psi_j)$, where $N = 14$. Time increases from (**a**) to (**f**). Without stimulation the oscillators would rotate in counterclockwise direction. Due to the stimulation the cluster straddles ψ_r, the phase repellor corresponding to the initial cluster phase. The straddling is indicated by the bent arrows. Both the advancing and the delaying subcluster are attracted by the phase attractor ψ_a corresponding to the final cluster phase. In (**c**) the cluster variable Z vanishes, and, thus, the cluster phase φ is not defined.

$$V(\psi) = -\Omega\psi - I \sin\psi \,. \tag{2.54}$$

Figure 2.5 displays V for subcritical, critical and supercritical stimulation intensity I.

Comparing the desynchronization mechanism in the ensemble model with that one described by Winfree (1980) for the deterministic model one encounters similarities and differences. Let us first consider a supercritical stimulation according to (2.52), i.e. $I > \Omega$. Moreover, we assume that at the beginning of the stimulation the oscillators are synchronized, so that the distribution function $f(\psi, t_\mathrm{B})$ has one peak. Figure 2.6 shows the straddling process which occurs provided the initial distribution is centered around the phase repellor ψ_r corresponding to the maximum of the potential $V(\psi)$ shown in Fig. 2.5: The peak splits into an advancing peak and a delaying peak. Both melt in the vicinity of the attractor ψ_a. The latter corresponds to the min-

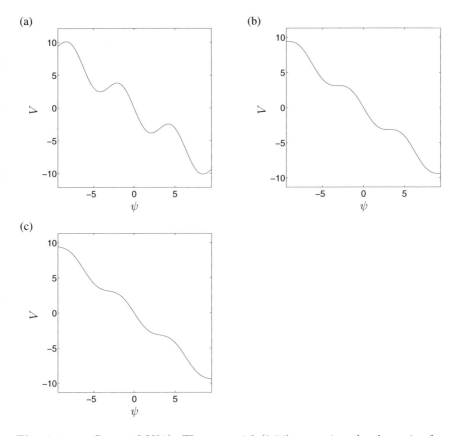

Fig. 2.5a–c. *Potential $V(\psi)$:* The potential (2.54) governing the dynamics for vanishing noise amplitude Q is plotted for supercritical ($I = 2 > \Omega = 1$, (**a**)), critical ($I = \Omega = 1$, (**b**)) and subcritical ($I = 0.8 < \Omega = 1$, (**c**)) stimulation intensity.

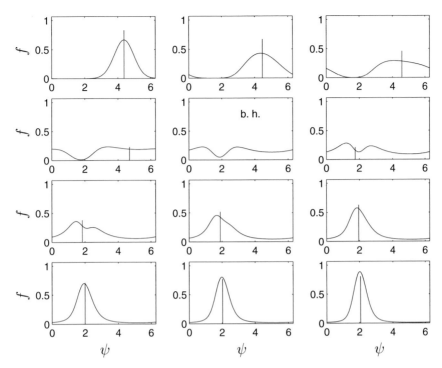

Fig. 2.6. *Supercritical stimulation* $(I = 2 > \Omega = 1)$*:* The dynamics of the density $f(\psi, t)$ was revealed by numerically integrating Fokker–Planck equation (2.30). The initial distribution $f(\psi, t_B)$ is a Gaussian distribution fitted in the interval $[0, 2\pi]$, so that boundary conditions (2.34) and (2.35) are fulfilled. $f(\psi, t_B)$ has standard deviation $\sigma = 0.6$, and its peak is located in $\varphi_B = 4.376$. Noise amplitude is $Q = 0.4$. $f(\psi, t)$ is plotted after equidistant time steps. Time increases from the left upper corner to the right lower corner. The ensemble variable Z is indicated by the thin vertical line which is located in $\psi = \varphi$ and is of length R, where φ and R are ensemble phase and amplitude, respectively. At the critical stimulation time $T_{\text{crit}} = 0.791$ the ensemble reaches the black hole (*as shown in the plot denoted by* '*b.h.*').

imum of the potential $V(\psi)$. The plot denoted by 'b.h.' shows the typical double peak-distribution in the black hole.

Figure 2.7 presents different aspects of the intimate relationship between desynchronization and phase singularity. For the simulation shown in Fig. 2.6 the motion of the ensemble variable $Z(t)$ is depicted in Fig. 2.7a. Starting at t_B ('begin') the center of mass approaches the origin of the Gaussian plane, i.e. the black hole. As a consequence of the double peak-distribution in the black hole the ensemble amplitude vanishes (Fig. 2.7b). In the vicinity of the black hole a jump of the ensemble phase φ occurs (Fig. 2.7c). The difference of the ensemble phase close before and close behind the black hole is $\approx \pi$.

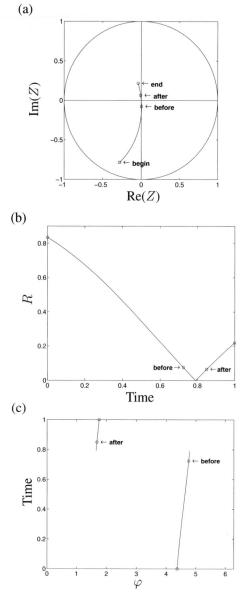

Fig. 2.7a–c. *Desynchronization and phase singularity:* Ensemble variable $Z(t)$ (**a**), ensemble amplitude $R(t)$ (**b**) and ensemble phase $\varphi(t)$ (**c**) are plotted for the same simulation as in Fig. 2.6. In (**a**) the trajectory of $Z(t)$, the center of mass of the ensemble's distribution $f(\psi, t)$, is plotted in the Gaussian plane (cf. (2.33)). In all plots small circles indicate four different timing points: begin and end of the stimulation as well as two timing points immediately before and after the phase singularity. The latter occurs at the critical stimulation time $T_{\text{crit}} = 0.791$ when Z and R vanish, and φ undergoes a jump.

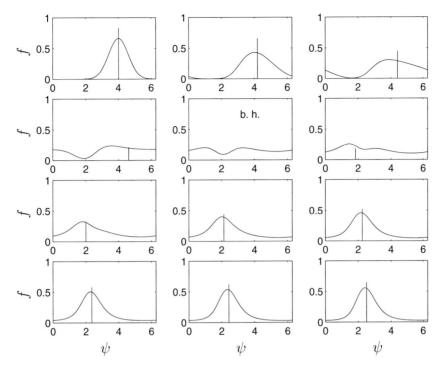

Fig. 2.8. *Critical stimulation* $(I = \Omega = 1)$*:* Plots display the dynamics of the density $f(\psi, t)$ during critical stimulation. The initial density $f(\psi, t_B)$ has standard deviation $\sigma = 0.6$, and its peak is located in $\varphi_B = 4.009$. All other parameters and format as in Fig. 2.6. At the critical stimulation time $T_{\text{crit}} = 1.553$ the ensemble's center of mass vanishes (*as shown in the plot denoted by 'b.h.'*).

As shown in Figs. 2.8 and 2.9 for critical $(I = \Omega)$ as well as subcritical $(I < \Omega)$ stimulation intensity a similar straddling process is observed. For a detailed comparison I refer to a former study (Tass 1996a).

Figures 2.10–2.12 show how the type of resetting depends on the stimulation duration. In particular, these figures illustrate why in the model under consideration a black hole is related with a transition from type 1 to type 0 resetting. A series of simulations of Fokker–Planck equation (2.30) was carried out with φ_B equally spaced in $[0, 2\pi]$, where φ_B is the location of the peak of the initial distribution $f(\psi, t_B)$. All other model parameters were kept constant throughout the series of simulations.

Figure 2.10 displays the trajectories of the ensemble variable Z of the series of simulations. The beginning of each trajectory is indicated by a small circle, where the filled circle belongs to the trajectory which passes through the black hole, i.e. the origin of the Gaussian plane. The ends of the trajectories are connected by means of a cubic spline. If one runs a simulation with the same model parameters and stimulation duration the trajectory ends on

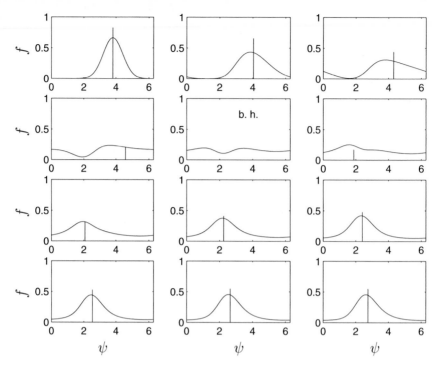

Fig. 2.9. *Subcritical stimulation* $(I = 0.8 < \Omega = 1)$: During stimulation the dynamics of the density $f(\psi, t)$ is governed by Fokker–Planck equation (2.30). $f(\psi, t_B)$ has standard deviation $\sigma = 0.6$, and $\varphi_B = 3.804$. Noise amplitude is $Q = 0.4$. $f(\psi, t)$ is plotted after equidistant time steps. All other parameters and format as in Fig. 2.6. The black hole is reached at the critical stimulation time $T_{crit} = 0.4875$ (*as shown in the plot denoted by 'b.h.'*).

this cubic spline, no matter which initial value φ_B was chosen. This was tested by integrating (2.30) numerically for 500 different values of φ_B. Hence, the cubic spline consists of the ending points of all possible simulations obtained by varying the initial ensemble phase φ_B.

At the end of Sect. 2.4.1 it was pointed out that $f_{st}(\psi)$ from (2.39) is a global attractor, so that with increasing time $f(\psi, t)$ tends towards $f_{st}(\psi)$, no matter from which $f(\psi, t_B)$ the dynamics of Fokker–Planck equation (2.30) starts. Accordingly, the ending points of all trajectories in the sixth plot of Fig. 2.10 coincide.

Let us consider the stimulation's impact on the ensemble phase. To this end two notations will be introduced: The ensemble phase at the beginning and at the end of the stimulation will be denoted by φ_B and φ_E, respectively. φ_B will be called old phase, and φ_E new phase. The relationship between old ensemble phase φ_B and new ensemble phase φ_E can topologically be described with the so-called *winding number* (cf. Milnor 1965, Winfree 1980).

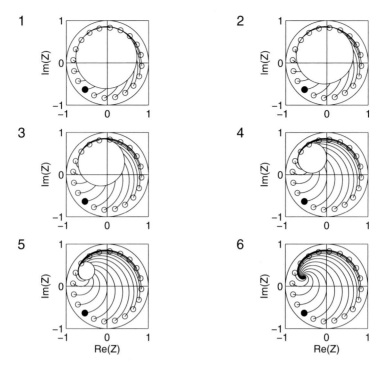

Fig. 2.10. *Winding numbers:* 40 simulations of Fokker–Planck equation (2.30) were performed with φ_B, i.e. the location of the peak of the intial distribution $f(\psi, t_B)$, equally spaced in $[0, 2\pi]$. All other parameters were the same as for the simulation of Fig. 2.8. This plot and Figs. 2.11 and 2.12 refer to these 40 simulations, where in this figure only half of the simulations are shown in order not to overload the plots. The simulations stopped after different stimulation durations T (increasing from plot 1 to 6). In each plot the ends of the trajectories are connected by means of a cubic spline to illustrate the corresponding winding number. The latter equals 1 in the first three plots, whereas in plots 4 to 6 it is equal to 0.

The winding number is the net number of times φ_E runs through a full cycle from 0 to 2π while the corresponding φ_B runs through one full cycle. In other words, the winding number tells us how often the trajectories' ends run around the origin as one follows the path given by the small circles starting and ending at the black dot in Fig. 2.10. So, the winding number is +1 if the cubic spline runs once around the origin of the Gaussian plane (with the same orientation as φ_B, i.e. the small cicles). If the cubic spline does not run around the origin, the winding number is 0. Negative signs of the winding number indicate that the cubic spline runs around the origin in the opposite direction. Obviously in the first three plots of Fig. 2.10 the winding number is +1, whereas in the other half of the plots it is 0. The transition of the resetting type takes place when the trajectory starting at the filled circle reaches the origin of the Gaussian plane, i.e. the black hole (cf. Fig. 2.7).

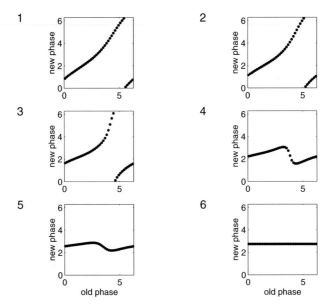

Fig. 2.11. *Phase resetting curves:* Plots show new phase, i.e. the ensemble phase at the end of the stimulation, plotted over old phase, i.e. the ensemble phase at the beginning of the stimulation, for the 40 simulations from Fig. 2.10. The old phase of a single simulation is the ensemble phase of the beginning of the corresponding trajectory of Z, whereas the new phase is the phase of the end of the trajectory. Stimulation durations and arrangement of the plots correspond to those in Fig. 2.10.

This transition can additionally be illustrated by means of the corresponding *phase resetting curves*, where φ_E is plotted over φ_B for given stimulation parameters, i.e. intensity and duration (Fig. 2.11). In this way we can easily determine the net number of times φ_E runs through a full cycle (modulo 2π) while φ_B runs through one full cycle. Clearly, between the third and the fourth plot there is a transition from type 1 resetting to type 0 resetting.

In Fig. 2.10 it was already shown that the transition of the type of resetting is connected with a singularity of the ensemble phase. At a critical stimulation duration T_{crit} the singularity occurs. For a stimulation duration which is smaller or larger than T_{crit} one observes type 1 or type 0 resetting, respectively. The phase singularity is additionally illustrated by plotting the evolution of the ensemble phase in time in Fig. 2.12.

Let us summarize the results of the investigation of the stimulation induced desynchronization of a synchronized ensemble of phase oscillators:

1. If a stimulus of a given stimulation intensity I is administered at a critical initial ensemble phase φ_B for a critical duration T_{crit} the ensemble's synchronization is annihilated.

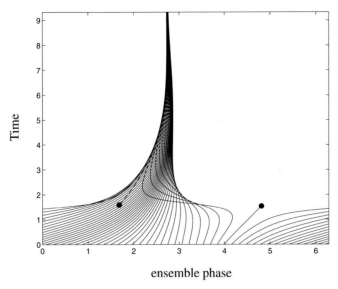

Fig. 2.12. *Ensemble phase:* Time is plotted over ensemble phase φ for the 40 simulations from Fig. 2.10. At the critical stimulation duration $T_{\mathrm{crit}} = 0.791$ a phase singularity occurs as indicated by the two black dots (cf. Fig. 2.7c).

2. When the stimulation duration exceeds T_{crit} there is a transition from type 1 resetting to type 0 resetting.

Finally it should be mentioned that for the stimulation mechanism (2.52) the plots of the trajectories of the ensemble variable Z and the ensemble phase φ as well as the phase resetting curves are qualitatively the same, respectively, no matter whether the stimulation intensity is supercritical $(I > \Omega)$, critical $(I = \Omega)$ or subcritical $(I < \Omega)$ (Tass 1996a).

2.7.2 Stimulation Induced Frequency Shift

Stimulation induced frequency shifts have to be analyzed in order to understand the relationship between the dynamics of a single oscillator and the ensemble's collective dynamics. According to (2.26) without stimulation the mean frequency of a single oscillator equals Ω. In this section the mean frequency of a single oscillator during stimulation will be determined.

The phase of a single oscillator can be considered as a particle moving in the potential $V(\psi)$ from (2.54) under the influence of noise (cf. Fig. 2.5). Thus, the oscillator's mean period is the time it takes the particle in the mean to move from $\psi = 0$ to $\psi = 2\pi$. This time is called *mean first passage time* (see, e.g., Haken 1977, 1983, Gardiner 1985, Risken 1989). To derive the formula for the mean first passage time appropriate boundary conditions have to be chosen. To this end a reflecting barrier at $\psi = 0$ and an absorbing barrier at

$\psi = 2\pi$ are assumed. Obviously we are allowed to assume a reflecting barrier in 0 provided the noise amplitude Q is small enough (i.e. $Q \ll \Omega$), so that the particle does not move upwards in the potential. This is not a drawback because one is mainly interested in small noise amplitudes as large noise amplitudes would quickly annihilate the ensemble's synchronized oscillation without any stimulation.

According to Gardiner (1985) ((5.2.160), p. 139) the mean first passage time for passing through one cycle reads

$$\bar{T}(I) = \frac{2}{Q} \int_0^{2\pi} \frac{\mathrm{d}\mu}{\Lambda(\mu)} \int_0^{\mu} \Lambda(\eta)\,\mathrm{d}\eta \ , \qquad (2.55)$$

with

$$\Lambda(\mu) = \exp\left[-\frac{2}{Q}V(\mu)\right] = \exp\left[\frac{2}{Q}\int_0^{\mu} K(\xi)\,\mathrm{d}\xi\right] \qquad (2.56)$$

and with $K(\psi)$ and $V(\psi)$ from (2.29) and (2.54). Accordingly, the mean first passage time given by (2.55) depends on the noise amplitude Q, the eigenfrequency Ω and stimulation parameters. For the simple stimulation mechanism (2.52) the intensity I is the only stimulation parameter. It should be stressed that $\bar{T}(I)$ is constant throughout the whole stimulation. As $\bar{T}(I)$ is nothing but the mean period of a single oscillator the *mean frequency of a single oscillator during stimulation* can be written in the form

$$\bar{\Omega}(I) = \frac{2\pi}{\bar{T}(I)} \ . \qquad (2.57)$$

Figure 2.13 illustrates how $\bar{\Omega}$ depends on I for given noise amplitude Q and eigenfrequency Ω. Of course, without stimulation there is no frequency shift. With increasing intensity the mean frequency decreases. As a consequence of the presence of noise there is no abrupt transition from subcritical to supercritical values of the stimulation intensity. However, for larger values of I the oscillators are practically stopped.

Figure 2.13 points out that we have to distinguish between the motion of the probability distribution $f(\psi, t)$ and the motion of a single phase oscillator. In Sect. 2.7.1 it was shown that a well-timed stimulus of suitable intensity induces a straddling process of $f(\psi, t)$, so that after a critical stimulation duration T_{crit} the ensemble is desynchronized. Straddling of the distribution occurs for subcritical, critical and supercritical intensity (cf. Figs. 2.6, 2.8 and 2.9). According to (2.57) for sufficiently high stimulation intensity in the mean the oscillators are stopped, two straddling subpopulations evolve in a similar way as described by Winfree (1980) for the deterministic model (cf. Fig. 2.4). One encounters a different scenario for subcritical intensity: The oscillators continue rotating with non-vanishing mean frequency $\bar{\Omega}(I)$. Nonetheless, stimulation gives rise to a rearrangement of the oscillators' mutual phase differences, so that at time T_{crit} in the black hole there is no net synchronization any more.

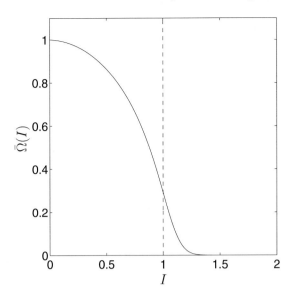

Fig. 2.13. *Stimulation induced frequency shift:* The mean frequency of a single oscillator $\bar{\Omega}(I)$ is plotted over stimulation intensity I for $\Omega = 1$ and $Q = 0.1$ according to (2.57). The vertical dashed line indicates the critical stimulation intensity $I = \Omega = 1$.

2.7.3 Stimulation Mechanism with Higher Harmonics

This section is about additional stimulation induced phenomena which are important from the experimentalist's point of view. Phenomena of this kind are due to harmonics of the stimulation mechanism S, a simple example of which is given by

$$S(\psi, I_1, I_2) = I_1 \cos \psi + I_2 \cos(2\psi) \tag{2.58}$$

(cf. (2.28)). In this case

$$V(\psi) = -\int_0^\psi K(\xi)\, \mathrm{d}\xi = -\Omega\psi - I_1 \sin\psi - \frac{I_2}{2}\sin(2\psi) \tag{2.59}$$

is the potential governing the dynamics in the noise-free setting. Inserting (2.58) into (2.47) one obtains the Fourier transformed Fokker–Planck equation

$$
\begin{aligned}
\frac{\partial \hat{f}(k,t)}{\partial t} = {}&- \left(\frac{Q}{2}k^2 + \mathrm{i}\Omega k \right) \hat{f}(k,t) \\
&- \mathrm{i}\frac{k}{2}\Big\{ I_1 \left[\hat{f}(k-1,t) + \hat{f}(k+1,t) \right] \\
&\qquad + I_2 \left[\hat{f}(k-2,t) + \hat{f}(k+2,t) \right] \Big\}.
\end{aligned}
\tag{2.60}
$$

Before the transient features of the dynamics of (2.60) will be investigated, let us first dwell on the attractor of this dynamics. The latter is determined by the stationary solution f_{st}. According to Sect. 2.4.1 with increasing time every distribution $f(\psi, t)$ finally decays to $f_{st}(\psi)$. Hence, inserting f_{st} into (2.33) yields the *attractor of the ensemble variable*

$$Z_{st} = R_{st} \exp\left(i\varphi_{st}\right) = \int_0^{2\pi} f_{st}(\psi) \exp(i\psi) \, d\psi . \qquad (2.61)$$

In order to point out the impact of harmonics of the stimulation S three different stimulation mechanisms will be compared. In all three cases Ω is set equal to 1, whereas the parameters I_1 and I_2 of (2.58) are chosen to be

1. $I_1 = 1, I_2 = 0$: This mechanism was already discussed in detail in the former section. For suitable initial ensemble phase and stimulation duration it causes a straddling mechanism of the distribution f. This straddling mechanism forces f into a black hole which is associated with a transition from type 1 resetting to type 0 resetting.
2. $I_1 = 0, I_2 = 1$,
3. $I_1 = I_2 = 1$.

Figures 2.14–2.16 show the potential V and the corresponding stationary solution f_{st} of the three stimulation mechanisms determined according to (2.59) and (2.39). Figure 2.14 displays f_{st} and V for critical stimulation intensity $I_1 = 1$ and $I_2 = 0$. f_{st} has one peak which is located in the neighbourhood of the inflexion point of the potential V. This peak is larger or smaller for supercritical or subcritical stimulation intensity ($I_1 > \Omega$ or $I_1 < \Omega$), respectively. For supercritical I_1 the potential V exhibits a minimum, whereas for subcritical I_1 the potential V has neither a minimum nor an inflexion point. The plots for super- and subcritical I_1 are not shown as they are only gradually different to that one for critical I_1. For $I_1 = 0, I_2 = 1$ one observes a symmetric double peak distribution with consequently vanishing ensemble amplitude R_{st} (Fig. 2.15). Figure 2.15 refers to the stimulation mechanism with $I_1 = 1, I_2 = 1$. As a result of the two minima of V (modulo 2π) f_{st} has two peaks. In contrast to the second stimulation mechanism in this case f_{st} is not symmetrical. For this reason the ensemble amplitude does not vanish: $Z_{st} \neq 0$.

The ensemble dynamics during stimulation was investigated by integrating (2.60) numerically. To analyze the impact of the initial ensemble phase on the outcome of the stimulation a series of simulations with equally spaced φ_B was performed as in Sect. 2.7 (cf. Fig. 2.10). To this end as initial distribution a Gaussian distribution with standard deviation 0.6 and mean φ_B was fitted in the interval $[0, 2\pi]$, so that conditions (2.34) and (2.35) were fulfilled. Simulations were carried out for 40 different values of φ_B. In Figs. 2.10 and 2.17 only half of these 40 simulations are plotted in order not to overload the presentation.

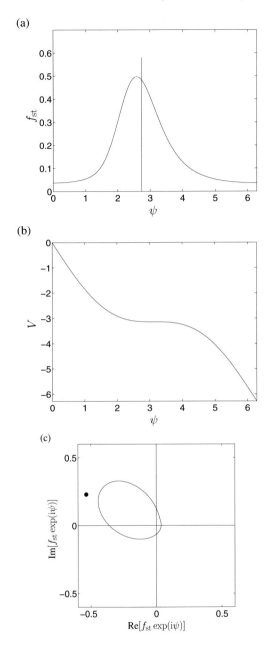

Fig. 2.14a–c. *Attractor of the distribution f for* $I_1 = 1, I_2 = 0$ $(\Omega = 1)$: Potential $V(\psi)$ **(b)** and corresponding stationary solution $f_{st}(\psi)$ **(a)** are plotted. In **(a)** Z_{st} from (2.61) is indicated by the vertical line which is located in $\psi = \varphi_{st}$ and is of length R_{st}. In **(c)** $f_{st} \exp(i\psi)$ is plotted in the Gaussian plane, and the attractor of the ensemble variable Z_{st} is indicated by the black dot.

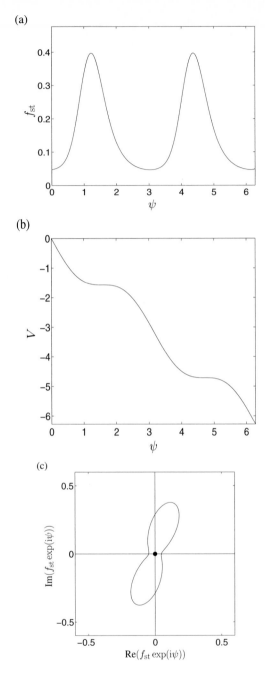

Fig. 2.15a–c. *Attractor of the distribution f for* $I_1 = 0, I_2 = 1$ $(\Omega = 1)$: Plots show potential $V(\psi)$ and corresponding stationary solution $f_{st}(\psi)$. Same format as in Fig. 2.14. Due to the symmetric distribution f_{st} the ensemble variable Z_{st} vanishes.

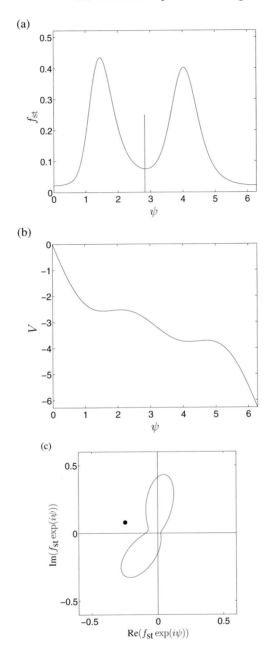

Fig. 2.16a–c. *Attractor of the distribution f for $I_1 = I_2 = 1$ $(\Omega = 1)$: Plots show potential $V(\psi)$ and corresponding stationary solution $f_{st}(\psi)$. Same format as in Fig. 2.14. f_{st} is not symmetric and, thus, Z_{st} does not vanish.*

The simulations with $I_1 = I = 1, I_2 = 0$ are displayed in Figs. 2.10–2.12. For the stimulation mechanisms with $I_1 = 1, I_2 = 0$ and $I_1 = I_2 = 1$ the trajectories' attractors Z_{st} are not located in the origin of the Gaussian plane, respectively (cf. Figs. 2.14 and 2.16). For this reason in both cases there is one trajectory which passes through the black hole, i.e. the origin of the Gaussian plane, after a critical stimulation duration T_{crit} (Figs. 2.10 and 2.17). Passing through the black hole a transition from type 1 to type 0 resetting occurs (Figs. 2.11 and 2.18). This transition is due to a singularity of the ensemble phase φ (Figs. 2.12 and 2.19).

In contrast to these stimulation mechanisms the mechanism given by $I_1 = 0, I_2 = 1$ gives rise to an attractor of the ensemble variable located in the origin of the Gaussian plane, i.e. $Z_{st} = 0$ (Fig. 2.15). With increasing time all trajectories tend to this attractor, but no trajectory passes through this attractor (Fig. 2.20). Consequently, this type of stimulation causes only type

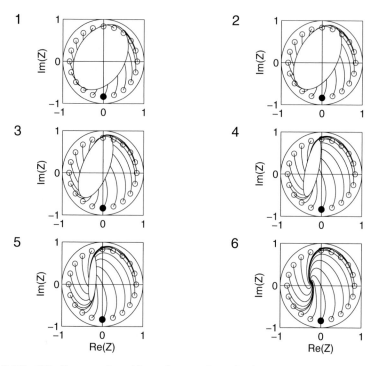

Fig. 2.17. *Winding numbers* $(I_1 = I_2 = 1, \Omega = 1)$: A series of simulations of the Fokker–Planck equation (2.30) was performed with φ_B, i.e. the location of the peak of the intial distribution $f(\psi, t_B)$, equally spaced in $[0, 2\pi]$. Same format as in Fig. 2.10. Z_{st}, the ensemble variable's global attractor is not located in the origin of the Gaussian plane (cf. Fig. 2.16). Correspondingly one trajectory passes through the origin. The winding number equals 1 in the first three plots, whereas in plots 4 to 6 it is equal to 0.

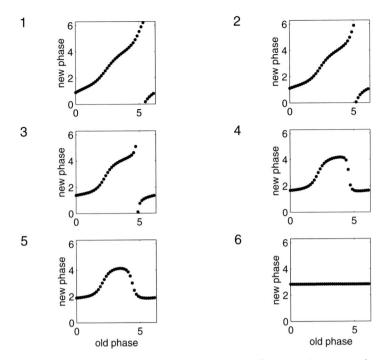

Fig. 2.18. *Phase resetting curves* ($I_1 = I_2 = 1, \Omega = 1$)*:* Plots show new phase, i.e. the ensemble phase at the beginning of the stimulation, plotted over old phase, i.e. the ensemble phase at the end of the stimulation, for the 40 simulations from Fig. 2.17. Same format as in Fig. 2.11.

1 resetting (Fig. 2.21) since the ensemble phase performs a continuous motion (modulo 2π) without any singularities (Fig. 2.22).

Comparing Figs. 2.10, 2.17, and 2.20 illustrates how the trajectories of Z of the 'combined' stimulation mechanism ($I_1 = I_2 = 1$) compromise on the two different (isolated) stimulation mechanisms ($I_1 = 1, I_2 = 0$ and $I_1 = 0, I_2 = 1$).

The stimulation mechanism with parameters $I_1 = I = 1, I_2 = 0$ gives rise to a straddling process as shown in Fig. 2.6: An initial peak of the distribution $f(\psi, t)$ straddles so that two smaller peaks evolve which finally melt. To understand the impact of a harmonic of the stimulation mechanism on the dynamics of the distribution one should recall how higher order stimulation influences f_{st}, the attractor of the distribution. As illustrated in Figs. 2.14–2.16 f_{st} may have more than one peak provided the stimulation mechanism contains higher order terms of sufficient strength (e.g., $I_2 \neq 0$). Accordingly due to a second order stimulation term an initial peak of $f(\psi, t)$ may split into two smaller peaks which do not melt again (Figs. 2.23 and 2.24). In order to stress that several peaks of the distribution f may arise due to higher order stimulation terms this type of desynchronization process was called a

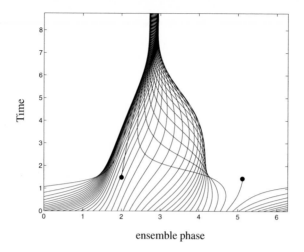

ensemble phase

Fig. 2.19. *Ensemble phase* ($I_1 = I_2 = 1, \Omega = 1$): Time is plotted over ensemble phase φ for the series of simulations from Fig. 2.10. At the critical stimulation duration $T_{\text{crit}} = 1.456$ a phase singularity shows up as indicated by the two black dots. Same format as in Fig. 2.12.

splitting mechanism in contrast to the straddling mechanism observed for $I_1 = I = 1, I_2 = 0$ (Tass 1996b).

2.8 Firing Patterns

The former sections revealed that apart from ensemble phase and ensemble amplitude the shape of the distribution $f(\psi, t)$ may change markedly during stimulation. In this section the consequences of these changes will be considered from the experimentalists' point of view. First it has to be clarified which quantity is observed in a typical stimulation experiment. To this end let us consider an ensemble of periodically spiking or bursting neurons. Approximating the dynamics of populations of spiking or bursting neurons with networks of phase oscillators will be explained in detail in Sect. 3.4.1. For the time being a neuron is modeled by means of a phase oscillator generating a spike (or burst) whenever its phase equals $\rho + k2\pi$, where ρ is a constant parameter, and k is an integer. As already mentioned in Sect. 2.3 the relative number of oscillators with a phase within the interval $\psi \dots \psi + \mathrm{d}\psi$ at time t is given by $f(\psi, t)\,\mathrm{d}\psi$. Therefore

$$p(t) = f(\rho, t) \tag{2.62}$$

provides us with the density of firing (or bursting) neurons at time t which will briefly be denoted as *firing density* (cf. Tass 1996a, 1996b). In this way $p(t)$ serves as a link between the experimentally observed firing of a population

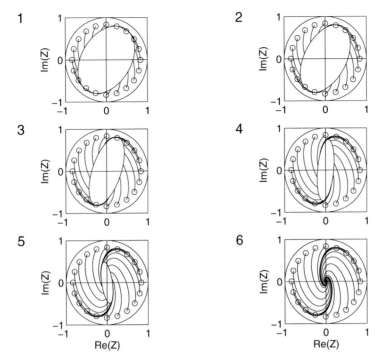

Fig. 2.20. *Winding numbers* $(I_1 = 0, I_2 = 1, \Omega = 1)$: Plots show a series of simulations of the Fokker–Planck equation (2.30) with φ_B equally spaced in $[0, 2\pi]$. Same format as in Fig. 2.10. All trajectories tend to the origin which is the global attractor Z_{st} (cf. 2.15). There is no trajectory passing through the origin. In all plots the winding number equals 1.

of neurons and the ensemble's dynamics as described, for instance, by means of the ensemble variable Z.

In order to avoid misunderstandings we should distinguish between a burst of a single neuron and an ensemble burst. A burst of a single (bursting) neuron occurs whenever the neuron's phase equals $\rho + k2\pi$. In contrast to this an *ensemble burst* denotes a group of coincident spikes and bursts generated by single spiking and bursting neurons. Therefore an ensemble burst corresponds to a peak of the firing density $p(t)$. In this chapter the ensemble burst will briefly be denoted as burst.

Below $\rho = 0$ will be chosen. For non-vanishing ρ one may introduce the shifted phase

$$\psi' = \psi - \rho \,, \tag{2.63}$$

so that the stimulation mechanism S from (2.28) takes the form

$$S(\psi', \{I_m\}, \{\gamma'_m\}) = \sum_{m=1}^{\infty} I_m \cos(m\psi' + \gamma'_m) \text{ with } \gamma'_m = \gamma_m + m\rho \,. \tag{2.64}$$

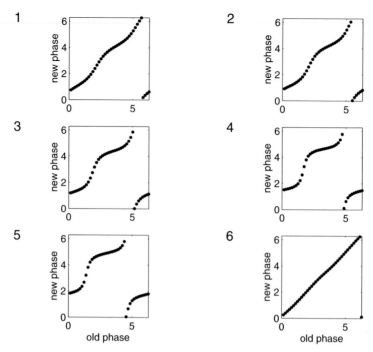

Fig. 2.21. *Phase resetting curves* ($I_1 = 0, I_2 = 1, \Omega = 1$)*:* Plots show new phase, i.e. the ensemble phase at the beginning of the stimulation, plotted over old phase, i.e. the ensemble phase at the end of the stimulation, for the simulations from Fig. 2.17. All plots exhibit type I resetting. Same format as in Fig. 2.11.

This means that for nonvanishing ρ one simply has to adjust the stimulation parameters by setting $\gamma_m \to \gamma'_m$. In this way taking into account $\rho \neq 0$ the model neuron generates a spike or burst whenever its transformed phase ψ' equals $k2\pi$ (with integer k).

Typical stimulation induced patterns of the firing density p are displayed in Figs. 2.25–2.27. For the three different stimulation mechanisms analyzed in the former sections a series of simulations of the Fokker–Planck equation (2.30) was performed. In these series the stimulus was administered at subsequent timing points, so that in all series the initial ensemble phase was equally spaced. Uppermost plots of Figs. 2.25 and 2.26 show suitably timed simulations which force the ensemble into a black hole. Correspondingly, the ensemble amplitude R vanishes at the end of the stimulation. According to (2.50) it remains equal to 0 after stimulation. Comparing both simulations illustrates how stimulation terms of higher order excite higher order Fourier modes of the distribution $f(\psi, t)$. In particular, the second order term ($I_2 = 1$) gives rise to a pronounced double-peak distribution (see uppermost plot in Fig. 2.26). Put otherwise: A *black hole* in the sense of a vanishing ensemble

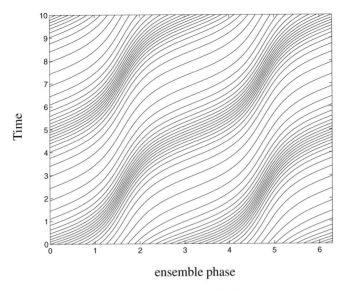

Fig. 2.22. *Ensemble phase* ($I_1 = 0, I_2 = 1, \Omega = 1$): Time is plotted over ensemble phase φ for the series of simulations from Fig. 2.10. No phase singularity occurs because no trajectory passes through the black hole, i.e. the origin of the complex plane (cf. Fig. 2.17). Same format as in Fig. 2.12.

variable (i.e. $Z = 0$) need not be connected with a fully desynchronized state, i.e. a uniform distribution $f(\psi, t) \approx 1/(2\pi)$.

As illustrated in Figs. 2.25 and 2.26 an additional severe consequence of higher order terms of the stimulation is *burst splitting* (Tass 1996b). The latter means that bursts of the firing density $p(t)$ are splitted into two or more smaller bursts. For this reason from the experimentalist's viewpoint the detection of the ensemble phase becomes involved. This fact should be discussed in more detail because it is of crucial importance for carrying out stimulation experiments. In order to find out appropriate stimulation parameters (stimulation intensity, duration, and initial ensemble phase) experimentalists perform series of stimulation experiments as in Figs. 2.25 and 2.26. In each series, for instance, stimulation intensity and duration are kept constant, while the initial ensemble phase is systematically varied (for a review see Winfree 1980). In all trials the ensemble phase before and after stimulation is detected, and the type of resetting is determined by plotting one over the other. According to the results of this resetting analysis one stimulation parameter, e.g. duration, is modified for running the subsequent series. In this way iteratively carrying out several series of experiments with modified stimulation duration the experimentalist may detect the crtical duration T_{crit} which is connected with a transition from type 1 to type 0 resetting (cf. Figs. 2.11, 2.18 and 2.21). The appropriate initial phase is given by the trial associated with the singularity in the phase resetting curve.

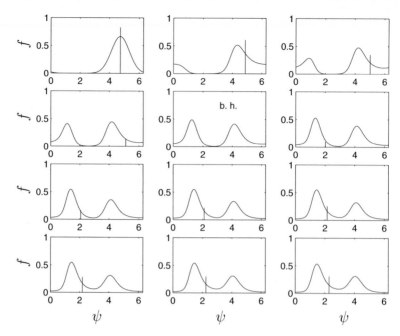

Fig. 2.23. *Splitting mechanism* ($I_1 = 1, I_2 = 1, \Omega = 1$): Plots show the dynamics of the distribution $f(\psi,t)$ during stimulation. $f(\psi,0)$ has mean $\varphi_B = 4.565$ and standard deviation $\rho = 0.6$. Same format as in Fig. 2.6. The vertical line indicates the ensemble variable as in Fig. 2.6. The initial peak splits into two peaks which do not melt after the ensemble has passed through the black hole.

The determination of the ensemble phase is typically based on simple algorithms detecting peak or onset of ensemble bursts. Unfortunately these algorithms fail when burst splitting occurs. This is illustrated in Fig. 2.26, where the thick vertical lines indicate the timing points when the ensemble phase φ equals $\rho(= 0)$ (modulo 2π), i.e. when the center of mass of the ensemble fires. Considering all trials within a single series one realizes that there is no fixed relation between these timing points and the burst peaks or onsets. However, the iterative experimental approach sketched above does only work provided it is possible to detect the ensemble phase in a reliable way. In Chap. 8 a different method for detecting phases of bursts will be presented.

In the former section it turned out that within a finite period of time the stimulation mechanism with $I_1 = 0, I_2 = 1$ cannot force the ensemble variable Z into the black hole, i.e. the origin of the Gaussian plane in Fig. 2.20. Accordingly in none of the trials of Fig. 2.27 the ensemble amplitude R vanishes at the end of the stimulation. However, as a consequence of the stimulation term of second order one encounters burst splitting.

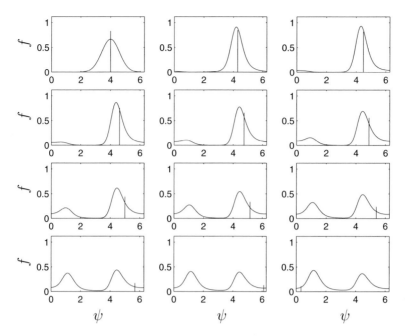

Fig. 2.24. *Splitting mechanism* $(I_1 = 0, I_2 = 1, \Omega = 1)$: The dynamics of the distribution $f(\psi, t)$ during stimulation is shown, where $\varphi_B = 4.565$ and $\rho = 0.6$. Same format as in Fig. 2.6. As a consequence of the stimulation term of second order two peaks of the distribution evolve. The ensemble does not pass through a black hole (cf. Fig. 2.20).

2.9 Summary and Discussion

A large number of theoretical studies addressed the issue of phase resetting (for a review see Winfree 1980). These studies refered to deterministic models where, in particular, the impact of stimulation on a single oscillator was investigated. However, noise is inevitable in physiological systems, and, additionally, in such systems one typically encounters populations of synchronously active oscillators (see, e.g., Cohen, Rossignol and Grillner 1988, Steriade, Jones, and Llinás 1990). For this reason in this chapter we investigate phase resetting of a population of oscillators in the presence of noise by means of an ensemble model (Tass 1996a, 1996b).

The ensemble consists of identical phase oscillators which do not interact via deterministic couplings. Thus, the ensemble's collective dynamics is exclusively determined by the interplay of stimulation and noise governed by the Fokker–Planck equation (2.30) for $f(\psi, t)$, which serves as a distribution function of the ensemble. One can describe the ensemble's state in a condensed way by focussing on the dynamics of the center of mass of the

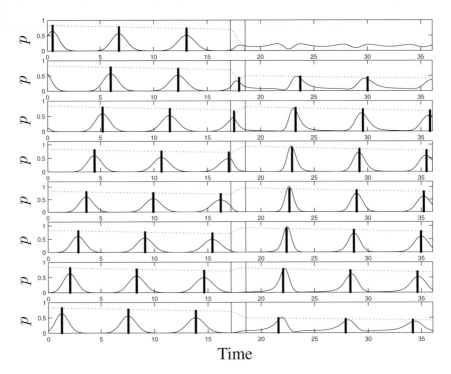

Fig. 2.25. *Firing patterns:* Fokker–Planck equation (2.30) was integrated numerically for $I_1 = 1, I_2 = 0, \Omega = 1$. The firing density $p(t) = f(0, t)$ is plotted over time. In all simulations the initial distribution $f(\psi, 0)$ is a Gaussian distribution with standard deviation $\sigma = 0.6$ fitted in the interval $[0, 2\pi]$. From the uppermost plot to the lowest plot the initial value of the ensemble phase is increased by steps of $\pi/4$ in order to illustrate the impact of a modification of the initial ensemble phase on the stimulation's outcome. In each plot begin and end of the stimulation are indicated by two vertical lines. The cluster amplitude R (*dotted line*) serves to estimate the extent of the ensemble's net synchronization. The thick vertical lines indicate the center of mass of the ensemble burst as they are located in timing points where $\varphi(t)$ equals $k2\pi$ (with integer k). The uppermost plot shows a desynchronizing stimulation.

distribution $f(\psi, t)$. With this aim in view macroscopic variables, namely the ensemble amplitude R and the ensemble phase φ, were introduced by (2.33).

The stimulus' impact on a single oscillator depends on the oscillator's phase and on stimulation parameters, for instance, the stimulation intensity (cf. (2.28)). Every single oscillator is affected by the stimulus. This leads to macroscopic changes of the ensemble's state of synchronization which are estimated with the macrovariables R and φ. The ensemble is forced into a black hole provided a stimulus of a given stimulation intensity is administered at a critical initial ensemble phase for a critical duration T_{crit}. Within the black hole the ensemble amplitude vanishes and the ensemble phase exhibits

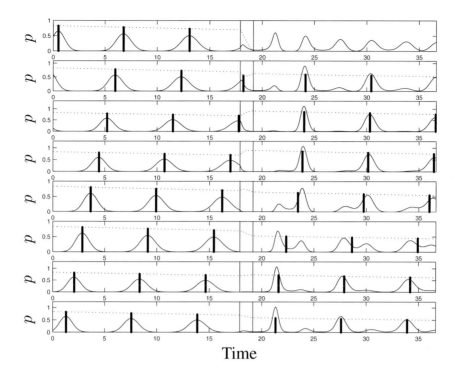

Fig. 2.26. *Firing patterns:* A series of simulations of (2.30) with equally spaced initial ensemble phase was performed for $I_1 = 1, I_2 = 1, \Omega = 1, \sigma = 0.6$. Format and arrangement of the plots as in Fig. 2.25. In the uppermost plot the stimulation forces the ensemble into the black hole. Due to the stimulation term of second order ($I_2 = 1$) burst splitting occurs.

a phase singularity. Furthermore, a transition from type 1 to type 0 resetting occurs when the stimulation duration exceeds T_{crit}. It is important to note that a black hole need not be associated with a fully desynchronized state, i.e. with a uniform distribution $f(\psi, t) = 1/(2\pi)$. As illustrated in Fig. 2.3 the distribution may exhibit two or more peaks, while its center of mass is located in the origin of the Gaussian plane, i.e. the black hole. In fact, stimulation mechanisms containing terms of higher order may excite higher order Fourier modes of the distribution $f(\psi, t)$ in this way giving rise to *burst splitting*. One has to take into account two important consequences of burst splitting: On the one hand algorithms detecting peak or onset of bursts commonly used for the analysis of experimental data are not appropriate for detecting the ensemble phase (cf. Fig. 2.26). On the other hand the ensemble's state of synchronization cannot sufficiently be described by only two variables, amplitude and phase.

Based on the investigation of one- and two-dimensional oscillator models it was claimed that one cannot obtain type 0 resetting curves from only a

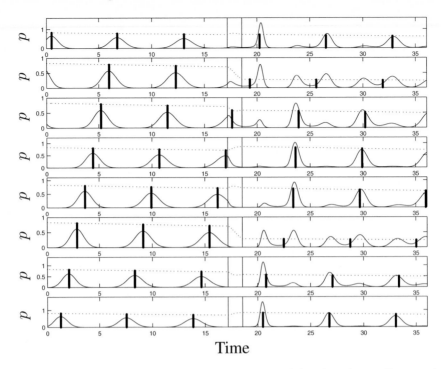

Time

Fig. 2.27. *Firing patterns:* A series of simulations of (2.30) with equally spaced initial ensemble phase was performed for $I_1 = 0, I_2 = 1, \Omega = 1, \sigma = 0.6$. Format and arrangement of the plots as in Fig. 2.25. Burst splitting occurs as a consequence of the stimulation term of second order ($I_2 = 1$). It is important to note that $p(t)$ is a firing *density* and, thus, may be larger than 1. However, the corresponding firing *probability* obtained by integration of $p(t)$ lies within the interval $[0, 1]$ since $f(\psi, t)$ is normalized according to (2.35).

phase stimulus (Murray 1989, p. 208). Actually, in contrast to the behavior of deterministic two-dimensional models (cf. Winfree 1980, Murray 1989) even subcritical stimulation of an ensemble of phase oscillators can cause type 0 resetting (Tass 1996a). Certainly, this raises the question as to whether synchronizing couplings might decrease the population's vulnerability to stimulation. In general, we have to expect that couplings of this kind may crucially determine in which way the population reacts to stimulation. Thus, the following chapters are dedicated to the patterns of synchrony arising due to oscillators' mutual interactions, and the stimulation induced changes of these patterns. In this way we shall become familiar with describing these patterns of synchrony by means of a few appropriate variables. To contribute to a better understanding of stimulation experiments Chap. 8 will be concerned with the issue of how to estimate these variables in experimental data.

3. Synchronization Patterns

3.1 Introductory Remarks

Chapter 2 was devoted to an ensemble of oscillators interacting via random forces. We studied the oscillators' spontaneous behavior and their reaction to stimulation. In this way we were able to investigate type 0 and type 1 resetting of an ensemble of oscillators. Moreover by encountering burst splitting we learned how difficult it may be to assign the ensemble's collective activity to a single phase value. However, in the ensemble scenario one important dynamical feature was completely missing: No self-organized patterns of synchronization occured. Actually, self-synchronized collective activity abounds in physiological systems (see, for instance, Steriade, Jones, Llinás 1988). For this reason it is not sufficient to model the oscillators' mutual interactions by means of random forces. Rather we have to take into account synchronizing couplings among the oscillators.

The main concern in this book is to study how stimulation affects synchronized oscillatory activity. With this aim in view in this chapter we become acquainted with several types of synchronized states emerging as a result of different coupling mechanisms. We shall emphasize the strong dependence of the synchronization frequency on the arrangement of the synchronized oscillators given by their mutual phase differences. This will prepare us for investigating transient synchronization processes which occur during and after stimulation.

To motivate the physiological relevance of different types of synchronization we shall dwell on synchronized neuronal activity in the visual cortex and its implications for pattern recognition and neural coding. Additionally, however, we shall discuss why the significance of complex synchronization patterns goes beyond the issue of neural coding.

3.2 Pattern Recognition

One of the central issues in neuroscience is the *binding problem*, i.e. the question of how widely distributed neuronal processing gets bound together to form a meaningful pattern of perception. In other words, during the act of perception the sensory input has to be separated into pieces which form patterns. For instance, single objects of the visual world can only be recognized if different objects are separated from each other and from the background. The process which joins together local information to form distinct objects is also called scene segmentation or figure-ground segregation (cf. Marr 1976, Treisman 1980, 1986, Singer 1989, Julesz 1991). As yet, the neurophysiological mechanism which realizes scene segmentation is open to debate. Several authors suggested a temporal neuronal firing code to be the physiological correlate of this segmentation process (cf. Hebb 1949, Milner 1974, Abeles 1982, Crick 1984, von der Malsburg and Schneider 1986, Shimizu, Yamaguchi, Tsuda, Yano 1985, Singer 1989). In particular, in-phase synchronization was suggested to be the neuronal firing code under discussion (Milner 1974, von der Malsburg and Schneider 1986). According to this hypothesis an object is recognized as a whole by synchronizing the firing activity of all neurons which encode local features belonging to this particular object. A variety of experimental findings in rabbit (Freeman 1975), cat (Gray and Singer 1987, Eckhorn et al. 1988), pigeon (Neuenschwander and Varela 1993) and monkey (Livingstone 1991, Kreiter and Singer 1992, Murthy and Fetz 1992, Eckhorn et al. 1993) supported this hypothesis.

Let me briefly sketch some of the experimental results. For more information I refer, for instance, to Singer and Gray (1995). Hubel and Wiesel (1959) found that local visual features are detected by single neurons with suitable receptive field properties. Neurons which are stimulated by similar local features form columns which are vertically arranged with respect to the surface of the cortex (Hubel and Wiesel 1962, 1963). When a column is stimulated by a moving light bar with suitable orientation and movement direction, a huge amount of neurons within this column synchronizes its firing activity with a frequency of about 40 Hz (Eckhorn et al. 1988, Gray and Singer 1989). Neurons with overlapping receptive fields tend to synchronize even if they are located in different areas of the cortex (Eckhorn et al. 1988, Gray and Singer 1989). Several experimental results indicate that the neurons' synchronization behavior reflects global stimulus features (Eckhorn et al. 1988, Gray and Singer 1989, Engel et al. 1990). For instance, synchronized neuronal firing is more efficiently induced by one long moving bar compared to two short bars with same orientation and movement direction (Gray et al. 1989).

It is important to note that synchronization in this context means that the neurons' oscillatory activity is in phase except for small mutual phase differences. Data analysis tools used in the experiments mentioned above are typically based on linear techniques such as auto-correlation and cross-correlation. Experimentalists use these tools in order to decide whether the oscillatory neuronal activity is synchronized or not. However, taking into account the behavior of populations of oscillators the idea suggests itself that besides in-phase synchronization other dynamical processes might play a role in scene segmentation, too. This will be discussed in the following section.

3.3 Clustering

The dynamics of networks of oscillators is the subject of a vast literature. A multitude of theoretical studies revealed that apart from in-phase synchronization and incoherent activity there is a variety of qualitatively different dynamical phenomena. Let us consider some of them:

1. *Splay-phase states:* The oscillators' phases are equally spaced in the interval $[0, 2\pi]$, i.e. the phase of the jth oscillator may be written as $\psi_j(t) = \phi(t + jT/N)$ for $j = 1, \ldots, N$ and some function $\phi(t + T) = \phi(t) + 2\pi$, where T is the period of the splay-phase state (Nicholis and Wiesenfeld 1992, Swift, Strogatz, Wiesenfeld 1992, Strogatz and Mirollo 1993).

2. *Attractor crowding:* Wiesenfeld and Hadley (1989) investigated an array of N coupled oscillators, where the number of stable limit cycles scaled as $(N-1)!$. For large enough N the basins of attraction crowd so tightly that a diffusive hopping among the many coexisting attractors occurs.

3. *Clustering:* A population of oscillators forms several groups, so-called clusters, in each of which the oscillators behave identically. This kind of synchronized state is called cluster state (Sakaguchi, Shinomoto, Kuramoto 1987, 1988, Strogatz and Mirollo 1988a, 1988b, Golomb, Hansel, Shraiman, Sompolinsky 1992, Hakim and Rappel 1992, Hansel, Mato, Meunier 1993a, Nakagawa, Kuramoto 1993, Okuda 1993, Tass and Haken 1996a, Tass 1997a). Clustering will be discussed below in great detail.

4. *Dephasing* and *bursting:* Han, Kurrer and Kuramoto (1995) analyzed coupled neural oscillators, where simple diffusive coupling gives rise to burstlike behavior: The oscillators switch between low and high oscillation amplitude.

5. *Collective chaotic behavior:* Due to the oscillators' mutual interactions collective chaotic dynamics emerges (see, for instance, Matthews and Strogatz 1990, Hakim and Rappel 1992, Nakagawa, Kuramoto 1993).

Hence from a theoretical point of view one has to take into account that apart from synchrony of periodic activity and incoherency the experimental data might display other dynamical states, too. Although the latter may not be detectable by means of the typically applied linear correlation techniques, they might play an important role, for instance, concerning the neuronal code and scene segmentation.

From a physiological standpoint one might expect some synchronized states to be more relevant than others. In particular, different physiological mechanisms may be related to different types of synchronization behavior. Let the problem of the neural coding guide us to the synchronized states which are worthwhile to be studied in detail. In particular, in the context of the binding problem increasing experimental evidence indicates that cortical neurons act as coincidence detectors (cf. von der Malsburg and Schneider 1986, König, Engel, Singer 1996). For this reason we focus on the cluster states because clusters consist of oscillators with (nearly) the same, i.e., mutually coinciding phases. Moreover, clustering appears to be important in the context of neurological diseases, too (Golomb, Wang, Rinzel 1996).

To compare cluster states with other synchronized states we should recall the phase oscillators' different levels of synchronization:

1. *Synchrony* or *in-phase synchronization:* This is the strongest form of synchronization as the phase difference between any two oscillators vanishes (modulo 2π).
2. *Phase locking* is a weaker form of synchronization: The oscillators' mutual phase differences are constant, but generally do not vanish.
3. *Frequency locking:* The synchronizing interactions are not strong enough to establish fixed phase relationships among the oscillators. Nevertheless, all oscillators have the same average frequency. *Quasientrainment* is a special case of frequency locking, where mutual phase differences diverge as time increases (Daido 1992a). Quasientrainment was found by Daido (1992a) in a population of phase oscillators with random and frustrated couplings.
4. *Clustering:* On the one hand clustering is a local type of synchronization which occurs in lattices of oscillators with local interactions (Sakaguchi, Shinomoto, Kuramoto 1987, 1988, Strogatz and Mirollo 1988a, 1988b). In this case a cluster typically consists of neighbouring oscillators. On the other hand clustering is observed in populations of globally coupled oscillators, too (Golomb, Hansel, Shraiman, Sompolinsky 1992, Hakim and Rappel 1992, Hansel, Mato, Meunier 1993a, Nakagawa, Kuramoto 1993, Okuda 1993, Tass 1997a). In both cases the oscillators form several clusters as a consequence of their mutual interactions. Within each cluster all oscillators have (nearly) the same phase. Between different clusters the phase differences do not vanish; often they are functions in time. The number of different clusters remains finite even if the number of oscillators tends towards infinity.

In this chapter we will become familiar with the synchronization behavior of a population of globally coupled and continuously interacting phase oscillators. The approach presented here is based on a former study addressing the impact of the clusters' sizes on the synchronization frequency and the phase differences between the clusters (Tass 1997a).

3.4 Populations of Neurons

Populations of neurons have been modeled in rather different ways (see, for instance, Reichardt and Poggio 1981, Koch and Segev 1989, Eggermont 1990, Müller and Reinhardt 1990, Edelman 1992, Aertsen 1993, Gerstner, Ritz, van Hemmen 1993, Arbib 1995, Omidvar 1995, Haken 1996, Hoppensteadt and Izhikevich 1997). Among other things these modelling approaches differ from each other as far as the level of description is concerned. Of course, the latter depends on the particular features of interacting neurons one is interested in. For instance, from an abstract point of view populations of neurons were considered as networks of information processing devices switching between two states, the resting state and an active state (McCulloch and Pitts 1943, cf. Hopfield 1982). McCulloch and Pitts showed that such a network is able to perform all the logical processes of a Boolean algebra provided the interactions among the model neurons are suitably chosen.

Along the lines of a different approach it turned out to be fruitful to model the spatially averaged behavior of neuronal populations (cf. Beurle 1956, Griffith 1963, 1965) by introducing field variables which describe the activities of excitatory and inhibitory neurons (Wilson and Cowan 1972, 1973, Nunez 1979, 1981, 1995, Jirsa and Haken 1996, 1997). Such models were used, e.g., in order to relate neuronal activity to the electromagnetic field of the brain (Nunez 1979, 1981, 1995, Jirsa and Haken 1996, 1997). Additionally this approach was used to model the cortical activity during visual hallucinations (Ermentrout and Cowan 1979, Cowan 1987, Tass 1995a, 1997b).

The mutual coordination and, in particular, synchronization of the firing of neurons can be studied in networks of integrate-and-fire oscillators (Mirollo and Strogatz 1990, Kuramoto 1991, Grossberg and Somers 1991, Chawanya et al. 1993) and in networks of limit cycle oscillators (Schuster and Wagner 1990a, 1990b, Sompolinsky, Golomb, Kleinfeld 1991, Schillen and König 1994, Tass and Haken 1996b). In this book we shall decide in favour of a network of limit cycle oscillators for two reasons: (a) In general, it is an appropriate model for the study of neuronal synchronization processes. Accordingly, networks of limit cycle oscillators were used to model synchronized neuronal activity in the visual cortex (cf. Schuster and Wagner 1990a, 1990b, Grossberg and Somers 1991, Sompolinsky, Golomb, Kleinfeld 1991, Chawanya et al. 1993, Schillen and König 1994, Tass and Haken 1996b). Actually, according to, for example, animal studies synchronization processes are also relevant for motor control and not just for scene segmentation (Roelfsema et al. 1997). For this reason we should avoid to restrict ourselves to analyze a highly specified model. Rather we shall choose a general model of interacting neuronal limit-cycle oscillators. (b) Networks of coupled limit cycle oscillators serve as models for collective processes in other branches of physiology, too. For example, the energy metabolism in a population of cells can be considered as a network of limit cycle oscillators (cf.Winfree 1980). Thus, insights concerning the emergence of different patterns of synchronization can also be applied in these fields.

3.4.1 Model Neuron

The dynamics of a single periodically firing or bursting neuron can be approximated by means of a limit cycle oscillator: The membrane potential of periodically firing neurons runs on a limit cycle (Hodgkin and Huxley

1952, FitzHugh 1961, Nagumo, Arimoto, Yoshizawa 1962), whereas the slow dynamics of the relaxation oscillation which generates the bursts displays a limit cycle, too (Plant 1978, 1981, Rinzel 1986, Murray 1990). For this reason we model a single neuron of both types by means of a limit cycle oscillator with time-dependent phase $\phi(t)$ and time-dependent amplitude $s(t)$. Phase and amplitude of an isolated neuron are governed by evolution equations of the form $\dot{s} = f(s, \phi)$ and $\dot{\phi} = \omega + g(s, \phi)$, where ω denotes the eigenfrequency. As illustrated in Fig. 3.1 the normal form theorem (cf. Elphik et al. 1987, Iooss 1987) enables us to perform a suitable transformation so that on the transformed limit cycle $\dot{r} = 0$ holds, where r denotes the transformed amplitude (Kuramoto 1984, Murray 1990). Whenever the transformed phase $\psi(t)$ equals a fixed value ρ (modulo 2π), the neuron fires or bursts, depending on whether it is a firing or bursting neuron (cf. Sect. 2.8). The eigenfrequency ω models, for instance, the retinal input of a neuron in the visual cortex (see, for instance, Tass and Haken 1996b).

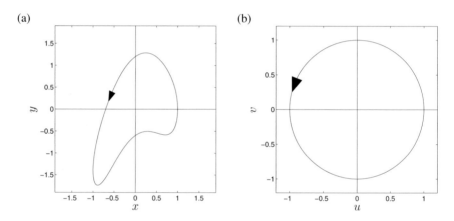

Fig. 3.1a,b. *Schematic illustration of the normal form theorem:* (**a**) shows a limit cycle with amplitude $s = \sqrt{x^2 + y^2}$ and phase ϕ, where $x = s\cos\phi$ and $y = s\sin\phi$. The transformation given by the normal form theorem (cf. Elphik et al. 1987, Iooss 1987) gives rise to a limit cycle with transformed coordinates u and v as shown in (**b**). $r = \sqrt{u^2 + v^2}$ is the amplitude and ψ is the phase of the transformed limit cycle, where $u = r\cos\psi$ and $v = r\sin\psi$. Clearly, $\dot{r} = 0$ holds on the transformed limit cycle. Hence, by means of the normal form theorem a bent limit cycle (**a**) becomes a radially symmetrical limit cycle (**b**). In both plots the arrow indicates the orientation of motion along the limit cycle.

3.4.2 Neuronal Interactions

In the cortex there are different types of neurons interacting via excitatory and inhibitory synapses (see, e.g., Creutzfeldt 1983, Braitenberg and Schüz

1991). Apart from the excitatory pyramidal cells there are at least two classes of inhibitory neurons: stellate cells and Martinotti cells (cf. Braitenberg and Schüz 1991).

Taking into account these three classes of neurons means modelling three populations of coupled limit cycle oscillators. As yet, the knowledge of the synchronization behavior of one population of coupled limit cycle oscillators is far from being complete. However, it is to be expected that a network of several populations of oscillators exhibits an even more complex synchronization behavior than a single population. If one is interested to understand the model's dynamics appropriately one may think it advisable to choose a more macroscopic level of description. Hence, instead of modelling every neuron as a limit cycle oscillator a different approach can be performed (cf. Tass and Haken 1996b): One takes into account an arbitrary number of excitatory neurons, each of which modeled by means of a limit cycle oscillator. For the sake of simplicity the mutual synaptic interactions of the excitatory neurons as well as the net effect of the inhibitory neurons on the excitatory neurons are taken into account as coupling mechanism between the limit cycle neurons.

Thus, a network of three interacting populations of neurons (pyramidal, stellate and Martinotti cells) is approximated by means of a population of limit cycle oscillators (pyramidal cells), where the effect of the inhibitory populations (stellate and Martinotti cells) is incorporated into the couplings among the limit cycle oscillators. As the coupling mechanism is not derived from microscopic equations, the oscillators' mutual interactions have to be formulated as general as possible in order to avoid unphysiological restrictions. Based on the modelling outlined in this book in future studies the next step can be taken in analyzing the dynamics of several interacting populations of oscillators.

3.5 Populations of Phase Oscillators

Now one encounters two questions: 1. Which model of coupled limit cycle oscillators should be chosen? 2. How should the model's dynamical behavior be analyzed? Obviously these two questions are intertwined. On the one hand the model should take into account as much details of the physiological problem under consideration as possible. On the other hand the model should allow one to perform an appropriate mathematical investigation providing real knowledge of the model's dynamics. Otherwise a modelling approach would run the risk of being as puzzling as a sum of incoherent experimental observations.

In this situation it is of benefit to recall that in 1967 Winfree proposed to approximate the dynamics of a population of limit cycle oscillators by means of the dynamics of a population of phase oscillators. His approach was based on the assumption that the coupling predominantly affects the motion of each oscillator around its limit cycle provided the coupling is weak compared to the attractiveness of the limit cycle. In the case of weak coupling Kuramoto (1984) developed a sound mathematical framework for this notion of phase oscillators. As far as phase and frequency shifts in the synchronized states are concerned one is allowed to neglect the amplitude dynamics for several networks of limit cycle oscillators also in the case of strong coupling (Tass and Haken 1996).

According to this approach with phase oscillators Hansel, Mato and Me-unier (1993b) derived an evolution equation for the phase dynamics of two spiking Hodgkin-Huxley neurons coupled by weak excitatory interactions. Our study should not be restricted to a pair of Hodgkin-Huxley neurons. Rather a typical and generic model equation for smoothly interacting phase oscillators will be considered. The latter is given by

$$\dot{\psi}_j = \omega_j - \sum_{k=1}^{N} \Gamma_{jk}(\psi_j - \psi_k) \quad (j = 1, \ldots, N) \ , \tag{3.1}$$

where ψ_j is the phase of the jth oscillator, and ω_j is the eigenfrequency of the jth oscillator (see Winfree 1967,1980, Ermentrout and Rinzel 1981, Kuramoto 1984). The couplings Γ_{jk} model the oscillators' mutual interactions, where $\Gamma_{jk}(x + 2\pi) = \Gamma_{jk}(x)$ holds. Thus, Γ_{jk} can be expanded in terms of Fourier modes. For globally coupled oscillators one is allowed to set $\Gamma_{jk}(x) = \Gamma(x)$. Of course, the oscillators' synchronization behavior critically relies on the couplings Γ_{jk}. If one takes into account only first order terms of $\Gamma(x)$, for instance $\Gamma(x) = K N^{-1} \sin(x)$, as the coupling strength K passes its critical value, a transition from incoherency to in-phase synchronization appears (Kuramoto 1984). Higher harmonics of the coupling Γ enrich the dynamical behavior of Eq. (3.1) considerably: Apart from macroscopic synchronization (cf. Daido 1994, 1996) one encounters clustering (cf. Hansel, Mato and Me-unier 1993a, Okuda 1993, Tass 1997a). Since we want to study clustering processes we have to take into account Fourier modes of higher order. Let us, thus, consider the model equation

$$\dot{\psi}_j = \omega_j + \frac{1}{N} \sum_{k=1}^{N} M(\psi_j - \psi_k) \quad (j = 1, \ldots, N) \tag{3.2}$$

with the mean-field-coupling

$$M(\psi_j - \psi_k) = - \sum_{m=1}^{4} \{ K_m \sin[m(\psi_j - \psi_k)] + C_m \cos[m(\psi_j - \psi_k)] \} \tag{3.3}$$

and with nonnegative coupling constants $K_1, \ldots, K_4$ (Tass 1997a). $C_1, \ldots, C_4$ are real constants which will be specified below. Above it was already mentioned that the jth model neuron fires (or bursts) whenever $\psi_j(t)$ equals ρ (modulo 2π). As in (2.63) for non-vanishing ρ one may introduce the shifted phase

$$\psi_j' = \psi_j - \rho \, . \tag{3.4}$$

With this transformation the jth model neuron fires (or bursts) whenever its shifted phase $\psi_j'(t)$ vanishes (modulo 2π). Obviously model equation (3.2) is invariant against transformation (3.4) because the argument of the coupling function M only contains differences $\psi_j - \psi_k$. Hence, in general, we may assume that ρ equals zero.

In this chapter we shall see that clustering occurs as a consequence of coupling terms with higher harmonics like $\sin[2(\psi_j - \psi_k)]$, $\cos[2(\psi_j - \psi_k)]$, $\sin[3(\psi_j - \psi_k)]$, $\cos[3(\psi_j - \psi_k)]$ etc. This stresses the importance of this type of coupling in models of phase oscillators. Since the latter are approximations of models of limit cycle oscillators it should be illustrated which interaction mechanisms in models of limit cycle oscillators cause higher harmonic couplings of the phase dynamics. With this aim in view we focus on a cluster of limit cycle oscillators which are exclusively interacting via phase coupling.

For this purpose let us first recall the simple and representative limit cycle model (2.5) discussed in Sect. 2.2.1. The dynamics of a cluster of N noninteracting limit cycle oscillators of this type is governed by

$$\dot{z}_j = (\alpha + \mathrm{i}\omega_j)z_j - \beta z_j^2 z_j^* \, , \tag{3.5}$$

where $j = 1, \ldots N$, and z_j is the complex variable of the jth oscillator consisting of the nonnegative amplitude r_j and the phase ψ_j according to

$$z_j = r_j \exp(\mathrm{i}\psi_j) \, . \tag{3.6}$$

z_j^* denotes the complex conjugate of z_j. ω_j is the oscillator's eigenfrequency, whereas α is a nonnegative parameter, and β is positive. Inserting (3.6) into (3.5) yields equations for the amplitudes and phases, respectively:

$$\dot{r}_j = \alpha r_j - \beta r_j^3 \, , \tag{3.7}$$

$$\dot{\psi}_j = \omega_j \, . \tag{3.8}$$

The dynamics of (3.7) and (3.8) was already discussed in Sect. 2.2.1: A stable fixed point with vanishing amplitude (i.e. $r_j = 0$) occurs for negative α, whereas a Hopf bifurcation takes place when α becomes positive. This means that for $\alpha > 0$ a stable limit cycle oscillation with amplitude $\sqrt{\alpha/\beta}$ and phase $\psi_j(t) = \omega_j t + \theta_j$ ($\theta_j = \mathrm{const}$) emerges (cf. Haken 1983).

Obviously, in (3.5) coupling terms are still missing. Typical synchronizing coupling terms are cubic terms like $z_j^2 z_k^*$. However, they lead to interaction terms of lowest order, namely $\sin(\psi_j - \psi_k)$ and $\cos(\psi_j - \psi_k)$. For this reason we additionally have to take into account higher order terms. Accordingly, an example of an evolution equation of a cluster of globally coupled limit cycle oscillators exclusively interacting via phase coupling reads

$$\dot{z}_j = (\alpha + \mathrm{i}\omega_j)z_j - \beta z_j^2 z_j^* - \frac{z_j}{N}\sum_{k=1}^{N}\sum_{m=1}^{4}\left[\mu_m(z_j z_k^*)^m - \mu_m^*(z_j^* z_k)^m\right] \tag{3.9}$$

with complex coupling constants

$$\mu_m = k_m + ic_m . \tag{3.10}$$

According to the normal form theorem (Elphik et al. 1987, Iooss and Adelmeyer 1992) the dynamics of numerous and more complex clusters of interacting limit cycle oscillators can be transformed onto or can be approximated by the dynamics given by (3.9). In this sense (3.9) may be considered as a minimal model.

Inserting (3.6) into (3.9) yields equations for the amplitudes r_j and phases ψ_j, respectively:

$$\dot{r}_j = \alpha r_j - -\beta r_j^3 , \tag{3.11}$$

$$\dot{\psi}_j = \omega_j - \frac{1}{N} \sum_{k=1}^{N} \sum_{m=1}^{4} \left\{ K_m^{(j,k)} \sin[m(\psi_j - \psi_k)] + C_m^{(j,k)} \cos[m(\psi_j - \psi_k)] \right\} ,$$
$$\tag{3.12}$$

where

$$K_m^{(j,k)} = 2k_m r_j^m r_k^m , \quad C_m^{(j,k)} = 2c_m r_j^m r_k^m . \tag{3.13}$$

Comparing (3.11) with (3.7) shows that for positive α the amplitudes of all oscillators relax towards the limit cycle amplitude $\sqrt{\alpha/\beta}$. In particular, the oscillators' interactions do not influence the dynamics of the amplitudes. According to (3.12) the oscillators interact via phase couplings where the amplitudes determine the strengths of the couplings as given by (3.13). Inserting $r_j = \sqrt{\alpha/\beta}$ into (3.13) provides us with the coupling strengths

$$K_m^{(j,k)} = 2k_m \left(\frac{\alpha}{\beta}\right)^m , \quad C_m^{(j,k)} = 2c_m \left(\frac{\alpha}{\beta}\right)^m , \tag{3.14}$$

which occur when all oscillators move along their limit cycles. In this case we end up with the phase dynamics given by (3.2) with the mean-field-coupling (3.3) with the coupling coefficients $K_m = K_m^{(j,k)}$ and $C_m = C_m^{(j,k)}$ from (3.14). In summary, this example illustrates that coupling terms of higher order in the limit cycle models lead to coupling terms with higher harmonics in phase oscillator models.

Before we shall come back to the study of the synchronization behavior of the model of phase oscillators (3.2) let us first focus on the mathematical tool which will be used for this.

3.6 Slaving Principle and Center Manifold

In the middle of this century an increasing number of natural scientists and mathematicians became interested in self-organization processes in nature. Consequently they faced the problem of how spatio-temporal patterns emerge without external driving force. Amazing features of self-organized dynamics became evident. For instance, as an increasing control parameter passes its critical value, a new and qualitatively different spatio-temporal behavior occurs without any specific external input. As a result of the pioneering work, in particular, in the late 1960s it turned out that many of these emergent patterns share one common feature: Though the system consists of a large number of subsystems, only a few variables govern the system's collective dynamics (for a review see Haken 1977, 1983, 1996). By means of the slaving principle (Haken 1975, Wunderlin and Haken 1975) it became possible to understand how these few variables, the so-called order parameters, determine the dynamics of the entire system. As self-organization is not at all confined to systems analyzed in physics Haken (1973, 1977) founded the interdisciplinary field of synergetics in order to promote the study of self-organization in physics, chemistry, biology and sociology (Haken and Graham 1971, cf. Nicolis and Prigogine 1977, Winfree 1980, Meinhardt 1982, Glass and Mackey 1988, Murray 1990).

Initially hardly recognized by the physicists a branch of mathematics arose which was dedicated to nonlinear dynamics, i.e. the dynamics of self-organizing systems (see, for instance, Thom 1972, Arnold 1983, Guckenheimer and Holmes 1990, Strogatz 1994). In the 1960s Pliss (1964) and Kelley (1967) proved the center manifold theorem. According to the latter under certain conditions the dynamics of a high dimensional system can be approximated by a low dimensional manifold. Though both slaving principle and center manifold theorem address similar issues, there are some differences:

1. The center manifold theorem is a mathematical tool which has been fruitfully applied, in particular, to the analysis of bifurcations of dynamical systems (Pliss 1964, Kelley 1967) and partial differential equations (Carr 1981, Kirchgässner 1982). For a review see, for instance, Aulbach (1984), Guckenheimer and Holmes (1990), Iooss and Adelmeyer (1992). In contrast to this from the very beginning the slaving principle aimed at the explanation of self-organizing processes in nature (Haken 1975, 1977, Wunderlin and Haken 1975). In other words, the slaving principle is not just a mathematical method. Rather its deep physical meaning and relevance already turned out in laser physics (Haken 1964) and, consequently, lead Haken (1973, 1977) to establish the interdisciplinary synergetic approach (cf. Haken and Graham 1971).

2. The reasoning of the two approaches is different: While the center manifold theorem is proved by means of the variation of constants (cf. Vanderbauwhede 1989) or by means of the graph transform (cf. Hirsch, Pugh and Shub 1976, Sandstede, Scheel and Wulff 1997) the slaving principle is based on the separation of time scales (Haken 1975, Wunderlin and Haken 1975).

3. The center manifold theorem refers to dynamical systems without any noise. From the standpoint of a natural scientist, however, it is inevitable

to take into account noise, provided a level of description is chosen more macroscopic than that of quantum mechanics. For this reason along the lines of the synergetic approach the impact of noise on emerging spatio-temporal patterns was studied in detail right from the beginning (Haken 1977, Haken and Wunderlin 1982). As a result of these efforts it was shown that noise gives rise to critical fluctuations and critical slowing down (Haken 1977,1983). Both phenomena were verified experimentally, e.g., in the context of laser physics (cf. Haken 1970) and bimanual coordination (Schöner, Haken and Kelso 1986).

A detailed comparison between slaving principle and center manifold theory was presented by Aulbach (1984). In spite of all possible differences in my opinion it is impressive that two different trends in science lead us to the same goal in a complementary way. Based on a deep principle the synergetic approach lightly encompasses an overwhelming plethora of dynamical phenomena. Attracted by the same beauty, mathematicians elucidate a large variety of sophisticated aspects of nonlinear dynamics, in this way contributing to a sound understanding of dynamical processes in nature. For me this is a lucky situation because we need both types of reasoning, and every rivalry between physicists and mathematicians should give way to mutual fertilization.

During our cluster states' analysis it will turn out that the slow variables are constant, and one can conveniently calculate the center manifold (cf. Tass 1997a). Hence, in the next section it will be explained how to apply the center manifold theorem. For a detailed explanation of the slaving principle including its applications to partial differential equations, stochastic differential equations and delay differential equations I refer to Haken (1975, 1977, 1983, 1988, 1996), Haken and Wunderlin 1982, Wunderlin and Haken 1975, 1981, Wischert et al. (1994).

3.6.1 Center Manifold Theorem

Phase and frequency shifts in four different types of cluster states will be determined below by explicitly calculating the center manifold. Therefore in this section we shall see how to apply the center manifold theorem in this particular case. For an existence proof of the center manifold I refer to the center manifold theorem of Pliss (1964) and Kelley (1967). The application of this theorem to the study of bifurcations is presented in detail, for instance, in Iooss and Adelmeyer (1992).

To derive the center manifold in Sects. 3.7.2–3.7.5 suitable transformations will be performed so that the center manifold theorem can be applied effectively. For every type of n-cluster state these transformations will be different. However, as a result of these transformations the state of the population of oscillators is described by two sets of variables: the center modes and the stable modes. The vectors of the center and the stable modes will be denoted by $\boldsymbol{x}_c$ and $\boldsymbol{x}_s$. With these notations the transformed system reads

$$\dot{\boldsymbol{x}}_c = B_c \boldsymbol{x}_c + \boldsymbol{N}_c(\boldsymbol{x}_c, \boldsymbol{x}_s) \tag{3.15}$$

$$\dot{\boldsymbol{x}}_s = B_s \boldsymbol{x}_s + \boldsymbol{N}_s(\boldsymbol{x}_c, \boldsymbol{x}_s) \ . \tag{3.16}$$

B_c and B_s are matrices, whereas $\boldsymbol{N}_c(\boldsymbol{x}_c, \boldsymbol{x}_s)$ and $\boldsymbol{N}_s(\boldsymbol{x}_c, \boldsymbol{x}_s)$ only contain nonlinear terms. The benefit of the above mentioned transformations is that the linear parts of the center and the stable modes are separated, i.e. in (3.15) and (3.16) there is no mutual linear coupling. Neglecting nonlinear terms in (3.15) and (3.16) yields Jordan's normal form of the linear problem (see, for instance, Hirsch and Smale 1974).

Note that $\boldsymbol{x}_c$ and $\boldsymbol{x}_s$ differ from each other as far as the eigenvalues of the linearized problem (i.e. $\boldsymbol{N}_c$ and $\boldsymbol{N}_s$ vanish) is concerned: The real part of all eigenvalues of the matrix B_c vanishes, whereas the eigenvalues of the matrix B_s have negative real parts. Actually, in this particular case all eigenvalues of B_c vanish. This makes the application of the center manifold theorem so fruitful. The slow variables, i.e. the center modes, finally determine the dynamics of the entire system. In particular, in a small but finite neighborhood of zero the stable modes are given by the center modes according to

$$\boldsymbol{x}_s = \boldsymbol{h}(\boldsymbol{x}_c) \ , \tag{3.17}$$

where $\boldsymbol{h}$ only contains terms of second and higher order. The *center manifold* is characterized by (3.17). The parameter range in which (3.17) is valid will be stated more precisely below. The dynamics on the center manifold is governed by

$$\dot{\boldsymbol{x}}_c = B_c \boldsymbol{x}_c + \boldsymbol{N}_c(\boldsymbol{x}_c, \boldsymbol{h}(\boldsymbol{x}_c)) \ . \tag{3.18}$$

This equation is the so-called reduced problem obtained by inserting (3.17) into (3.15) (cf. Iooss and Adelmeyer 1992). Independent of the approach with the center manifold, in synergetics a corresponding equation, the so-called *order parameter equation*, was derived in the more extensive setting of stochastic differential equations (Haken 1975, Wunderlin and Haken 1975).

From (3.18) one immediately reads off that in order to analyze the dynamics on the center manifold, first the center manifold itself, i.e. $\boldsymbol{h}$, has to be determined. To this end (3.17) is differentiated with respect to time yielding

$$\dot{\boldsymbol{x}}_s = D\boldsymbol{h}(\boldsymbol{x}_c) \, \dot{\boldsymbol{x}}_c \ , \tag{3.19}$$

where D denotes the differentiation operator. Inserting (3.15), (3.16) and (3.17) into (3.19) yields an implicit equation for $\boldsymbol{h}$:

$$B_s \boldsymbol{h}(\boldsymbol{x}_c) + \boldsymbol{N}_s(\boldsymbol{x}_c, \boldsymbol{h}(\boldsymbol{x}_c)) = D\boldsymbol{h}(\boldsymbol{x}_c)\big[B_c \boldsymbol{x}_c + \boldsymbol{N}_c(\boldsymbol{x}_c, \boldsymbol{h}(\boldsymbol{x}_c))\big] \ . \tag{3.20}$$

To determine $\boldsymbol{h}$, in general, the latter is expanded as a polynomial, inserted into (3.20), and the coefficients are determined up to the appropriate order (see, for instance, Haken 1975, Wunderlin and Haken 1975, Iooss and

Adelmeyer 1992). Obviously the determination of h may be rather complicated. Moreover the resulting reduced problem may be so complex, that it would be impossible to carry out an analytic investigation.

Aware of these obstacles the power of the center manifold theorem can easily be unfolded by performing a *nonlinear renormalization procedure* introduced by introduced by Tass (1997a). This procedure will be explained in detail in Sect. 3.7.2. It has a specific aim: $B_c \boldsymbol{x}_c$ as well as $\boldsymbol{N}_c(\boldsymbol{x}_c, \boldsymbol{h}(\boldsymbol{x}_c))$ vanish on the center manifold as a consequence of the nonlinear renormalization. In this way the analysis is decisively simplified: (a) Conveniently the reduced problem becomes trivial as it reads $\dot{\boldsymbol{x}}_c = \boldsymbol{0}$. (b) On the other hand (3.20) takes the simple form

$$\boldsymbol{h}(\boldsymbol{x}_c) = -B_s^{-1}\boldsymbol{N}_s(\boldsymbol{x}_c, \boldsymbol{h}(\boldsymbol{x}_c)) \ . \tag{3.21}$$

Anyhow, it may still be impossible or at least complicated to invert the matrix B_s analytically. Therefore (a) an additional variable is (rigorously) introduced, and (b) some of the coupling constants are assumed to obey certain smallness conditions. As a result of these manipulations the matrix of the stable part of the transformed model equation is given by

$$B_s = -\gamma E \ , \tag{3.22}$$

where E denotes the identity matrix, and γ is a positive constant. Inserting (3.22) into (3.21) one ends up with

$$\boldsymbol{h}(\boldsymbol{x}_c) = \frac{1}{\gamma}\boldsymbol{N}_s(\boldsymbol{x}_c, \boldsymbol{h}(\boldsymbol{x}_c)) \ . \tag{3.23}$$

In this way h will easily be determined in lowest order. According to Pliss (1964) and Kelley (1967) the center manifold given by (3.17) is a local attractor. On the other hand the nonlinear renormalization procedure guarantees that $\dot{\boldsymbol{x}}_c = \boldsymbol{0}$ holds. Hence, (3.17) determines a *stable fixed point*. Due to the above mentioned suitable transformations the latter is nothing but the synchronized state under consideration. In summary, the n-cluster states analyzed below are stable.

The center manifold theorem is a local theorem. That means, it only holds provided $\|\boldsymbol{x}_c\|$ is small as compared with $|\text{Re}(\lambda_s)|$, where λ_s denotes the eigenvalues of B_s, and $\text{Re}(\xi)$ denotes the real part of ξ (cf. Pliss 1964 and Kelley 1967). So,

$$\|\boldsymbol{x}_c\| \ll \gamma \tag{3.24}$$

has to be fulfilled. Obviously smallness condition (3.24) restricts the parameter range for which the analysis is valid. Consequently the deviations $\eta_1, \ldots, \eta_N$ and some of the coupling constants have to be small compared to γ. This will be discussed in detail below. For larger values of $\|\boldsymbol{x}_c\|$, violating condition (3.24), the synchronized states will be investigated numerically.

3.6.2 Strategy

Below it will turn out that model equation (3.2) is impressive because of its rich synchronization behavior. Apart from synchrony the oscillators may form different types of cluster states which have to be analyzed separately. Nevertheless, the study's strategy remains the same in all four cases. In the next four sections a lot of analytic calculations will be performed. In order to be prevented from losing track in the heat of the moment, in this section the main features of the analysis are briefly sketched.

1. First a rotating coordinate system is introduced. The suitable choice of the rotation frequency in the rotating coordinate system will guarantee that the synchronized state under consideration is nothing but a stable fixed point. The proper rotation frequency is the synchronization frequency denoted by Ω^*.
2. The system's dynamics acts on two time scales, where the motion of the fast variables is governed by the slow variables. In particular, in all cluster states under consideration the slow variables are constant. Because of that it is convenient to determine phase and frequency shifts explicitly by means of the center manifold theorem. Moreover the latter enables us to prove the stability of the different synchronized states.
3. The coordinate system has to rotate with a suitable frequency because otherwise the center manifold theorem cannot be applied. A nonlinear renormalization procedure guarantees that the proper rotation frequency is chosen. Renormalization by linear counterterms is a common tool for the determination of frequency shifts of coupled oscillators (cf. Haken 1983). In previous studies phase and frequency shifts were derived in several oscillator models by means of a combination of linear renormalization and center manifold theorem (Tass 1995b, Tass and Haken 1996a). However, as far as model equation (3.2) is concerned a linear renormalization procedure won't fly (Tass 1997a). This will be discussed in Sect. 3.7.2.
4. Finally phase and frequency shifts are calculated by determining the center manifold in lowest order.

3.7 n-Cluster States

3.7.1 Configuration of Cluster States

This chapter is devoted to cluster states which are phase-locked states, i.e. all oscillators run at the same frequency. Denoting the phase of the jth oscillator by ψ_j, in the synchronized state one, thus, obtains $\dot{\psi}_j = \Omega^*$ for $j = 1, \ldots, N$, where N is the number of all oscillators, and Ω^* is called synchronization frequency.

Let me anticipate some of the model's dynamical features to motivate the way cluster states will be described: The intial conditions and the model parameters determine whether all oscillators form one giant cluster or whether they break into two, three or four phase-locked clusters of arbitrary size. Within one cluster the phase difference between any two oscillators vanishes or is small compared to the phase difference between different clusters. Figure 3.2 shows the different types of cluster states. Both synchronization frequency and shifts of the mutual phase differences critically depend on the number of oscillators within each cluster. For this reason the term *configuration of the synchronized state* or briefly *configuration* was introduced (Tass 1997a). The latter is defined by the number of oscillators within each cluster, no matter, which particular oscillator belongs to a certain cluster. Denoting the number of oscillators of the νth cluster by N_ν, the configuration of an n-cluster state can be written as $(N_1, \ldots, N_n)$. Below the configuration's impact on the synchronization frequency will be investigated in detail.

Before we turn to the analytical analysis let us first introduce some notations refering to the clusters' eigenfrequencies. Mean and variance of all eigenfrequencies will be denoted by

$$\Omega = \frac{1}{N} \sum_{k=1}^{N} \omega_k \quad \text{and} \quad V = \frac{1}{N} \sum_{k=1}^{N} (\omega_k - \Omega)^2 \ . \tag{3.25}$$

By means of deviations η_k the eigenfrequencies may be written as

$$\omega_k = \Omega + \eta_k \quad (k = 1, \ldots, N) \ . \tag{3.26}$$

The *cluster frequency* of the νth cluster, i.e. the mean of the eigenfrequencies of the oscillators in the νth cluster will be introduced by

$$\Omega_\nu = \frac{1}{N_\nu} \sum_{k(\nu)} \omega_k \ . \tag{3.27}$$

$\sum_{k(\nu)}$ denotes summation over all k, where the kth oscillator is in the νth cluster. N_ν denotes the number of oscillators in the νth cluster. Deviations of the cluster frequencies Λ_ν are introduced by setting

$$\Omega_\nu = \Omega + \Lambda_\nu \ . \tag{3.28}$$

The variance of the eigenfrequencies of the νth cluster are denoted by

$$V_\nu = \frac{1}{N_\nu} \sum_{k(\nu)} (\omega_k - \Omega_\nu)^2 = \frac{1}{N_\nu} \sum_{k(\nu)} (\eta_k - \Lambda_\nu)^2 \ . \tag{3.29}$$

The third moment of the distribution of all eigenfrequencies reads

$$T = \frac{1}{N} \sum_{k=1}^{N} (\omega_k - \Omega)^3 = \frac{1}{N} \sum_{k=1}^{N} \eta_k^3 \ . \tag{3.30}$$

(a)

(b)

(c)

(d)

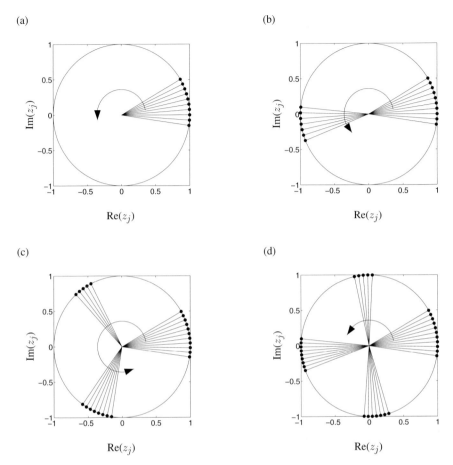

Fig. 3.2a–d. *Schematic illustration of the different synchronized states:* A single oscillator is represented by a dot (connected with the origin) on the unit circle in the complex plane according to $z_j = \exp(i\psi_j)$. A synchronized state consists of one (**a**), two (**b**), three (**c**), or four (**d**) clusters in each of which the oscillators have constant small (or vanishing) mutual phase differences. All oscillators run at the same frequency. The latter depends on the model parameters as well as on the synchronization configuration, i.e. the number of oscillators within the different clusters. Different configurations are typically associated with different synchronization frequencies as indicated by arrows of different length. The synchronized states will be called one-cluster state (**a**), two-cluster state (**b**), three-cluster state (**c**), four-cluster state (**d**).

3.7.2 One Cluster

In this section we will deal with a one-cluster state, i.e. a synchronized state where all oscillators have the same frequency and small mutual phase differences as shown in Fig. 3.2a. For this reason it is favourable to introduce *relative phases*

$$\phi_j(t) = \psi_j(t) - \Omega^* t - \theta \ , \tag{3.31}$$

where θ is a constant phase shift which will be determined below. Ω^* denotes the *synchronization frequency* which is the oscillators' frequency in the synchronized state. The latter need not be equal to Ω, the mean of the eigenfrequencies. This is taken into account by setting

$$\Omega^* = \Omega - \Delta \tag{3.32}$$

with Ω as introduced in Sect. 3.5. The *counterterm* Δ is expanded in terms of ε, where $0 < \varepsilon \ll 1$ holds:

$$\Delta = \sum_{\nu=0}^{\infty} \Delta^{(\nu)} \varepsilon^\nu \ . \tag{3.33}$$

Δ will contribute to a *nonlinear renormalization* of the eigenfrequencies. Accordingly $\Delta^{(0)}, \Delta^{(1)}, \Delta^{(2)}, \ldots$ will be determined appropriately below. With (3.31) the model equation (3.2) reads

$$\dot\phi_j = \eta_j + \Delta - \frac{1}{N} \sum_{k=1}^{N} \sum_{m=1}^{4} \left\{ K_m \sin\left[m(\phi_j - \phi_k)\right] + C_m \cos[m(\phi_j - \phi_k)] \right\} \ . \tag{3.34}$$

For the sake of brevity some shortforms are used:

$$a = \sum_{m=1}^{4} m\,K_m \ , \quad b = \frac{1}{2} \sum_{m=1}^{4} m^2\,C_m \ , \quad c = \frac{1}{6} \sum_{m=1}^{4} m^3\,K_m \ . \tag{3.35}$$

Moreover an additional variable is introduced by setting

$$\rho(t) = \frac{1}{N} \sum_{k=1}^{N} \phi_k(t) \ . \tag{3.36}$$

This (rigorous) trick will turn out to simplify the analysis decisively. Actually, the matrix of the stable modes (B_s) will simply be a diagonal matrix which can be inverted straightforwardly (cf. (3.22)). In this way the center manifold theorem can be applied in a convenient way.

In what follows two important features will be shown: (a) The equation governing the dynamics of the center modes simply reads $\dot{\boldsymbol{x}}_c = \boldsymbol{0}$ (cf. (3.15)). (b) All other modes are stable. Thus, according to the center manifold theorem the one-cluster state is nothing but a stable fixed point which lies within a small neighborhood around zero. Actually, this fixed point will be calculated by means of the center manifold theorem in order to determine phase and frequency shifts.

Let us first dwell on the first point concerning the dynamics of the center modes. As ε and $\eta_1, \ldots, \eta_N$ are constant they can (rigorously) be treated as variables which do not change in time, i.e. $\dot{\varepsilon} = \dot{\eta}_1 = \ldots = \dot{\eta}_N = 0$. Hence, $\varepsilon, \eta_1, \ldots, \eta_N$ are center modes (cf. Sect. 3.6.1). In order to cast (3.15) into a most suitable form we postulate that

$$\dot{\rho} = 0 \quad \text{on the center manifold} \tag{3.37}$$

has to be fulfilled. This can easily be achieved by choosing the counterterms $\Delta^{(0)}, \Delta^{(1)}, \Delta^{(2)}, \ldots$ from (3.33) appropriately. Condition (3.37) yields $\dot{\boldsymbol{x}}_c = \boldsymbol{0}$.

To illustrate the importance of condition (3.37) let us assume that the latter is not fulfilled. In this case we obtain $\dot{\rho} = G(\varepsilon, \eta_1, \ldots, \eta_N)$, where G does not depend on ρ. Thus, $\rho(t) = Gt + \rho_0$ ($\rho_0 = \text{const}$). Accordingly ρ increases or decreases monotonously, thereby removing the system from the small neighborhood around zero. As a matter of fact, in this case the center manifold cannot be applied. That is exactly why one has to guarantee that condition (3.37) is fulfilled by means of proper counterterms ($\Delta^{(0)}, \Delta^{(1)}, \Delta^{(2)}, \ldots$). Moreover our analysis will show that for $C_j \neq 0$ condition (3.37) is not fulfilled if a linear renormalization is performed by merely fitting counterterms $\Delta^{(0)}$ and $\Delta^{(1)}$. It is important to note that in a generic case condition (3.37) can only be fulfilled provided a *nonlinear* renormalization is carried out.

Of course, we still have to address the second point concerning the modes' stability. Moreover the counterterms $\Delta^{(0)}, \Delta^{(1)}, \ldots$ are not yet determined. Fortunately both issues can be addressed at the same time by going on as follows:

1. Equations for $\dot{\phi}_1, \ldots, \dot{\phi}_N$ and condition (3.37) are expanded according to Taylor. $\Delta^{(0)}, \Delta^{(1)}$ are determined by taking into account constant terms and terms of first order, where the order of terms refers to the center modes. Thus, terms of nth order are $\sim \|\boldsymbol{x}_c\|^n$.
2. Next, the transformation $\varphi_j = \phi_j - \eta_j/a - \rho$ separates the linear parts of center and stable modes which are given by $\boldsymbol{x}_c = (\varepsilon, \rho, \eta_1, \ldots, \eta_N)$ and $\boldsymbol{x}_s = (\varphi_1, \ldots, \varphi_N)$, respectively. As a consequence of this transformation the matrix of the stable modes is of the form presented by (3.22), where $\gamma = a$ (cf. (3.35)). In particular, the modes given by $\boldsymbol{x}_s$ are stable modes.
3. According to (3.23) $\boldsymbol{h}$ is determined in lowest order.
4. Still the counterterms of higher order, i.e. $\Delta^{(2)}, \Delta^{(3)}, \Delta^{(4)}, \ldots$, remain to be determined. To this end one has to take into account nonlinear terms by inserting $\boldsymbol{h}$ into condition (3.37). In principle the determination of these counterterms is trivial. However, it is quite tedious. Therefore the investigation will be restricted to the determination of frequency shifts of second order, that is $\Delta^{(2)}$. Actually, below it will turn out that it is sufficient to determine nonlinear counterterms of lowest order to illustrate the complex relationship between configuration and synchronization frequency. Note that this *nonlinear renormalization* is only possible because ε is treated as variable.

A little calculation shows that condition (3.37) is *always* fulfilled (not only on the center manifold) if $C_1, \ldots, C_4$ vanish. In this case ρ is a *con-*

served quantity, no matter whether or not the system is synchronized, and correspondingly Δ vanishes. But in the generic case, i.e. if at least one of the coefficients $C_1, \ldots, C_4$ does not vanish, the counterterms $\Delta^{(0)}, \Delta^{(1)}, \Delta^{(2)}, \ldots$ have to be chosen appropriately so that condition (3.37) is fulfilled. One obtains

$$\Delta^{(0)} = \sum_{m=1}^{4} C_m \ , \quad \Delta^{(1)} = 0 \ , \quad \Delta^{(2)} = -\frac{2b}{\varepsilon^2 a^2} V \ , \tag{3.38}$$

where V denotes the variance of all eigenfrequencies defined by (3.25). These counterterms describe the frequency shifts in the synchronized state, which occur due to the oscillators' mutual interactions.

As yet, the phase shifts $\phi_1, \ldots, \phi_N$ in the synchronized state (cf. (3.31)) are not determined. To this end the stable modes in the synchronized state are derived by means of the center modes according to (3.17): $\varphi_j = h_j(\boldsymbol{x}_c)$, where $\boldsymbol{h}(\boldsymbol{x}_c) = (h_1(\boldsymbol{x}_c), \ldots, h_N(\boldsymbol{x}_c))$. ρ vanishes on the center manifold provided θ is suitably chosen, i.e. $\theta = N^{-1} \sum_{k=1}^{N} \psi_k(0)$. Taking into account (3.31) one finally ends up with the synchronized state

$$\psi_j(t) = \Omega^* t + \psi_j^{\text{stat}} \ , \quad \text{where} \quad \psi_j^{\text{stat}} = \theta + \frac{\eta_j}{a} + \chi_j \tag{3.39}$$

is a constant phase shift, and $\chi_j = \eta_j a^{-1} + h_j(\boldsymbol{x}_c)$. Higher order phase shifts read $\chi_j = (c\eta_j^3 + 3c\eta_j V - cT)/a^4 + O(\|\boldsymbol{x}_c\|^4)$ if $C_1 = \ldots = C_4 = 0$, else one gets $\chi_j = b(\eta_j^2 - V)/a^3 + O(\|\boldsymbol{x}_c\|^3)$, where T is the third moment defined by (3.30). From (3.39) one immediately reads off that in the synchronized state all oscillators have the same frequency

$$\Omega^* = \Omega - \sum_{m=1}^{4} C_m + \frac{\sum_{m=1}^{4} m^2 C_m}{\left(\sum_{m=1}^{4} m K_m\right)^2} V + O(\|\boldsymbol{x}_c\|^3) \ , \tag{3.40}$$

(cf. (3.32), (3.33), and (3.38)), where V denotes the variance of the eigenfrequencies (cf. Sect. 3.5). According to condition (3.24)

$$|\eta_j| \ll \sum_{m=1}^{4} K_m \tag{3.41}$$

has to be fulfilled for $j = 1, \ldots, N$. This means that the distribution of the eigenfrequencies $\omega_1, \ldots, \omega_N$ is narrow as compared with the coupling strength. According to (3.39) for randomly distributed $\eta_1, \ldots, \eta_N$ one encounters small mutual phase differences within the cluster, whereas for vanishing $\eta_1, \ldots, \eta_N$ all oscillators are perfectly synchronized, i.e. $\psi_j(t) = \psi(t)$ for $j = 1, \ldots, N$.

The stability of the synchronized state still remains to be elucidated. To this end it is important to note that $\dot{\boldsymbol{x}}_c = \boldsymbol{0}$ holds as a consequence of condition (3.37). Moreover one has to take into account that apart from the center modes there are only stable modes. For this reason the center manifold is actually nothing but a point. On the other hand the center manifold is a *local attractor* as shown by Pliss (1964) and Kelley (1967). Thus, the center manifold, that means the investigated one-cluster state, is a stable fixed point.

3.7.3 Two Clusters

To study antiphase synchronization we analyze a two-cluster state with oppositely arranged clusters as shown in Fig. 3.2b. Let the oscillators be labeled appropriately so that the first cluster consists of oscillators $1, \ldots, N_1$, whereas the second cluster consists of oscillators $N_1 + 1, \ldots, N$, where N_ν denotes the number of oscillators in the νth cluster, and $N = N_1 + N_2$. The normalized size of the νth cluster will be denoted by $n_\nu = N_\nu / N$. The ansatz

$$\phi_j(t) = \psi_j(t) - \Omega^* t - \theta - \vartheta_j \tag{3.42}$$

reflects the clusters' antiphase arrangement, where $\vartheta_j = 0$ for the oscillators of the first cluster (i.e. $j = 1, \ldots, N_1$) and $\vartheta_j = \pi$ for the oscillators of the second cluster (i.e. $j = N_1 + 1, \ldots, N$).

According to condition (3.24) our analytical approach outlined in Sect. 3.6.1 cannot be performed for arbitrary values of $K_1, \ldots, K_4$ and $C_1, \ldots, C_4$. We, thus, have to assume that $K_1, K_3, C_1, \ldots, C_4$ are small in comparison with the synchronizing coupling constants K_2 and K_4. For this reason we set

$$K_1 \longrightarrow \varepsilon K_1 \ , \quad K_3 \longrightarrow \varepsilon K_3 \ , \quad C_m \longrightarrow \varepsilon C_m \ \ (m = 1, \ldots, 4) \ , \tag{3.43}$$

where the smallness parameter ε is of the same order of magnitude as $\eta_1,$ $\ldots, \eta_N$. This assumption guarantees that B_s is simply a diagonal matrix (cf. (3.22)) which can be inverted straigthforwardly according to (3.21) and (3.23). As in the former section we intend to cast the system under consideration into the form given by (3.15) and (3.16). To this end the nonlinear renormalization is performed, which guarantees that condition (3.37) is fulfilled. Unlike the one-cluster state due to the proper choice of $\Delta^{(1)}$ we encounter new quantities, which will be denoted by

$$\eta_j^\dagger = \varepsilon(C_1 + C_3)\left[n_\mu - n_\nu + (n_\mu - n_\nu)^2\right] \ , \tag{3.44}$$

where the jth oscillator is in the νth cluster, and $\mu \neq \nu$ (i.e. $\nu = 1, \mu = 2$ or vice versa). These quantities cause a shift of the deviations $\eta_1, \ldots, \eta_N$ and, thus, a shift of the eigenfrequencies, too. Let us set

$$\xi_j = \eta_j + \eta_j^\dagger \ , \tag{3.45}$$

in order to introduce *shifted eigenfrequencies* $\omega_j^\dagger$ according to

$$\omega_j^\dagger = \omega_j + \eta_j^\dagger = \Omega + \eta_j + \eta_j^\dagger \ . \tag{3.46}$$

The linear parts of center and stable modes are separated by the transformation $\varphi_j = \phi_j - \xi_j/(2K_2 + 4K_4) - \rho$, yielding

$$\boldsymbol{x}_c = (\varepsilon, \rho, \xi_1, \ldots, \xi_N) \ , \quad \boldsymbol{x}_s = (\varphi_1, \ldots, \varphi_N) \ . \tag{3.47}$$

In contrast to the one-cluster state in the two-cluster state shifted eigenfrequency deviations ξ_j serve as center modes. As explained in Sect. 3.7.2 condition (3.37) is taken into account in order to determine the frequency shift Δ in the synchronized state. In this way, we obtain

$$\Delta^{(0)} = \Delta^{(2)} = 0 \ , \quad \Delta^{(1)} = C_2 + C_4 + (C_1 + C_3)(n_1 - n_2)^2 \ , \tag{3.48}$$

$$\Delta^{(3)} = \delta_{(3)} + \delta_{(3)}^\dagger \tag{3.49}$$

with $\delta_{(3)}$ and $\delta_{(3)}^\dagger$ as in (B.1) and (B.2) in the appendix. The dagger indicates all quantities depending on $\eta_1^\dagger, \ldots, \eta_N^\dagger$ (such as $\delta_{(3)}^\dagger$). For the interpretation of the phase and frequency shifts it is important to note that these quantities only depend on the configuration. In contrast to these labeled variables the corresponding variables without dagger (such as $\delta_{(3)}$) additionally depend on the eigenfrequency distribution $\{\eta_j\}$.

According to the instructions of Sect. 3.7.2 the phase shifts are determined, where in this case $\gamma = 2K_2 + 4K_4$. With the appropriately chosen $\theta = N^{-1} \sum_{k=1}^N \psi_k(0) - n_2\pi$ on the center manifold ρ vanishes. Finally we obtain the stable synchronized state which is given by

$$\psi_j(t) = \Omega^* t + \psi_j^{\text{stat}} \ , \quad \text{where} \quad \psi_j^{\text{stat}} = \theta + \chi_j + \chi_j^\dagger + \vartheta_j \tag{3.50}$$

is the constant phase shift of the jth oscillator. χ_j and $\chi_j^\dagger$, the shifts of higher order, are listed as (B.6) and (B.7) in the appendix. χ_j and $\chi_j^\dagger$ influence the arrangement of the oscillators' phases in the two-cluster state in a different way. χ_j gives rise to mutual phase differences within a single cluster. By virtue of condition (3.24) these mutual phase differences are small in comparison with the coupling strength $(2K_2 + 4K_4)$. In particular, within a single cluster the oscillators are perfectly synchronized provided all oscillators have the same eigenfrequency, i.e. $\eta_1 = \cdots = \eta_N = 0$. In contrast to this $\chi_j^\dagger$ does not influence the oscillators arrangement within a single cluster. Rather by inducing a systematic phase shift of a whole cluster it disturbs the opposite arrangement of both clusters, where the latter is determined by ϑ_j according to (3.42) and (3.50) (cf. Fig. 3.2b). This deformation of the clusters' opposite arrangement will be illustrated numerically in Sect. 3.8.

According to (3.32), (3.33), (3.48), and (3.49) in the generic case, i.e. for non-vanishing coupling constants $C_1, \ldots, C_4$, the synchronization frequency Ω^* critically depends on the configuration of the synchronized state.

3.7.4 Three Clusters

Let us focus on a three-cluster state with three equally spaced clusters (cf. Fig. 3.2c). We can only perform our analytical analysis if we pay tribut to the choice of the model parameters: B_s, the matrix of the stable modes, becomes a simple diagonal matrix provided

$$K_\nu \longrightarrow \varepsilon K_\nu \;, \quad C_m \longrightarrow \varepsilon C_m \quad (\nu = 1, 2, 4 \text{ and } m = 1, \ldots, 4) \qquad (3.51)$$

holds. ε is the smallness parameter already used in the former section. Assumption (3.51) guarantees that B_s, the matrix of the stable modes, is a simple diagonal matrix as in (3.22). The clusters' equal spacing is taken into account by means of the ansatz

$$\phi_j(t) = \psi_j(t) - \Omega^* t - \theta - \vartheta_j \;, \qquad (3.52)$$

where $\vartheta_j = 0$ for the oscillators of the first cluster (i.e. $j = 1, \ldots, N_1$) and $\vartheta_j = 2\pi/3$ for the oscillators of the second cluster (i.e. $j = N_1 + 1, \ldots, N_1 + N_2$), and $\vartheta_j = 4\pi/3$ for the oscillators of the third cluster (i.e. $j = N_1 + N_2 + 1, \ldots, N$). Again, the renormalization and, in particular, the proper choice of $\Delta^{(1)}$, forces us to introduce shifts of the eigenfrequencies according to (3.46). The shift terms are listed in the appendix (cf. (B.10)).

The transformation $\varphi_j = \phi_j - \xi_j/(3K_3) - \rho$ separates the linear parts of center and stable modes, with x_c and x_s given by (3.47). As in the former sections the determination of the frequency shift Δ is based on condition (3.37). In this way, we end up with

$$\Delta^{(1)} = C_3 + \frac{1}{2}(C_1 + C_2 + C_4)\left[(n_1 - n_2)^2 + (n_2 - n_3)^2 + (n_3 - n_1)^2\right] \;, \quad (3.53)$$

$$\Delta^{(0)} = 0 \;, \quad \Delta^{(2)} = \delta_{(2)} + \delta_{(2)}^\dagger \;, \qquad (3.54)$$

where $\delta_{(2)}$ and $\delta_{(2)}^\dagger$ are listed in the appendix ((B.11) and (B.12)). The phase shifts are determined in a similar way as in the case of the two-cluster state. With $\gamma = 3K_3$ we immediately obtain $h_j(x_c) = H_j(x_c) + H_j^\dagger(x_c)$ according to (3.23). $H_j(x_c)$ and $H_j^\dagger(x_c)$ are presented in appendix B.2. ρ vanishes on the center manifold provided $\theta = N^{-1}\sum_{k=1}^N \psi_k(0) - (n_2 + 2n_3)2\pi/3$ is chosen. The stable synchronized state is obtained by transforming back to the phases. The former is given by the phases

$$\psi_j(t) = \Omega^* t + \psi_j^{\text{stat}} \;, \quad \psi_j^{\text{stat}} = \theta + \vartheta_j + \frac{\eta_j + \eta_j^\dagger}{3K_3} + H_j(x_c) + H_j^\dagger(x_c) \quad (3.55)$$

with $H_j(x_c)$ and $H_j^\dagger(x_c)$ from (B.8) and (B.9). As a consequence of condition (3.24) Δ, η_j, $\eta_j^\dagger$, H_j and $H_j^\dagger$ are small in comparison with the coupling strength $\gamma = 3K_3$. If this condition is not fulfilled our approach is no longer valid, and we have to restrict ourselves to a numerical investigation, which will be presented below.

In the synchronized state determined by (3.55) all oscillators have the same frequency Ω^* which deviates from the mean of the eigenfrequencies Ω according to (3.32), (3.33), (3.53), and (B.11). Generically, i.e. for non-vanishing coupling constants $C_1, \ldots, C_4$, the synchronization frequency crucially depends on the configuration of the synchronized state given by n_1, n_2, n_3. Figure 3.2c illustrates the oscillators' arrangement in the synchronized state: We encounter three clusters consisting of N_1, N_2, and N_3 oscillators. Within a single cluster the oscillators display small mutual phase differences according to (3.55). The equal spacing of the clusters is disturbed if $H_j^\dagger$ and $\eta_j^\dagger$ do not vanish. In particular, clusters of equal size are equally spaced.

3.7.5 Four Clusters

Let us finally investigate a four-cluster state consisting of four synchronized clusters which are equally spaced (cf. Fig. 3.2d). Hence, we choose the ansatz

$$\phi_j(t) = \psi_j(t) - \Omega^* t - \theta - \vartheta_j \ , \tag{3.56}$$

where $\vartheta_j = 0$ for the first cluster (i.e. $j = 1, \ldots, N_1$), $\vartheta_j = \pi/2$ for the second cluster (i.e. $j = N_1+1, \ldots, N_1+N_2$), $\vartheta_j = \pi$ for the third cluster (i.e. $j = N_1 + N_2 + 1, \ldots, N_1 + N_2 + N_3$), and $\vartheta_j = 3\pi/2$ for the fourth cluster (i.e. $j = N_1 + N_2 + N_3 + 1, \ldots, N$), where $N = N_1 + N_2 + N_3 + N_4$.

In order to achieve that B_s is of the simple form given by (3.22), we have to assume that

$$K_\nu \longrightarrow \varepsilon K_\nu \ , \quad C_m \longrightarrow \varepsilon C_m \quad (\nu = 1, 2, 3 \text{ and } m = 1, \ldots, 4) \tag{3.57}$$

holds. As in Sects. 3.7.3 and 3.7.4, the determination of $\Delta^{(1)}$ gives rise to shifts of the eigenfrequencies (cf. (B.13) in the appendix). The separation of the linear parts of the center and the stable modes is realized by the transformation $\varphi_j = \phi_j - \xi_j/(4K_4) - \rho$ with ξ_j from (3.45). In this way, we obtain $\boldsymbol{x}_c$ and $\boldsymbol{x}_s$ from (3.47). Based on condition (3.37) the frequency shift Δ is derived as in the former sections. Some calculations finally yield

$$\begin{aligned}
\Delta^{(0)} &= 0 \ , \\
\Delta^{(1)} &= C_4 + C_2(n_1 + n_3 - n_2 - n_4)^2 \\
&\quad + (C_1 + C_3)\left[(n_1 - n_3)^2 + (n_2 - n_4)^2\right] \ , \\
\Delta^{(2)} &= \delta_{(2)} + \delta_{(2)}^\dagger
\end{aligned} \tag{3.58}$$

with $\delta_{(2)}$ and $\delta_{(2)}^\dagger$ from (B.14) and (B.15). The phase shifts still have to be derived. To this end first $h_j(\boldsymbol{x}_c) = H_j(\boldsymbol{x}_c) + H_j^\dagger(\boldsymbol{x}_c)$ is determined by means of

(3.23) with $\gamma = 4K_4$, where $H_j(\boldsymbol{x}_c)$ and $H_j^\dagger(\boldsymbol{x}_c)$ are listed in appendix B.3. ρ vanishes on the center manifold if $\theta = N^{-1}\sum_{k=1}^N \psi_k(0) - (n_2 + 2n_3 + 3n_4)\pi/2$ is chosen. Finally one arrives at the stable synchronized state

$$\psi_j(t) = \Omega^* t + \psi_j^{\text{stat}} , \quad \text{where} \quad \psi_j^{\text{stat}} = \theta + \vartheta_j + \frac{\eta_j + \eta_j^\dagger}{4K_4} + H_j(\boldsymbol{x}_c) + H_j^\dagger(\boldsymbol{x}_c)$$
(3.59)

is a constant phase shift, with $H_j(\boldsymbol{x}_c)$ and $H_j^\dagger(\boldsymbol{x}_c)$ from appendix B.3. From condition (3.24) it follows that Δ, η_j, $\eta_j^\dagger$, H_j and $H_j^\dagger$ are small in comparison with $\gamma = 4K_4$.

In the four-cluster state given by (3.55) all oscillators have the same frequency Ω^*. The latter critically depends on the configuration $(n_1, \ldots, n_4)$. Within a single cluster the oscillators diplay small mutual phase differences due to χ_j, whereas $\chi_j^\dagger$ perturbs the clusters' equal spacing (cf. Fig. 3.2).

3.8 Complexity of Synchronized States

3.8.1 Hierarchy of Frequency Levels

Before we discuss the different influences of the configuration and the eigenfrequency distribution on the synchronization frequency, let us consider a homogenous population of oscillators, where all oscillators have the same eigenfrequency (i.e. $\eta_j = 0$). Obviously we encounter a *multiplicity of attractors* (modulo 2π) which depends on the configuration:

1. In the one-cluster state there is only one attractor because all oscillators are in phase (Fig. 3.2a).

2. In the two-cluster state (Fig. 3.2b) for fixed configuration (N_1, N_2) there are different combinations, depending on which particular oscillators belong to the first and second cluster, respectively. For instance, for $N_1 = 2, N_2 = 3$ there are 10 different possible combinations: The first cluster may consist of oscillators $\{1,2\}$, $\{1,3\}$, $\{1,4\}$, $\{1,5\}$, $\{2,3\}$, $\{2,4\}$, $\{2,5\}$, $\{3,4\}$, $\{3,5\}$ or $\{4,5\}$. In general, the number of different combinations is given by

$$n_{\text{comb}}(N_1, N_2) = \binom{N}{N_1} = \frac{N!}{N_1! \, N_2!} ,$$
(3.60)

where $N_1 + N_2 = N$ (cf. Fig. 3.3). Analogously we obtain the number of combinations in the three and four-cluster state. A little calculation shows that in the n-cluster state the number of combinations for fixed configuration $(N_1, \ldots, N_n)$ is given by

$$n_{\text{comb}}(N_1, \ldots, N_n) = \frac{N!}{\prod_{j=1}^n N_j!} , \quad \text{where} \quad N_1 + \ldots + N_n = N .$$
(3.61)

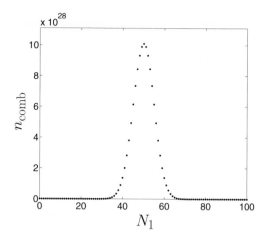

Fig. 3.3. *Multiplicity of attractors:* Number of attractors corresponding to different configurations (N_1, N_2) in the two-cluster state of 100 oscillators with eigenfrequencies $\omega_1 = \cdots = \omega_{100} = \Omega$ (cf. (3.60) and Fig. 3.2b). N_ν denotes the number of oscillators of the νth cluster.

Let us now assume that the eigenfrequencies are randomly distributed, where the distribution's width is small as compared with the coupling strength. In this case the smallness condition (3.24) of the center manifold is fulfilled, and the results of our analytical analysis are valid. In the one-cluster state the deviations $\eta_1, \ldots, \eta_N$ give rise to small mutual phase differences according to (3.39) and a frequency shift of second order (cf. (3.40)).

If the synchronized state consists of two or more clusters we encounter a qualitatively different situation: A variety of frequency levels occurs depending on both, the configuration and the eigenfrequency distribution. To illustrate this some notations will be introduced. From (3.48), (3.53), and (3.58) it follows, that $\Delta^{(1)}$ consists of two parts:

$$\Delta^{(1)} = \Delta_{\mathrm{I}}^{(1)} + \Delta_{\mathrm{II}}^{(1)} \ . \tag{3.62}$$

Only $\Delta_{\mathrm{II}}^{(1)}$ depends on the configuration, whereas $\Delta_{\mathrm{I}}^{(1)}$ does not depend on the configuration. Both $\Delta_{\mathrm{I}}^{(1)}$ and $\Delta_{\mathrm{II}}^{(1)}$ are independent of the eigenfrequency distribution. With these notations the lower order frequency shifts in the n-cluster state read:

$$\Delta_{\mathrm{I}}^{(1)} = \begin{cases} C_2 + C_4 & : & n = 2 \\ C_3 & : & n = 3 \\ C_4 & : & n = 4 \end{cases} \ . \tag{3.63}$$

On the other hand we obtain

$$\Delta_{\mathrm{II}}^{(1)} = (C_1 + C_3)(n_1 - n_2)^2 \tag{3.64}$$

for $n = 2$,

$$\Delta_{\text{II}}^{(1)} = \frac{1}{2}(C_1 + C_2 + C_4)[(n_1 - n_2)^2 + (n_2 - n_3)^2 + (n_3 - n_1)^2] \quad (3.65)$$

for $n = 3$, and

$$\Delta_{\text{II}}^{(1)} = C_2(n_1 + n_3 - n_2 - n_4) + (C_1 + C_3)[(n_1 - n_3)^2 + (n_2 - n_4)^2] \quad (3.66)$$

for $n = 4$. Obviously the configuration dependent shift $\Delta_{\text{II}}^{(1)}$ reflects the configuration's asymmetries.

Equations (B.1), (B.2), (B.11), (B.12), (B.14), and (B.15) of Appendix B provide us with the frequency shifts of lowest nonlinear order. While $\delta_{(2)}$ and $\delta_{(3)}$ depend on the eigenfrequency distribution $\{\eta_j\}$, $\delta_{(2)}^\dagger$ and $\delta_{(3)}^\dagger$ only depend on the configuration. Taking into account nonlinear terms of lowest order for $n = 2$ the synchronization frequency may, thus, be written as

$$\Omega_{\text{III}} = \underbrace{\overbrace{\Omega - \Delta_{\text{I}}^{(1)}\varepsilon}^{\Omega_{\text{I}}} - \Delta_{\text{II}}^{(1)}\varepsilon - \delta_{(3)}^\dagger\varepsilon^3}_{\Omega_{\text{II}}} - \delta_{(3)}\varepsilon^3 . \quad (3.67)$$

For $n = 3, 4$ one has to replace $\delta_{(3)}^\dagger\varepsilon^3$ and $\delta_{(3)}\varepsilon^3$ by $\delta_{(2)}^\dagger\varepsilon^2$ and $\delta_{(2)}\varepsilon^2$. The different terms of (3.67) reflect a hierarchy of frequency levels as illustrated in Fig. 3.4:

1. Ω_{I} refers to frequency shifts of first order, which are independent of both, configuration as well as eigenfrequency distribution.

2. Ω_{II} summarizes frequency shifts of first and higher order which exclusively depend on the configuration. Note that Ω_{II} does not depend on the eigenfrequency distribution. Above we already saw that every Ω_{II} level is associated with $n_{\text{comb}}(N_1, \ldots, N_n)$ different synchronized states (cf. (3.61)). That means, we encounter a configuration dependent multiplicity of attractors. For this reason a *degeneracy of Ω_{II}-frequency levels* occurs. The latter is additionally increased as a consequence of symmetry properties of Ω_{II} (cf. (3.62) and (3.67)): In the two- and three-cluster state Ω_{II} is invariant with respect to the interchange $N_j \leftrightarrow N_k$, and in the four-cluster state Ω_{II} is invariant with respect to the interchange $N_2 \leftrightarrow N_4$ for $N_1 = N_3$ and $N_1 \leftrightarrow N_3$ for $N_2 = N_4$.

3. Apart from configuration dependent effects Ω_{III} additionally takes into account the influence of the eigenfrequencies' detuning. Terms of lowest nonlinear order ($\delta_{(3)}\varepsilon^3$ for $n = 2$ and $\delta_{(2)}\varepsilon^2$ for $n = 3, 4$) are determined by the eigenfrequency distribution $\{\eta_j\}$. These terms vanish provided the eigenfrequencies are homogenously distributed ($\omega_j = \Omega$). Otherwise, e.g., for randomly distributed $\{\omega_k\}$, every degenerated Ω_{II} level splits into a multitude of slightly different Ω_{III}-levels as illustrated in Fig. 3.4 (cf. (B.1), (B.11), and (B.14)).

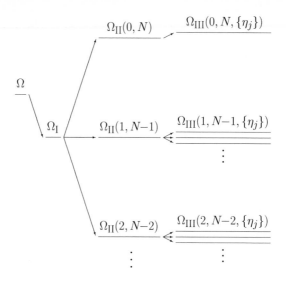

Fig. 3.4. *Hierarchy of frequency levels:* Schematic illustration of the different frequency levels corresponding to different configurations (N_1, N_2) of the two-cluster state. The frequency shift described by $\Omega_{\mathrm{I}} = \Omega - \Delta_{\mathrm{I}}^{(1)}\varepsilon$ is independent of the configuration as well as the eigenfrequency distribution. Lower order frequency splitting depends on the configuration: $\Omega_{\mathrm{II}}(N_1, N_2) = \Omega - (\Delta_{\mathrm{I}}^{(1)} + \Delta_{\mathrm{II}}^{(1)})\varepsilon - \delta_{(3)}^{\dagger}\varepsilon^3$. The eigenfrequency distribution determines higher order frequency splitting according to $\Omega_{\mathrm{III}}(N_1, N_2, \{\eta_j\}) = \Omega - (\Delta_{\mathrm{I}}^{(1)} + \Delta_{\mathrm{II}}^{(1)})\varepsilon - \delta_{(3)}^{\dagger}\varepsilon^3 - \delta_{(3)}\varepsilon^3$. Note that frequency splitting only occurs in a synchronized state with more than one cluster.

In summary, a coarse $(\sim \varepsilon)$ and a fine $(\sim \varepsilon^2$ or $\sim \varepsilon^3)$ splitting of Ω_{II} levels occurs. The former is due to the configuration, whereas the latter is due to the oscillators' (weak) detuning.

3.8.2 Phase and Frequency Shifts

Our analytical investigation is based on the center manifold theorem which is an essentially local tool. In order to apply this theorem according to condition (3.24) we have to assume that the eigenfrequency deviations η_j as well as the smallness parameter ε are small in comparison with the coupling strength. If condition (3.24) is not fulfilled strictly speaking we are not allowed to apply the center manifold theorem. Nevertheless, also in this case our results might be accurate at least qualitatively. For this reason this section is devoted to numerically checking our formulas for phase and frequency shifts, in particular, for model parameters not fulfilling the smallness condition (3.24).

First we examine phase shifts in two-cluster states of 20 oscillators, where $\eta_1 = \cdots = \eta_N = 0$, $K_2 = 0.5$, $K_1 = K_3 = K_4 = 0.01$. As all oscillators have the same eigenfrequency within one cluster all share the same phase (modulo 2π, cf. (B.6)). Model equation (3.2) was integrated for 19 different coexisting stable synchronization configurations ($N_1 = 1, \ldots, 19$). To this end in each simulation the system was started in the neighborhood of one of the 19 different coexisting stable synchronized states. Figures 3.5 and 3.6 display the analytically derived *normalized phase difference*

$$\Delta\psi^{(1,2)} = \frac{\psi_j^{\text{stat}} - \psi_k^{\text{stat}}}{2\pi} \tag{3.68}$$

for two oscillators (j and k) from two different clusters with ψ_j^{stat} and ψ_k^{stat} from (3.50). $\Delta\psi^{(1)}$ only takes into account terms of first order ($\sim \varepsilon$), whereas $\Delta\psi^{(2)}$ also takes into account terms of second order ($\sim \varepsilon^2$) (cf. (B.7)). Figures 3.5 and 3.6 additionally present $\Delta\psi^{\text{num}}$, the numerically determined normalized phase difference between both clusters (performed by means of a 4th order Runge-Kutta algorithm).

While in each cluster the oscillators are perfectly synchronized, the phase difference between the two clusters depends on the configuration (i.e. the number of oscillators within each cluster). A perfect anti-phase state ($\Delta\psi = 0.5$) does only occur if both clusters are of equal size ($n_1 = n_2$). For $n_1 \neq n_2$ the anti-phase arrangement is perturbed due to the quantities $\eta_1^\dagger, \ldots, \eta_N^\dagger$, which cause a shift of the eigenfrequencies as explained in Sect. 3.7.3.

For the same simulations Figs. 3.7 and 3.8 show the numerically determined mean synchronization frequency

$$\Omega_{\text{num}}^* = \frac{1}{20} \sum_{k=1}^{20} \omega_k^* , \tag{3.69}$$

where ω_k^* denotes the numerically determined synchronization frequency of the kth oscillator. In Figs. 3.7 and 3.8 the analytically derived synchronization frequency is additionally plotted. To illustrate the effect of terms of different order terms of first order ($\Omega^{(1)}$) and terms of third order ($\Omega^{(3)}$) are taken into account by setting

$$\Omega^{(1)} = \Omega - \Delta^{(1)}\varepsilon \quad , \quad \Omega^{(3)} = \Omega - \Delta^{(1)}\varepsilon - \Delta^{(3)}\varepsilon^3 \tag{3.70}$$

according to (3.32), (3.33), (3.48), and (3.49).

If the smallness condition (3.24) is fulfilled, the application of the center manifold theorem is justified, and the numerical results perfectly agree with the theoretical results (Figs. 3.5a–3.8a). There is a good agreement between theory and numerical results even if the parameters $C_1, \ldots, C_4$ are of the same order of magnitude as the synchronizing coupling (K_2) (Figs. 3.5b–3.8b). However, for larger values of $C_1, \ldots, C_4$ terms of first and third order are no longer sufficient to achieve a good agreement between theory and simulation (Figs. 3.5c,d–3.8c,d). In this case one has to take into account higher order terms $\Delta^{(4)}, \Delta^{(5)}, \ldots$, too. The latter can be derived in a straightforward way along the given lines of our approach.

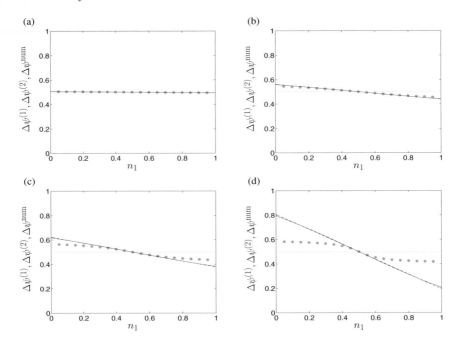

Fig. 3.5a–d. *Normalized phase difference of two clusters:* The plots compare analytically derived and numerically revealed phase shifts in the synchronized states of two clusters, where $N = 20$, $\eta_1 = \ldots = \eta_N = 0$, $K_2 = 0.5$, $K_1 = K_3 = K_4 = 0.01$. To this end $\Delta\psi^{\mathrm{num}}$ (*dots, surrounded by small circles*), $\Delta\psi^{(1)}$ (*dashed line, covered by solid line*) and $\Delta\psi^{(2)}$ (*solid line*) are determined and plotted according to the explanation in the text. Plots show normalized phase difference for different values of $C_1, \ldots, C_4$: $C_1 = C_3 = -0.01, C_2 = C_4 = 0.01$ (**a**) (fulfilling smallness condition (3.24), i.e. $|C_j| \ll \gamma = 2K_2 + 4K_4$), $C_1 = C_3 = -0.1, C_2 = C_4 = 0.1$ (**b**), $C_1 = C_3 = -0.2, C_2 = C_4 = 0.2$ (**c**), $C_1 = C_3 = -0.5, C_2 = C_4 = 0.5$ (**d**). In order to illustrate differences between analytically derived and numerically revealed phase differences $\Delta\psi^{(1)}$ and $\Delta\psi^{(2)}$ are plotted as continuous functions of n_1.

Note that for $|n_1 - n_2| \ll 1$ and for $n_1 = 1$ or $n_2 = 1$ there is a good agreement between analytical and numerical results even for large values of the model parameters (Figs. 3.5c,d–3.8c,d). As $|K_1|$ and $|K_3|$ are small compared to $\gamma = 2K_2 + 4K_4$, terms of second order contribute to $\Delta\psi$ only marginally (cf. (B.7)). Depending on the model parameters the synchronization frequency in the anti-phase state may be both larger or smaller than in the in-phase state.

3.8.3 Cluster Variables

In the next two sections we dwell on transitions between different cluster states. To this end we describe the state of synchronization by means of *cluster variables*

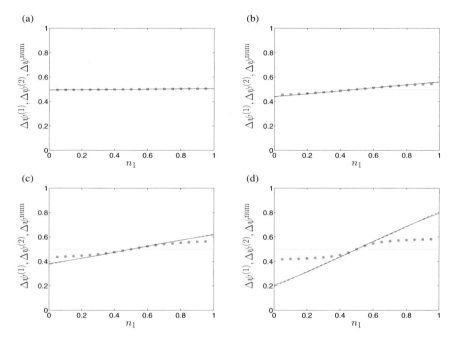

Fig. 3.6a–d. *Normalized phase difference of two clusters:* Plots show $\Delta\psi^{\mathrm{num}}$ (*dots, surrounded by small circles*), $\Delta\psi^{(1)}$ (*dashed line, covered by solid line*) and $\Delta\psi^{(2)}$ (*solid line*) for the same model parameters as in Fig. 3.5 except for the values of $C_1,\ldots,C_4$: $C_1 = C_3 = 0.01, C_2 = C_4 = -0.01$ (**a**) (fulfilling smallness condition (3.24), i.e. $|C_j| \ll \gamma = 2K_2 + 4K_4$), $C_1 = C_3 = 0.1, C_2 = C_4 = -0.1$ (**b**), $C_1 = C_3 = 0.2, C_2 = C_4 = -0.2$ (**c**), $C_1 = C_3 = 0.5, C_2 = C_4 = -0.5$ (**d**).

$$Z_m(t) = R_m(t)\,\exp[\mathrm{i}\varphi_m(t)] = \frac{1}{N}\sum_{k=1}^{N}\exp[\mathrm{i}m\psi_k(t)]\quad (m=1,\ldots,4)\,,\quad (3.71)$$

where ψ_k is the phase of the kth oscillator, and N is the number of oscillators. The real quantities R_m and φ_m will be denoted as mth *cluster amplitude* and mth *cluster phase*. For $m = 1$ (3.71) was used to describe in-phase synchronization (see, for instance, Aizawa 1976, Winfree 1980, Kuramoto 1984, Yamaguchi and Shimizu 1984, Shiino and Frankowicz 1989, Matthews and Strogatz 1990, Tass and Haken 1996a). Daido (1992b) used cluster variables in order to analyze the onset of entrainment by means of his order function. Hansel, Mato and Meunier (1993a) used Z_1 and Z_2 in order to investigate oscillations of two-cluster states. In this book $Z_1,\ldots,Z_4$ will be used to study the dynamics of n-cluster states, where $n = 1,\ldots,4$.

Apart from in-phase synchronization and incoherency the cluster variables enable us to detect complex synchronization patterns: While in the incoherent state $R_1,\ldots,R_4$ vanish, in a perfectly synchronized n-cluster state $R_1 = R_n = 1$ holds. In particular, in a perfectly synchronized one-cluster state

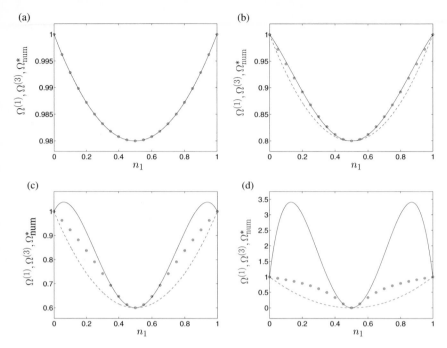

Fig. 3.7a–d. *Frequency shifts of two clusters:* Plots compare analytically derived and numerically revealed frequency shifts in the synchronized states of two clusters with parameters as in Fig. 3.5. Ω^*_{num} (*dots, surrounded by small circles*), $\Omega^{(1)}$ (*dashed line*) and $\Omega^{(3)}$ (*solid line*) are determined and plotted according to the explanation in the text. Plots show frequency shifts for different values of $C_1, \ldots, C_4$: $C_1 = C_3 = -0.01, C_2 = C_4 = 0.01$ (**a**) (fulfilling smallness condition (3.24), i.e. $|C_j| \ll \gamma = 2K_2 + 4K_4$), $C_1 = C_3 = -0.1, C_2 = C_4 = 0.1$ (**b**), $C_1 = C_3 = -0.2, C_2 = C_4 = 0.2$ (**c**), $C_1 = C_3 = -0.5, C_2 = C_4 = 0.5$ (**d**). In order to illustrate differences between analytically derived and numerically revealed frequency shifts $\Omega^{(1)}$ and $\Omega^{(3)}$ are plotted as continuous functions of n_1 (cf. Fig. 3.5).

$R_\nu = 1$ holds for $\nu = 1, 2, \ldots$. Less distinct n-cluster states correspond to smaller cluster amplitudes. Based on the slaving principle in the following chapters we will allocate a deeper meaning to the cluster variables.

3.8.4 Frozen States

In Sect. 3.8.2 we saw that due to the oscillators' mutual interactions pronounced shifts of the synchronization frequency may occur. To illustrate the strong impact of the configuration on the synchronization frequency the term *frozen state* was suggested to denote a stable n-cluster state with vanishing (or at most small) synchronization frequency (Tass 1997a). In Fig. 3.7d an example of a frozen state (for $n_1 = 0.5$) is plotted.

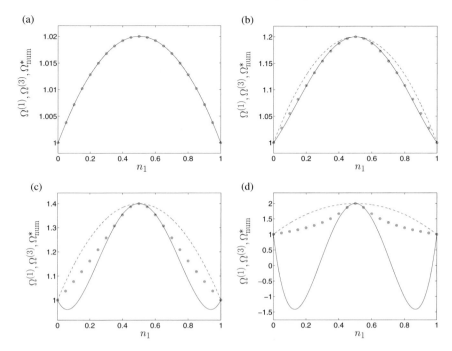

Fig. 3.8a–d. *Frequency shifts of two clusters:* Plots show Ω^*_{num} (*dots, surrounded by small circles*), $\Omega^{(1)}$ (*dashed line*) and $\Omega^{(3)}$ (*solid line*) for the same model parameters as in Fig. 3.5 except for the values of $C_1, \ldots, C_4$: $C_1 = C_3 = 0.01, C_2 = C_4 = -0.01$ (**a**) (fulfilling smallness condition (3.24), i.e. $|C_j| \ll \gamma = 2K_2 + 4K_4$), $C_1 = C_3 = 0.1, C_2 = C_4 = -0.1$ (**b**), $C_1 = C_3 = 0.2, C_2 = C_4 = -0.2$ (**c**), $C_1 = C_3 = 0.5, C_2 = C_4 = -0.5$ (**d**).

Transitions from an unstable n-cluster state to a frozen one-cluster state are shown in Fig. 3.9 for $n = 2$ and Fig. 3.10 for $n = 3$. Besides the oscillators' phase dynamics (illustrated by $\sin \psi_1, \sin \psi_2, \ldots$) the plot shows the cluster amplitudes R_1, R_n and the mean frequency of all oscillators given by

$$F = \frac{1}{N} \sum_{k=1}^{N} \dot{\psi}_k \ . \tag{3.72}$$

In Fig. 3.9 the oscillators start in an unstable antiphase arrangement. Correspondingly at first R_1 vanishes, whereas R_2 equals 1. With increasing time the oscillators join into one giant cluster, i.e. R_1 and R_2 tend to 1. As the one-cluster state is established all oscillators freeze, and F vanishes. Likewise in Fig. 3.10 an unstable three-cluster state ($R_1(0) = 0$, $R_3(0) = 1$) vanishes, making room for a frozen one-cluster state: R_1 and R_3 tend to 1, and F vanishes while time increases. Remarkably, not only the synchronization frequencies of the frozen states but also those of the two unstable initial states

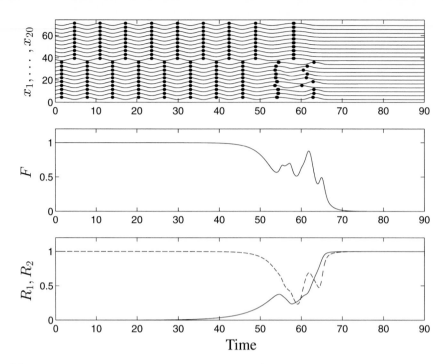

Fig. 3.9. *Frozen state:* Plots show a transition from an unstable two-cluster state to a stable frozen one-cluster state. Numerical integration of model equation (3.2) for $N = 20, K_1 = 0.2, K_2 = K_3 = K_4 = 0.01, C_1 = C_3 = 0.5, C_2 = C_4 = 0$. Initial phases of first and second cluster (i.e. $\psi_1(0), \ldots, \psi_{10}(0)$ and $\psi_{11}(0), \ldots, \psi_{20}(0)$) are normally distributed around 0 and π with standard deviation 0.001. In the upper plot the oscillators' phase dynamics is illustrated by $x_k = \sin\psi_k + 3.5k$ $(k = 1, \ldots, 20)$. Local maxima of $x_1, \ldots, x_{20}$ are indicated by black dots in order to emphasize the synchronization pattern. Middle plot shows mean frequency of all oscillators F according to (3.72). Lower plot displays the dynamics of the cluster amplitudes R_1 (*solid line*) and R_2 (*dashed line*).

totally agree with our analytically derived results (cf. (3.33), (3.48), (3.49), (3.53)) and (3.54)).

3.8.5 Transient Behavior

Cluster states of different configurations are generically (i.e. for $C_j \neq 0$) associated with different synchronization frequencies. For this reason transitions between different configurations are typically connected with transitions between different synchronization frequencies. The term synchronization frequency in the context of an unstable synchronized state denotes nothing but the oscillators' mean frequency in this particular state (cf. Tass 1997a).

We illustrate the intimate relationship between transitions of configurations and transitions of synchronization frequencies by considering a syn-

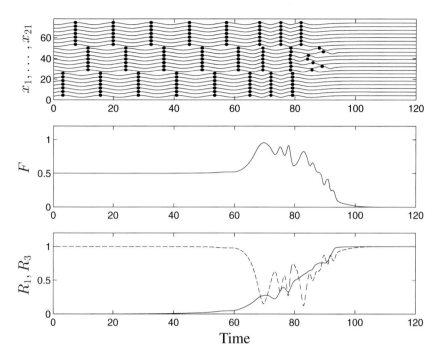

Fig. 3.10. *Frozen state:* An unstable three-cluster state vanishes while a stable frozen one-cluster state evolves. Numerical integration of model equation (3.2) for $N = 21, K_1 = 0.2, K_2 = K_3 = K_4 = 0.01, C_1 = C_3 = 0.5, C_2 = C_4 = 0$. Initial phases of clusters 1, 2, 3 (i.e. $\psi_1(0), \ldots, \psi_7(0), \psi_8(0), \ldots, \psi_{14}(0), \psi_{15}(0), \ldots, \psi_{21}(0)$) are normally distributed around $0, 2\pi/3, 4\pi/3$ with standard deviation 0.001. Upper plot shows $x_k = \sin \psi_k + 3.5k$ ($k = 1, \ldots, 21$). Middle plot shows mean frequency of all oscillators F. Lower plot displays cluster amplitudes R_1 (*solid line*), and R_3 (*dashed line*). Analogous format as in Fig. 3.9.

chronized state which is an unstable fixed point. Let the system start in a neighborhood of an unstable fixed point so that it will be attracted by a stable fixed point, i.e. a stable cluster state. While approaching the stable state, the model parameters $C_1, \ldots, C_4$ crucially determine whether the oscillators' mean frequency F increases (Fig. 3.11), decreases (Fig. 3.12) or remains unchanged (Fig. 3.13). The model parameters of the simulations illustrated in Figs. 3.11–3.13 differ only as far as $C_1, \ldots, C_4$ is concerned. For vanishing $C_1, \ldots, C_4$ the synchronization frequency remains unchanged because ρ and F are conserved quantities as shown in Sect. 3.7.3.

Interestingly, the numerically revealed synchronization frequencies of the unstable synchronized states of Figs. 3.11–3.13 perfectly agree with the analytical results derived in Sect. 3.7.3.

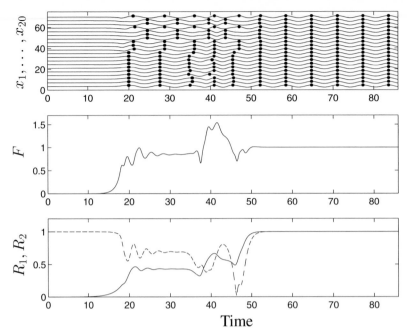

Fig. 3.11. *Transient behavior:* Unstable frozen two-cluster state vanishes, and stable one-cluster state evolves. Numerical integration of model equation (3.2) for $N = 20, K_1 = 0.5, K_2 = K_3 = K_4 = 0.01, C_1 = 0, C_3 = -1, C_2 = C_4 = 0.5$. Initial phases of clusters 1 and 2 (i.e. $\psi_1(0), \ldots, \psi_{10}(0)$ and $\psi_{11}(0), \ldots, \psi_{20}(0)$) are normally distributed around 0 and π with standard deviation 0.001. Upper plot shows $x_k = \sin\psi_k + 3.5k$ for all oscillators ($k = 1, \ldots, 20$). Local maxima of $x_1, \ldots, x_{20}$ are indicated by black dots. Middle plot shows mean frequency of all oscillators F (*solid line*) and amplitude of all oscillators R (*dashed line*). Lower plot displays amplitudes of cluster 1 (R_1, *solid line*), cluster 2 (R_2, *dashed line*), and cluster 3 (R_3, *dotted line, covered by solid line*).

3.8.6 Coupling Mechanism

During our analysis we encountered the intimate relationship between coupling mechanism and evolving cluster states. Phase-locked cluster states, such as two-, three- and four-cluster states, only occur provided the coupling contains higher harmonics. The majority of studies mentioned in Sect. 3.3 was not dedicated to phase-locked cluster states. As yet the latter were addressed by only few studies:

Okuda (1993) analyzed globally coupled phase oscillators, where the coupling contained higher harmonics, and the coupling coefficients where attractive as well as repulsive. His study revealed a rich dynamical behavior, for instance, besides in-phase states he observed anti-phase states, oscillating cluster states and chaotic long transients. Okuda did not study the dependence of the frequency shifts on the cluster state configuration and on the

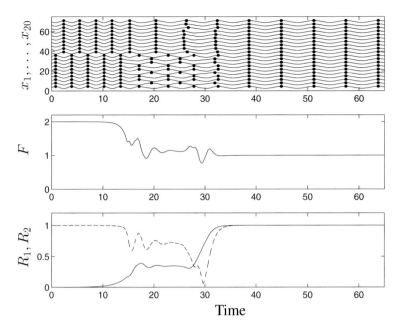

Fig. 3.12. *Transient behavior:* Transition from an unstable two-cluster state to a stable one-cluster state. Numerical integration and model parameters as in Fig. 3.11 except for $C_1 = 0, C_3 = 1, C_2 = C_4 = -0.5$. The same format as in Fig. 3.11 is used.

oscillators' detuning for two reasons: (a) He restricted his analysis to 'symmetric' cluster states where each cluster contains the same number of oscillators. (b) In his model all oscillators had the same eingenfrequency ($\omega_j = \Omega$).

Hansel, Mato and Meunier (1993a) focussed on the issue of phase-locked cluster states in a network of globally coupled oscillators, too. The coupling in their model contained Fourier modes of first and second order. They observed three types of dynamics: a one-cluster state, a totally incoherent state, and a pair of unstable two-cluster states connected by a homoclinic orbit. In the last regime adding noise causes oscillations between the two two-cluster states. As the oscillators in their model had uniform eigenfrequencies ($\omega_j = \Omega$), their study was not devoted to the influence of eigenfrequency detuning and configuration on the synchronization frequency of stable n-cluster states, too.

In a model for synchronized oscillations in the visual cortex Tass and Haken (1996a, 1996b) investigated how repulsive coupling gives rise to antiphase synchronization. In this model the synchronized state corresponds to one attractor (modulo 2π), so that no multiplicity of attractors occurred. In particular, the configuration only depends on the distribution of repulsive and attractive coupling.

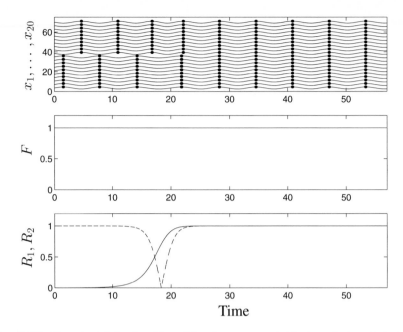

Fig. 3.13. *Transient behavior:* Transition from an unstable two-cluster state to a stable one-cluster state. Parameters as in Fig. 3.11 except for $C_1 = C_2 = C_3 = C_4 = 0$. The same format as in Fig. 3.11. Note that F is a conserved quantity.

The analysis presented in this chapter is in essence based on a former study (Tass 1997a), which served to work out the strong dependence of the synchronization frequency on the configuration. While analyzing the stimulation of synchronized activity we shall meet this dependence time and again.

Of course, in a realistic setting not all neurons interact with one another with equal strength. Hence, one has to take into account randomly distributed coupling coefficients as sketched in Tass (1997a). The scattering of the coupling constants will give rise to additional terms of phase and frequency shifts. For a detailed analysis of the impact of randomly distributed couplings on the synchronization frequency in one-cluster states I refer, for instance, to Tass and Haken (1996a). However, in order to point out the relationship between synchronization frequency and configuration it is sufficient to restrict the investigation to global coupling.

In addition to the investigations of phase-locked clusters a study addressing the influence of the coupling mechanism on the oscillators' arrangement within a single synchronized group should be mentioned. Daido (1993) focussed on a coupling where higher harmonics play a prominent role as compared with sinusoidal coupling. To this end he analyzed phase oscillators globally coupled via a periodic step function. He observed partial perfect synchrony, that means a part of the population is perfectly synchronized although the eigenfrequencies are randomly distributed. It would be very interesting to understand the impact of time delays on step function-like coupling in order to judge whether or not this type of coupling may be a candidate for interneuronal coupling giving rise to in-phase synchronization as discribed in Sect. 3.2.

3.9 Neural Coding

In this section we do not focus on the coding properties of single neurons. About this I refer, for instance, to Perkel and Bullock (1968), Orban (1984), Rieke et al. (1997). Rather let us come back to the question asked in Sect. 3.2: What is the neural mechanism realizing scene segmentation? Before the concept of von der Malsburg and Schneider (1986) will be augmented, let me first stress an important feature of scene segmentation.

3.9.1 Information Compression

In 1979 Haken pointed out the analogy between pattern recognition and pattern formation (cf. Haken 1991). According to his notion neural activity related to local features contributes to a dynamical pattern recognition process which is governed by the order parameters. According to the slaving principle the number of order parameters is typically much smaller as compared with the number of active neurons (Haken 1977, 1983). Consequently pattern recognition is intimately related to information compression (Haken 1979, 1988, 1991). As a consequence of the in-phase synchronization of the firing activity of all neurons encoding one particular object the degrees of freedom of these neurons would decrease drastically. In this way in-phase synchronization could serve as a means of information compression (cf. Shimizu 1985). The same holds for n-cluster states, where the dynamics of a huge amount of neurons can be described by only a few variables, namely the phase of one of the clusters and the configuration.

3.9.2 Coding by Clusters

Aware of the abundance of synchronized neuronal activity one may expect n-cluster states to occur in the firing patterns of neuronal populations, too. This conjecture may guide us to some speculations concerning the cluster states' physiological function. First we consider sensory segmentation. The original concept of the coding of sensory information by *von der Malsburg and Schneider* (1986) is based on in-phase synchronization of all neurons encoding one particular object. This concept was augmented in the following way (Tass 1997a): All neurons, which encode local features belonging to one particular object, may also join into a cluster state consisting of two or more phase-locked clusters. This type of encoding opens up additional effective properties: Apart from the object as a whole additional attributes of the object can be encoded by (a) the type of the cluster state (e.g. two-cluster state or three-cluster state) and (b) the configuration of the cluster state.

Besides the pattern related emergence of a single cluster state one might also expect a particular object to be encoded dynamically by means of (a) transitions between different configurations with fixed type of cluster state, and (b) transitions between different types of cluster states, e.g., transition from a two-cluster state to a three-cluster state. The dynamics of these transitions open up further encoding properties: On the one hand transitions from an unstable cluster state to a stable cluster state as analyzed in Sect. 3.8.5 might occur. On the other hand, for instance, oscillations between different unstable cluster states might appear as investigated in a population of phase oscillators by Hansel, Mato and Meunier (1993a).

We saw that transitions between different cluster states are typically associated with transitions between different synchronization frequencies (cf. Tass 1997a). For this reason it is difficult or even impossible to detect transient synchronization processes of this kind by means of linear correlation techniques. Obviously, the question as to whether or not phase-locked clusters play an important role in neural coding can only be answered if experimentalists are provided with appropriate data analysis tools.

The synchronization behavior of model equation (3.2) is not only interesting for the understanding of scene segmentation. Rather synchronized firing is a common feature of neuronal populations which is not at all limited to the visual system (see, for instance, Vaadia et al. 1995, Roelfsema et al. 1997). Hence, (3.2) may be a starting point for the analysis of other physiologically relevant synchronization processes, too. Let me sketch two examples:

(a) The first example refers to *bimanual movement coordination*: In an experiment by Kelso (1981, 1984) a subject was asked to perform parallel movements of both index fingers paced by a metronome. As the metronome's frequency exceeds a critical value an entirely involuntary transition to an antiparallel movement occured. In 1985 Haken, Kelso, and Bunz presented a paradigmatic macroscopic model for this transition from an antiphase coordination mode to an in-phase coordination mode. Along the lines of a top-down approach their model describes the oscillatory dynamics of both index fingers. We may suggest that model (3.2) can be used to develop a mesoscopic version of the Haken–Kelso–Bunz-model describing the activity of populations of neurons located, e.g., in the left and right motor cortex: For simplicity set $K_3 = K_4 = C_j = 0$. Let us assume that the remaining coupling parameters depend on the movement frequency. Furthermore we assume that, for instance, with increasing movement frequency K_1 increases and K_2 decreases giving rise to a transition from a two-cluster state to a one-cluster state. The two-cluster state consists of two neuronal populations firing in antiphase and located in different motor cortices, while in the one-cluster state both populations fire in unison.

However, such a model would merely be a first minimal model because it is well known that during bimanual coordination tasks besides the primary motor cortices several other brain areas are active, too (see, for instance, Stephan et al. 1997). Hence, in accordance with experimental results revealed by imaging techniques several neuronal populations have to be taken into account.

(b) The second example is about *visuomotor integration* : Langenberg et al. (1992) investigated how delayed visual feedback changes the performance during sinusoidal forearm tracking. In their experiment subjects were asked to track a visually presented target with a manipulandum. The matching of proprioceptive and visual feedback was perturbed by inserting an artificial delay into the visual feedback loop. To study how both types of feedback information contribute to movement control during this tracking task Tass, Wunderlin and Schanz (1995) developed a macroscopic model for the oscillatory forearm movement. This model predicted several qualitatively different tracking movement patterns, namely fixed point and oscillatory tracking behavior, cycle slipping and drift. All of them were verified experimentally (Tass et al. 1996).

In order to develop a mesoscopic version of the model of Tass, Wunderlin and Schanz one has to take into account several interacting neuronal populations located in different brain areas. As in the former case of bimanual coordination this can only be realized based on experimental data stemming from imaging techniques. In this way one may hope to understand how the oscillatory activity of several neuronal populations is coordinated during well-defined tracking movement patterns.

The issue of coordination of neuronal activity between different brain areas during a visuomotor integration task was addressed in an experiment by Roelfsema et al. (1997). In that experiment awake cats had to respond to visual stimuli by pressing and releasing a lever with their forepaw. Among other things Roelfsema et al. measured the local field potential by means of implanted electrodes. They found synchronization with zero time-lag between areas of the visual and the parietal cortex, and between the parietal and the motor cortex.

If one wants to model interactions of remote populations, and, moreover, if artificial time delays occur, actually one ends up with networks of limit-cycle oscillators interacting via time-delayed coupling. Networks of this kind may be expected to exhibit an overwhelming variety of synchronization phenomena. For instance, the time delay may additionally give rise to shifts of the synchronization frequency as it was shown for a network of phase oscillators with time-delayed nearest-neighbour coupling by Niebur, Schuster and Kammen (1991). In general, the analytical investigation of networks of time-delayed coupled oscillators is rather demanding. However, to understand, in particular, low-dimensional sychronization processes in such networks one may, e.g., profit from the extension of the slaving principle to delay differential equations by Wischert et al. (1994).

Finally, from the neurophysiological as well as theoretical point of view it is important to investigate the impact of noise on the synchronization behavior of (3.2). For instance, due to the multiplicity of attractors in particular in the thermodynamic limit $(N \gg 1)$ one might expect similar effects as observed by Wiesenfeld and Hadley (1989) in the context of attractor crowding. In other words, do the cluster states survive in spite of the presence of noise? This is a crucial point because in this case one has to expect the cluster states to occur in a natural setting. We will come back to this question below.

3.10 Summary

This chapter was about the synchronization behavior of a population of continuously interacting and globally coupled phase oscillators. Networks of this kind are used to model oscillatory activity, e.g., observed in different branches of physiology. Apart from in-phase synchronization one encounters clustering provided the oscillators' coupling contains higher harmonics, too (cf. Hansel, Mato and Meunier 1993a, Okuda 1993, Tass 1997a).

On the one hand one may observe stationary stable synchronized states: According to the intial conditions and the model parameters all oscillators form one giant cluster, or two, three or four phase-locked clusters of arbitrary size occur. As illustrated in Fig. 3.2 within one cluster the phase difference between any two oscillators is small compared to the phase difference between different clusters. These cluster states are phase-locked states, so that all oscillators run at the same frequency. The configuration of the cluster state, i.e. the number of oscillators within each cluster, critically determines synchronization frequency and shifts of the oscillators' mutual phase differences. In particular a hierarchy of synchronization frequency levels occurs, displaying a coarse and a fine splitting of frequency levels. The coarse splitting is due to the configuration, whereas the fine splitting is due to the oscillators' (weak) detuning (Tass 1997a). Transitions between different configurations are typically connected with changes of the synchronization frequency. For fixed model parameters the configuration may even give rise to a frozen state, i.e. a synchronized cluster state with vanishing synchronization frequency (Tass 1997a).

On the other hand different model parameters may give rise to dynamical synchronization phenomena such as oscillating cluster states and chaotic long transients (Okuda 1993) or oscillations between two two-cluster states (Hansel, Mato and Meunier 1993a).

Both stationary and dynamical synchronization patterns may be important for different physiological processes exhibiting synchronized oscillatory activity. For instance, in the context of visual pattern recognition in contrast to coding by synchrony (von der Malsburg and Schneider 1986) phase-locked clusters open up additional coding facilities. An object as a whole could, e.g., be encoded by means of neurons joined into a phase-locked cluster state, whereas additional information could be encoded by means of the configuration of this particular cluster state (Tass 1997a).

4. Stochastic Model

4.1 Introductory Remarks

The goal of this chapter is to present a model equation which describes the behavior of a population of interacting phase oscillators subjected to stimulation and random forces. With this aim in view first a stochastic differential equation, a so-called Langevin equation will be derived which describes how the oscillators' phase dynamics is influenced by their mutual interactions, by the stimulus and by the random forces. As a consequence of the presence of noise the cluster of oscillators does not move along a trajectory as it is known from systems without noise. Rather the system is permanently kicked by the random forces while its dynamics evolves in time. Hence, we have to deal with a stochastic description of the phase dynamics.

To this end we shall determine the Fokker–Planck equation which belongs to the system's Langevin equation. The Fokker–Planck equation is a partial differential equation providing us with the time course of the probability density $f(\{\psi_l\}, t)$, where $\{\psi_l\}$ is a shortform of $(\psi_1, \ldots, \psi_N)$, the vector of the phases of the cluster's oscillators. $f(\{\psi_l\}, t)\, d\psi_1 \ldots d\psi_N$ gives us the probability of finding the phases of the oscillators at time t in the intervals $\psi_j \ldots \psi_j + d\psi_j$ $(j = 1, \ldots, N)$. In other words, due to the random forces the time courses of the phases cannot be given with arbitrary precision. Nevertheless, this probabilistic description yields a tremendous amount of information especially if the cluster consists of a large number of oscillators.

Unfortunately, for large N it is difficult to analyze the Fokker–Planck equation, in particular, analytically. However, such an investigation is important to judge whether there are order parameters governing the dynamics of the cluster. Using the order parameter concept enables us to understand the principles which determine the behavior of high-dimensional and putatively unfathomable systems (Haken 1977, 1983). In the case under consideration it is important to understand the role of the order parameters during synchronization processes and, moreover, to understand how the stimulus acts on the order parameters. In this way we can approach the issue as to how a stimulus changes the cluster's state of synchronization.

For this reason we choose a level of description which is a bit more macroscopic: We introduce an average number density $n(\psi, t)$ which in a stochastic sense provides us with the number of the cluster's oscillators having phase ψ

at time t. This means that instead of focusing on the phase of every single oscillator we look at the cluster as a whole and ask *how many* oscillators most likely have phase ψ at time t. Using this level of description will enable us to study collective processes like synchronization and desynchronization in an effective way. We shall derive a partial differential equation for the average number density which can be analyzed analytically to a large extent so that we shall obtain important insights, e.g., into the cluster's synchronization behavior and into the cluster's endeavours to resynchronize after a desynchronizing stimulation.

Based on the model equation for the average number density the subsequent chapters will address different aspects of the collective dynamics: As a first step Chap. 5 is concerned with the network's spontaneous behavior. That means, in Chap. 5 the stimulator is turned off, so that the synchronization patterns can be studied which emerge due to the interaction of synchronizing couplings and random forces. In this way we shall reveal different types of synchronized states, namely noisy cluster states. The effects of pulsatile stimuli on these states will be analyzed in Chap. 6 whereas in Chap. 7 we shall study entrainment procceses caused by periodic stimuli.

4.2 Derivation of the Model Equation

The modelling approaches from Chap. 2 and from Chap. 3 have to be combined in order to derive a stochastic model which includes all facets of stimulation processes we are interested in. On the one hand in Chap. 2 the dynamics of a single oscillator within the ensemble was governed by

$$\dot{\psi}_j = \Omega + S(\psi_j) + F_j(t) \ , \tag{4.1}$$

where ψ_j denotes the phase of the jth oscillator (cf. (2.27)). All oscillators have the same eigenfrequency, namely Ω. The 2π-periodic function $S(\psi_j)$ stands for the stimulation mechanism taking into account that the effect of the stimulation depends on the phase of only this particular oscillator. In other words, the oscillator's phase determines its vulnerability as it is well known, for instance, from single neurons (Best 1979). Of course, $S(\psi_j)$ depends on additional parameters, such as stimulation intensity, which will be discussed below in detail. $F_j(t)$ models additive Gaussian white noise.

On the other hand in Chap. 3 the dynamics of a population of interacting phase oscillators was determined by

$$\dot{\psi}_j = \omega_j + \frac{1}{N} \sum_{k=1}^{N} M(\psi_j - \psi_k) \ , \tag{4.2}$$

where ω_j is the eigenfrequency of the jth oscillator, and $M(\psi_j - \psi_k)$ models the impact of the kth on the jth oscillator (cf. (3.2)).

All mechanisms, namely the mutual coupling and the impact of stimulation as well as random forces, can be summarized by adding the corresponding terms of the equations' right-hand sides. In this way combining (4.1) and (4.2) yields the model equation

$$\dot{\psi}_j = \underbrace{\Omega}_{\text{I}} + \underbrace{\frac{1}{N} \sum_{k=1}^{N} M(\psi_j - \psi_k)}_{\text{II}} + \underbrace{S(\psi_j)}_{\text{III}} + \underbrace{F_j(t)}_{\text{IV}} \ . \tag{4.3}$$

Based on the preparatory studies in Chaps. 2 and 3 the terms on the right hand side of (4.3) find clear interpretations:

Term I: For the sake of simplicity let us assume that all oscillators have the same eigenfrequency ($\omega_j = \Omega$).

Term II: The oscillators' mutual interactions are modeled by a mean-field-coupling, i.e. as in (3.3) every oscillator is coupled to all others with equal strength:

$$M(\psi_j - \psi_k) = -\sum_{m=1}^{4} \{ K_m \sin\left[m(\psi_j - \psi_k)\right] + C_m \cos\left[m(\psi_j - \psi_k)\right]\} \ . \tag{4.4}$$

Term III: The stimulation is modeled by:

$$S(\psi_j) = \sum_{m=1}^{4} [X_m \sin(m\psi) + Y_m \cos(m\psi)] = \sum_{m=1}^{4} I_m \cos(m\psi_j + \gamma_m) \ , \tag{4.5}$$

where

$$I_m = \sqrt{X_m^2 + Y_m^2} \ , \quad \cos\gamma_m = \frac{Y_m}{I_m} \ , \quad \sin\gamma_m = -\frac{X_m}{I_m} \tag{4.6}$$

(cf. (2.4) to (2.16)). Note that terms II and III consist of 2π-periodic functions, viz M and S. For this reason one can expand them in terms of Fourier modes. However, for the sake of simplicity let the analysis be restricted to an expansion of M and S up to terms of fourth order.

Term IV: For the sake of simplicity the random forces are modeled by Gaussian white noise which is delta-correlated with zero mean value:

$$\langle F_j(t) \rangle = 0 \ , \quad \langle F_j(t) F_k(t') \rangle = Q\delta_{jk}\delta(t - t') \ . \tag{4.7}$$

By introducing the abbreviation

$$\Gamma(\psi_j, \psi_k) = \Omega + M(\psi_j - \psi_k) + S(\psi_j) \tag{4.8}$$

(4.3) is cast into the concise form

$$\dot{\psi}_j = \frac{1}{N} \sum_{k=1}^{N} \Gamma(\psi_j, \psi_k) + F_j(t) \ . \tag{4.9}$$

The dynamics of the Langevin equation (4.9) can be investigated by means of the corresponding *Fokker–Planck equation* which is a partial differential equation for the probability density $f(\{\psi_l\}, t)$, where $\{\psi_l\}$ denotes the vector $(\psi_1, \ldots, \psi_N)$. $f(\{\psi_l\}, t)\, d\psi_1 \ldots d\psi_N$ provides us with the probability of finding the phases of the oscillators at time t in the intervals $\psi_j \ldots \psi_j + d\psi_j$ $(j = 1, \ldots, N)$. The Fokker–Planck equation belonging to (4.9) reads

$$\frac{\partial f(\{\psi_l\}, t)}{\partial t} + \sum_{k=1}^{N} \frac{\partial}{\partial \psi_k} \left[\frac{1}{N} \sum_{j=1}^{N} \Gamma(\psi_k, \psi_j) f(\{\psi_l\}, t) \right]$$

$$- \frac{Q}{2} \sum_{k=1}^{N} \frac{\partial^2 f(\{\psi_l\}, t)}{\partial \psi_k^2} = 0 \qquad (4.10)$$

with Q from (4.7) (cf. Haken 1983, Gardiner 1985, Risken 1989). Investigating a partial differential equation like (4.10) is not trivial. Fortunately the analysis of the oscillators' dynamics can be crucially simplified by profiting from a trick used by Kuramoto (1984) in a former study. By introducing an average number density he analyzed (4.10) for $S = 0$, i.e. in his model no stimulation occured. Though for nonvanishing S our model is quite different, we can apply this trick in our investigation, too.

To this end let us consider the stimulation scenario from a slightly different point of view. Instead of analyzing the phase of every single oscillator, we restrict ourselves to the question of *how many* oscillators most probably have phase ψ at time t. In this way we introduce the number density $\tilde{n}$ of the oscillators which have phase ψ

$$\tilde{n}(\{\psi_l\}; \psi) = \frac{1}{N} \sum_{k=1}^{N} \delta(\psi - \psi_k) \ . \qquad (4.11)$$

The stochastic aspect of the dynamics is taken into account by introducing the *average number density* n of the oscillators which have phase ψ according to

$$n(\psi, t) = \langle \tilde{n}(\{\psi_l\}; \psi) \rangle_t = \int_0^{2\pi} \cdots \int_0^{2\pi} d\psi_1 \cdots d\psi_N\ \tilde{n}(\{\psi_l\}; \psi) f(\{\psi_l\}; t)$$
$$(4.12)$$

(cf. Kuramoto 1984). An evolution equation for the dynamics of the average number density n can now be derived in a similar way as in *Kuramoto's* study. First $n(\psi, t)$ is differentiated with respect to time. With (4.10) and (4.12) one obtains

$$\frac{\partial n(\psi, t)}{\partial t} = \int_0^{2\pi} \cdots \int_0^{2\pi} d\psi_1 \cdots d\psi_N\ \tilde{n}(\{\psi_l\}; \psi)$$

$$\times \left\{ -\sum_{k=1}^{N} \frac{\partial}{\partial \psi_k} \left[\frac{1}{N} \sum_{j=1}^{N} \Gamma(\psi_k, \psi_j) f(\{\psi_l\}; t) \right] + \frac{Q}{2} \sum_{k=1}^{N} \frac{\partial^2 f(\{\psi_l\}; t)}{\partial \psi_k^2} \right\} .$$

$$(4.13)$$

Performing a partial integration and taking into account

$$\frac{\partial}{\partial \psi_k} \tilde{n}(\{\psi_l\}; \psi) = -\frac{1}{N} \frac{\partial}{\partial \psi} \delta(\psi - \psi_k) \qquad (4.14)$$

and

$$\sum_{j=1}^{N} \Gamma(\psi_k, \psi_j) = \sum_{j=1}^{N} \int_0^{2\pi} d\psi' \, \Gamma(\psi_k, \psi') \delta(\psi' - \psi_j) \qquad (4.15)$$

yields

$$\frac{\partial n(\psi, t)}{\partial t} = -\frac{1}{N} \int_0^{2\pi} \cdots \int_0^{2\pi} d\psi_1 \cdots d\psi_N \sum_{j,k=1}^{N} \left[\frac{1}{N} \frac{\partial}{\partial \psi} \delta(\psi - \psi_j) \right]$$

$$\times \int_0^{2\pi} d\psi' \, \Gamma(\psi_j, \psi') \delta(\psi' - \psi_k) f(\{\psi_l\}; t)$$

$$+ \frac{Q}{2N} \sum_{j=1}^{N} \int_0^{2\pi} \cdots \int_0^{2\pi} d\psi_1 \cdots d\psi_N$$

$$\times \left[\frac{\partial^2}{\partial \psi^2} \delta(\psi - \psi_j) \right] f(\{\psi_l\}; t) . \qquad (4.16)$$

The differentiation with respect to ψ and the integrations commute. Hence, by rearranging (4.16) one gets

$$(4.16) = -\frac{\partial}{\partial \psi} \int_0^{2\pi} d\psi' \, \Gamma(\psi, \psi') \int_0^{2\pi} \cdots \int_0^{2\pi} d\psi_1 \cdots d\psi_N$$

$$\times \left[\frac{1}{N} \sum_{j=1}^{N} \delta(\psi - \psi_j) \right] \left[\frac{1}{N} \sum_{k=1}^{N} \delta(\psi' - \psi_k) \right] f(\{\psi_l\}; t)$$

$$+ \frac{Q}{2} \frac{\partial^2}{\partial \psi^2} \int_0^{2\pi} \cdots \int_0^{2\pi} d\psi_1 \cdots d\psi_N$$

$$\times \left[\frac{1}{N} \sum_{j=1}^{N} \delta(\psi - \psi_j) f(\{\psi_l\}; t) \right] . \qquad (4.17)$$

Taking into account definition (4.12) the vector field can be written as

$$(4.17) = -\frac{\partial}{\partial \psi} \int_0^{2\pi} d\psi' \, \Gamma(\psi, \psi') \langle \tilde{n}(\{\psi_l\}; \psi) \tilde{n}(\{\psi_l\}; \psi') \rangle_t + \frac{Q}{2} \frac{\partial^2}{\partial \psi^2} n(\psi, t) .$$

$$(4.18)$$

As $\tilde{n}(\{\psi_l\}; \psi)$ is a macrovariable, one can expect the fluctuation given by

$$\delta\tilde{n}(\{\psi_l\}; \psi) = \tilde{n}(\{\psi_l\}; \psi) - n(\{\psi_l\}; \psi) \tag{4.19}$$

to be of order $1/\sqrt{N}$, where N is the number of oscillators. Because of that one is allowed to introduce the approximation

$$\langle \tilde{n}(\{\psi_l\}; \psi)\tilde{n}(\{\psi_l\}; \psi') \rangle_t \approx n(\psi, t)n(\psi', t) \tag{4.20}$$

(cf. Kuramoto 1984). Inserting (4.8) and (4.20) into (4.18) yields the evolution equation for the average number density

$$\frac{\partial n(\psi, t)}{\partial t} = \underbrace{-\frac{\partial}{\partial\psi}\left\{ n(\psi, t)\int_0^{2\pi} d\psi'\, M(\psi - \psi')n(\psi', t)\right\}}_{\text{II}}$$

$$\underbrace{-\frac{\partial}{\partial\psi}n(\psi, t)\,S(\psi)}_{\text{III}} \quad \underbrace{-\Omega\frac{\partial}{\partial\psi}n(\psi, t)}_{\text{I}} \quad \underbrace{+\frac{Q}{2}\frac{\partial^2 n(\psi, t)}{\partial\psi^2}}_{\text{IV}}. \tag{4.21}$$

The labeling of the terms in this equation corresponds to that in (4.3). The subsequent chapters concern synchronization and stimulation processes described by (4.21). It will turn out that synchronization patterns occur because the mutual interactions (term II) overcome the dissipative influence of the random forces (term IV). These patterns move as travelling waves (modulo 2π) due to the drift term (I). And it will be particularly interesting to study how stimulation changes the synchronization patterns (term III). However, I do not want to anticipate too much at the moment. Rather let our investigation start from the very beginning.

In order to investigate the partial differential equation (4.21) one has to take into account two *boundary conditions:*

1. ψ is a phase. For this reason one is allowed to identify 0 and 2π, and, thus, the boundary condition

$$n(0, t) = n(2\pi, t) \tag{4.22}$$

has to be fulfilled for all times t.
2. As n is an average number density,

$$\int_0^{2\pi} n(\psi, t)\, d\psi = 1 \tag{4.23}$$

has to be fulfilled for all times t. This means that the number of oscillators does not change in time.

To come back to the ensemble scenario of Chap. 2 one simply has to set $M = 0$. Through this (4.21) is of the same form as the evolution equation for the ensemble's distribution (2.30). Clearly, this indicates the consistency of both modeling approaches.

4.3 Fourier Transformation

The bifurcation analysis performed in the next chapter will be based on the
Fourier transformed model equation. So, let us expand n in terms of Fourier
modes according to condition (4.22):

$$n(\psi, t) = \sum_{k \in \mathbb{Z}} \hat{n}(k, t)\, \mathrm{e}^{\mathrm{i}k\psi} \ . \tag{4.24}$$

Analogously we are allowed to expand M and S. For this the shortforms

$$\hat{M}(k) = \begin{cases} \pi(-C_k + \mathrm{i}K_k) & : & k = 1, 2, 3, 4 \\ -\pi(C_{-k} + \mathrm{i}K_{-k}) & : & k = -1, -2, -3, -4 \\ 0 & : & \text{otherwise} \end{cases} \tag{4.25}$$

and

$$\hat{S}(k) = \begin{cases} I_k \exp(\mathrm{i}\gamma_k)/2 & : & k = 1, 2, 3, 4 \\ I_k \exp(-\mathrm{i}\gamma_k)/2 & : & k = -1, -2, -3, -4 \\ 0 & : & \text{otherwise} \end{cases} \tag{4.26}$$

are introduced. With (4.25) and (4.26) M and S take the form

$$M(\psi - \psi') = \sum_{k \in \mathbb{Z}} \hat{M}(k)\, \mathrm{e}^{\mathrm{i}k(\psi - \psi')} \ , \quad S(\psi) = \sum_{k \in \mathbb{Z}} \hat{S}(k)\, \mathrm{e}^{\mathrm{i}k\psi} \ . \tag{4.27}$$

Additionally introducing the abbreviation

$$\sum_{\pm m=1}^{4} \xi_m = \xi_1 + \xi_2 + \xi_3 + \xi_4 + \xi_{-1} + \xi_{-2} + \xi_{-3} + \xi_{-4} \tag{4.28}$$

and inserting (4.24) and (4.27) into (4.21) finally yields the Fourier transformed equation for the average number density

$$\frac{\partial \hat{n}(k, t)}{\partial t} = -\mathrm{i}k \sum_{\pm m=1}^{4} \hat{M}(m)\hat{n}(k - m, t)\hat{n}(m, t) - \mathrm{i}k \sum_{\pm m=1}^{4} \hat{n}(k - m, t)\hat{S}(m)$$

$$- \mathrm{i}k\Omega\hat{n}(k, t) - \frac{Q}{2}k^2\hat{n}(k, t) \ . \tag{4.29}$$

Setting $k = 0$ in (4.29) one immediately reads off that $\partial \hat{n}(0, t)/\partial t = 0$ is
fulfilled for all times t. For this reason (4.21) conserves $\int_0^{2\pi} n(\psi, t)\, d\psi$ in
accordance with condition (4.23), and the mean of n reads

$$\hat{n}(0, t) = \frac{1}{2\pi} \quad \text{for all times } t \ . \tag{4.30}$$

Note that $\hat{n}(-k, t) = \hat{n}^*(k, t)$ holds for all wave numbers k and for all times
t because $n(\psi, t)$ is a real function.

4.4 Summary and Discussion

Model equation (4.3) derived in this chapter enables us to investigate the reactions of a population of interacting phase oscillators to stimulation and random forces. The state of the oscillators is described in a convenient way by introducing an average number density $n(\psi, t)$ which gives the number of oscillators most likely having phase ψ at time t. We determined an evolution equation for the average number density (cf. (4.21)). The Fourier transformed version of this equation, (4.29), will serve as a basis for the study of both self-synchronization processes in the absence of a stimulator as well as stimulation induced desynchronization processes. A comparison with the modeling in Chap. 2 showed that for vanishing coupling strength, i.e. $M = 0$, the stochastic cluster model (4.21) is equivalent to the ensemble model (2.30), in this way confirming the consistency of both models.

Since our theoretical investigation is essentially based on the introduction of the average number density, that means on choosing a more macroscopic level of description, it is worth to discuss this aspect from the neurophysiological point of view. In Sect. 3.4.1 we already saw that a single neuron can be approximated by means of a phase oscillator: The jth model neuron fires or produces a burst whenever its phase ψ_j equals ρ modulo 2π, where ρ is a fixed parameter. Thus, according to (4.12), the definition of the averaged number density $n(\psi, t)$, the number of model neurons which are most probably firing at time t is given by $n(\rho, t)$. The latter is a macrovariable which corresponds to typical observables in neuroscience such as MEG or EEG signals. The latter reflect the dynamics of large clusters of synchronously active neurons. For instance, MEG signals measured in stimulation experiments require that at least a million synapses are active in a synchronized way (Hämäläinen et al. 1993). In summary, introducing the average number density means that we do not ask which but how many neurons fire at time t.

5. Clustering in the Presence of Noise

5.1 Introductory Remarks

This chapter presents different patterns of synchronized collective activity occurring provided there is no stimulation. Mathematically speaking, we analyze special solutions of (4.21) which arise provided the stimulation term S vanishes. All of these patterns can only emerge because the oscillators' mutual synchronizing interactions resist the influence of noise. As the coupling strength is increased and passes its critical value, noisy cluster states appear which can be considered as stochastic versions of the cluster states analyzed in Chap. 3.

Which order parameters occur and how the emerging patterns of synchrony look like is the subject of the following sections. With this aim in view order parameter equations are derived for different coupling mechanisms. The corresponding synchronization patterns are investigated analytically and numerically. Additionally the very difference between a population of coupled oscillators and an ensemble lacking any synchronizing interactions is pointed out by comparing both models.

From the neurophysiological standpoint in this chapter we consider the neurons' patterns of self-synchronized activity occurring if there is no stimulation. We shall study how the neurons' interactions determine the type of their synchronized state and in this way we shall encounter an inventory of noisy cluster states. Finally the latter will be discussed in the context of the neural code.

5.2 Modelling Emerging Synchronization

First it should be mentioned which form the model takes if there is no stimulation. Turning off the stimulator means setting $S = 0$ in (4.3) and (4.21). In this way one obtains the Langevin equation

$$\dot{\psi}_j = \Omega + \frac{1}{N} \sum_{k=1}^{N} M(\psi_j - \psi_k) + F_j(t) \qquad (5.1)$$

with coupling term M and noise term F_j as defined by (4.4) and (4.7), respectively. Correspondingly the evolution equation for the average number density reads

$$\frac{\partial n(\psi, t)}{\partial t} = -\Omega \, \frac{\partial n(\psi, t)}{\partial \psi} + \frac{Q}{2} \frac{\partial^2 n(\psi, t)}{\partial \psi^2}$$

$$- \frac{\partial}{\partial \psi} \left[n(\psi, t) \int_0^{2\pi} d\psi' \, M(\psi - \psi') n(\psi', t) \right] . \qquad (5.2)$$

While investigating the dynamics of (5.2) below, the impact of the different terms on the right-hand side will be clarified: The first term on the right-hand side of (5.2) is a drift term giving rise to travelling waves. The latter occur when the density n rotates on the circle obtained by identifying 0 and 2π as illustrated in Fig. 5.1. Travelling waves of this kind will be discussed and illustrated in the following sections in detail. The second term is a dissipative

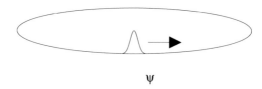

ψ

Fig. 5.1. *Schematic plot of a travelling density wave:* One obtains a phase circle by identifying 0 and 2π of the interval $[0, 2\pi]$. A travelling wave of the density n runs on that circle provided n simply rotates along that circle.

term counteracting the synchronizing coupling modeled by the third term. Before analyzing the dynamical behavior of (5.2) let us compare it with the ensemble model of Chap. 2.

5.3 Comparison with the Ensemble's Dynamics

As already discussed in Sect. 4.2 the stochastic cluster model (5.2) and the ensemble model (2.30) are equivalent for vanishing coupling strength. Hence for $M = S = 0$ one can derive an explicit solution of (5.2) as in Chap. 2:

$$n(\psi, t) = \sum_{k \in \mathbb{Z}} \underbrace{\hat{n}(k, 0)}_{\text{I}} \underbrace{\exp\left(-\frac{Q}{2} k^2 t \right)}_{\text{II}} \underbrace{\exp\left[ik(\psi - \Omega t) \right]}_{\text{III}} . \qquad (5.3)$$

The initial average number density $n(\psi, 0)$ is given by the Fourier coefficients for $t = 0$ (term I). Initially peaked density patterns drift as travelling waves (term III). Due to the random forces (term II) these patterns melt and converge to the uniform distribution (4.30). Hence, without synchronizing interactions n undergoes a stereotyped relaxation: n approaches the fully desynchronized state, i.e. $n(\psi, t) \to 1/2\pi$ for $t \to \infty$, irrespective of the initial density $n(\psi, 0)$.

5.4 Noisy Cluster States

A totally different situation occurs when the oscillators interact with synchronizing couplings. To understand how self-synchronized activity emerges a bifurcation analysis of (5.2) will be performed based on the Fourier transformed system. The latter is easily derived by inserting (4.25) in (4.29) and taking into account that there is no stimulation, i.e. $\hat{S}(k)$ vanishes for all k. So, the Fourier transformed model equation reads

$$\frac{\partial \hat{n}(k,t)}{\partial t} = \left[\frac{1}{2}\left(|k|K_{|k|} - k^2 Q\right) - ik\left(\Omega - \frac{C_{|k|}}{2}\right)\right]\hat{n}(k,t)$$

$$- ik \sum_{\substack{\pm m=1 \\ m \neq k}}^{4} \hat{M}(m)\hat{n}(k-m,t)\hat{n}(m,t) \tag{5.4}$$

for $k = \pm 1, \pm 2, \pm 3, \pm 4$ and

$$\frac{\partial \hat{n}(k,t)}{\partial t} = -\left(\frac{Q}{2}k^2 + ik\Omega\right)\hat{n}(k,t) - ik \sum_{\substack{\pm m=1 \\ m \neq k}}^{4} \hat{M}(m)\hat{n}(k-m,t)\hat{n}(m,t) \tag{5.5}$$

for $|k| > 4$, whereas the dynamics for $k = 0$ is simply given by (4.30): $\hat{n}(0,t) = 1/(2\pi)$ for all times t. $n(\psi, t)$ is a real function. Thus,

$$\hat{n}(-k,t) = \hat{n}^*(k,t) \tag{5.6}$$

holds for all wave numbers k and for all times t, where ξ^* denotes the complex conjugate of ξ. To apply the center manifold theorem as outlined in Sect. 3.6.1 an infinite dimensional vector $\boldsymbol{x}$ of all Fourier modes with non-vanishing wave number is introduced. With (5.6) the latter is given by

$$\boldsymbol{x} = \begin{pmatrix} \vdots \\ \hat{n}(2,t) \\ \hat{n}(1,t) \\ \hat{n}^*(1,t) \\ \hat{n}^*(2,t) \\ \vdots \end{pmatrix}. \tag{5.7}$$

Additionally the matrix $B = (b_{jk})$ is introduced by

$$
b_{jk} = \begin{cases} \left(|k|K_{|k|} - k^2 Q\right)/2 - ik\left(\Omega - C_{|k|}/2\right) & : & j = k \text{ and } |k| \leq 4 \\ -Qk^2/2 - ik\Omega & : & j = k \text{ and } |k| > 4 \\ 0 & : & \text{otherwise} \end{cases}
$$

$$(5.8)$$

Nonlinear terms of (5.4) are summarized by the vector

$$
\mathbf{N} = \begin{pmatrix} \vdots \\ N_2(\boldsymbol{x}) \\ N_1(\boldsymbol{x}) \\ N_{-1}(\boldsymbol{x}) \\ N_{-2}(\boldsymbol{x}) \\ \vdots \end{pmatrix} ,
$$

$$(5.9)$$

where

$$
N_k = -ik \sum_{\substack{\pm m = 1 \\ m \neq k}}^{4} \hat{M}(m)\hat{n}(k - m, t)\hat{n}(m, t) .
$$

$$(5.10)$$

With (5.7) to (5.10) the Fourier transformed model equation (5.4) can be written in the form

$$
\dot{\boldsymbol{x}} = B\boldsymbol{x} + \mathbf{N}(\boldsymbol{x}) .
$$

$$(5.11)$$

To apply the center manifold theorem first the linearized form of (5.11) has to be analyzed (cf. Sect. 3.6.1).

5.4.1 Linear Problem

By omitting nonlinear terms in (5.4) and (5.5) one easily obtains the linear problem:

$$
\frac{\partial \hat{n}(k, t)}{\partial t} = \left[\frac{1}{2}\left(|k|K_{|k|} - k^2 Q\right) - ik\left(\Omega - \frac{C_{|k|}}{2}\right)\right] \hat{n}(k, t)
$$

$$(5.12)$$

for $k = \pm 1, \pm 2, \pm 3, \pm 4$, and

$$
\frac{\partial \hat{n}(k, t)}{\partial t} = \left(-\frac{Q}{2}k^2 + ik\Omega\right) \hat{n}(k, t)
$$

$$(5.13)$$

for $|k| > 4$. The linear stability of the modes depends on the real parts of the eigenvalues of the linear problem (see, for instance, Pliss 1964, Kelley 1967, Hirsch and Smale 1974, Haken 1983). A mode is stable, unstable or a center mode provided the real part of its eigenvalue is negative, positive or vanishes. From (5.13) it immediately follows that Fourier modes with wave numbers fulfilling $|k| > 4$ are stable because the real parts of their eigenvalues read $-Qk^2/2$. Obviously this is a consequence of the fact that the coupling (3.3) only consists of Fourier modes up to fourth order.

According to (5.12) the real and imaginary parts of the eigenvalues λ_k of the modes with wave numbers $k = \pm 1, \pm 2, \pm 3, \pm 4$ read

$$\text{Re}(\lambda_k) = \frac{1}{2}\left(|k||K_{|k|}| - k^2 Q\right) \quad \text{and} \quad \text{Im}(\lambda_k) = k\left(\frac{C_{|k|}}{2} - \Omega\right) . \qquad (5.14)$$

Thus, the eigenvalues' real parts crucially depend on $K_1, \ldots, K_4$, while $C_1, \ldots, C_4$ exclusively act on the imaginary part. As a consequence of (5.14) the Fourier modes $\hat{n}(\pm j, t)$ become unstable when K_j exceeds its critical value

$$K_j^{\text{crit}} = jQ . \qquad (5.15)$$

Analogously for $K_j < K_j^{\text{crit}}$ both $\hat{n}(j, t)$ and $\hat{n}(-j, t)$ are stable modes. Interestingly the critical coupling strength scales with the order of the critical Fourier mode. In this book a pair of Fourier modes $\hat{n}(j, t)$ and $\hat{n}(-j, t)$ will briefly be denoted as jth mode.

In Chap. 3 it was shown that the synchronized states decisively depend on the coupling pattern. Therefore in the following sections the order parameter equations belonging to different coupling patterns will be derived.

5.4.2 First-Mode Instability

In this section we assume that K_1 is strong enough so that the first Fourier mode overcomes the noise induced damping. To this end we set

$$K_1 = Q + 2\varepsilon , \qquad (5.16)$$

where ε is a control parameter, also called bifurcation parameter. We assume that $|\varepsilon| \ll 1$ is fulfilled. That means we analyze the dynamics in the neighbourhood of the bifurcation occurring when K_1 exceeds its critical coupling strength $K_1^{\text{crit}} = Q$. Accordingly we assume that

$$K_j \ll jQ \qquad (5.17)$$

holds for $j = 2, 3, 4$. As a consequence of (5.17) the Fourier modes with wave vectors $k = \pm 2, \pm 3, \pm 4$ are stable. Thus, when the increasing ε changes its sign from negative to positive, the conjugate complex pair of stable eigenvalues $\lambda_{\pm 1}$ crosses the imaginary axis, and the real part of this pair becomes positive. Correspondingly a Hopf bifurcation occurs, where $\hat{n}(1, t)$ and $\hat{n}(-1, t)$ become order parameters, whereas all other modes remain stable.

To derive the order parameter equation one has to insert (5.16) into (5.4). Taking into account (4.25) one obtains

$$\begin{aligned}
\frac{\partial \hat{n}(1, t)}{\partial t} &= \left[\varepsilon + i\left(\frac{C_1}{2} - \Omega\right)\right]\hat{n}(1, t) \\
&\quad + \pi[(K_2 - Q - 2\varepsilon) + i(C_1 + C_2)]\hat{n}^*(1, t)\hat{n}(2, t) \\
&\quad + \pi[(K_3 - K_2) + i(C_2 + C_3)]\hat{n}^*(2, t)\hat{n}(3, t) \\
&\quad + \pi[(K_4 - K_3) + i(C_3 + C_4)]\hat{n}^*(3, t)\hat{n}(4, t) \\
&\quad + \pi(-K_4 + iC_4)\hat{n}^*(4, t)\hat{n}(5, t) .
\end{aligned} \qquad (5.18)$$

Because of (5.6) the corresponding equation for $\hat{n}(-1, t)$ is given by

$$\frac{\partial \hat{n}(-1, t)}{\partial t} = \left(\frac{\partial \hat{n}(1, t)}{\partial t}\right)^*. \tag{5.19}$$

According to the linear stability analysis one can distinguish between center modes $\boldsymbol{x}_{\mathrm{C}}$ and stable modes $\boldsymbol{x}_{\mathrm{S}}$:

$$\boldsymbol{x}_{\mathrm{C}} = \begin{pmatrix} \hat{n}(1, t) \\ \hat{n}^*(1, t) \end{pmatrix}, \quad \boldsymbol{x}_{\mathrm{S}} = \begin{pmatrix} \vdots \\ \hat{n}(3, t) \\ \hat{n}(2, t) \\ \hat{n}^*(2, t) \\ \hat{n}^*(3, t) \\ \vdots \end{pmatrix}. \tag{5.20}$$

On the center manifold the stable modes are given by the center modes according to

$$\hat{n}(j, t) = h_j\left(\hat{n}(1, t), \hat{n}^*(1, t)\right) \quad (j = \pm 2, \pm 3, \pm 4 \ldots) \tag{5.21}$$

(cf. Pliss 1964 and Kelley 1967). In order to derive the reduced problem in lowest order it is sufficient to determine h_2. This can easily be achieved by differentiating (5.21) with respect to time for $j = 2$. Consequently one obtains

$$\frac{\partial \hat{n}(2, t)}{\partial t} = \frac{\partial h_2}{\partial \hat{n}(1, t)} \frac{\partial \hat{n}(1, t)}{\partial t} + \frac{\partial h_2}{\partial \hat{n}^*(1, t)} \frac{\partial \hat{n}^*(1, t)}{\partial t}. \tag{5.22}$$

According to the center manifold theorem h_2 only contains nonlinear terms of the order parameters $\hat{n}(1, t)$ and $\hat{n}^*(1, t)$. For this reason both $\partial h_2/\partial \hat{n}(1, t)$ and $\partial h_2/\partial \hat{n}(1, t)$ are terms of first order, i.e. $= O(\|\boldsymbol{x}_{\mathrm{C}}\|)$. On the other hand, due to the second order term $\varepsilon \hat{n}(1, t)$ occurring on the right-hand side of (5.18) $\partial \hat{n}(1, t)/\partial t$ is of second order, i.e. $= O(\|\boldsymbol{x}_{\mathrm{C}}\|^2)$. Likewise $\partial \hat{n}^*(1, t)/\partial t$ is of second order. In summary, the right-hand side of (5.22) is of third order. Thus, based on (5.4) for $k = 2$ from the left-hand side of (5.22) one immediately reads off that

$$h_2\left(\hat{n}(1, t), \hat{n}^*(1, t)\right) = \zeta \hat{n}(1, t)^2 + O(\|\boldsymbol{x}_{\mathrm{C}}\|^3) \tag{5.23}$$

holds, where the coefficient ζ is listed in Appendix C.1. By inserting (5.23) into (5.18) we finally obtain the reduced problem which governs the dynamics of the order parameter on the center manifold:

$$\frac{\partial \hat{n}(1, t)}{\partial t} = \left[\varepsilon + \mathrm{i}\left(\frac{C_1}{2} - \Omega\right)\right] \hat{n}(1, t)$$
$$- (\alpha_1 + \mathrm{i}\beta_1)\, \hat{n}(1, t)\, |\hat{n}(1, t)|^2 + O(\|\boldsymbol{x}_{\mathrm{C}}\|^4) \tag{5.24}$$

with coefficients α_1 and β_1 in Appendix C.1. Because of (5.19) it is sufficient to restrict ourselves to the analysis of the order parameter equation (5.24). However, for the determination of h_2 it was necessary to take into account the pair of order parameters $\hat{n}(1,t)$ and $\hat{n}^*(1,t)$ (cf. (5.21) and (5.22)). As $\hat{n}(1,t)$ is a complex quantity it is favourable to choose the ansatz

$$\hat{n}(1,t) = A_1(t)\exp[i\phi_1(t)] \,, \tag{5.25}$$

where the real quantities $A_1(t)$ and $\phi_1(t)$ are amplitude and phase of the order parameter. Neglecting terms of fourth and higher order we obtain

$$\dot{A}_1 = \varepsilon A_1 - \alpha_1 A_1^3 \,, \tag{5.26}$$

$$\dot{\phi}_1 = \frac{C_1}{2} - \Omega - \beta_1 A_1^2 \,. \tag{5.27}$$

Equation (5.26) has a potential

$$V(A_1) = -\frac{\varepsilon}{2} A_1^2 + \frac{\alpha_1}{4} A_1^4 \,, \tag{5.28}$$

where

$$\dot{A}_1 = -\frac{dV(A_1)}{dA_1} \,. \tag{5.29}$$

According to (5.28) the order parameter's amplitude A_1 approaches stable states which are the potential's local minima (see, for instance, Hirsch and Smale 1974, Haken 1983). As illustrated in Fig. 5.2 for $\varepsilon < 0$ the stable state is given by $A_1^{\text{stat}} = 0$. For $\varepsilon > 0$ the latter becomes unstable, and two stable bifurcating solutions with amplitude

$$A_1^{\text{stat}} = \pm\sqrt{\frac{\varepsilon}{\alpha_1}} \tag{5.30}$$

appear. Inserting (5.30) into (5.27) one can immediately derive the order parameter's phase in the bifurcating stable states. It reads

$$\phi_1(t) = -\Omega_1 t + \theta_1 \,, \quad \Omega_1 = \Omega - \frac{C_1}{2} + \frac{\varepsilon\beta_1}{\alpha_1} \,, \tag{5.31}$$

where θ_1 is a constant, and α_1 and β_1 are listed in Appendix C.1. Ω_1 is the frequency in the stationary state. In comparison with Ω, the oscillators' eigenfrequency, Ω_1 is shifted due to the couplings ($\{C_j\}$ and $\{K_j\}$) and due to the random forces (Q).

The bifurcating patterns of synchronization which emerge for supercritical coupling strength ($\varepsilon > 0$) can easily be determined in lowest order by inserting (4.30), (5.30) and (5.31) into (4.24). This yields

$$n(\psi,t) = \frac{1}{2\pi} + 2\sqrt{\frac{\varepsilon}{\alpha_1}}\cos(\psi - \Omega_1 t + \theta_1) + O(\varepsilon) \,, \tag{5.32}$$

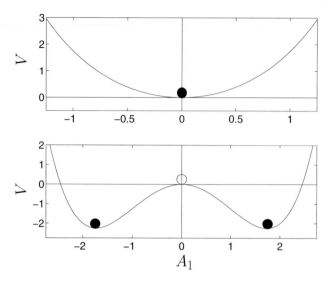

Fig. 5.2. *Potential of the order parameter's amplitude:* $V(A_1)$ is plotted versus A_1 according to (5.28). The order parameter A_1 behaves as a ball sliding down the hills of the potential (cf. Haken 1983). Hence, maxima and minima of V correspond to stable and unstable fixed points, respectively. The upper plot illustrates the potential's shape for $\varepsilon < 0$. In this case there is one stable state which is given by $A_1^{\text{stat}} = 0$ and indicated by the black ball. For supercritical coupling strength, i.e. for $\varepsilon > 0$, this fixed point becomes unstable as indicated by the white ball in the lower plot. Due to the system's symmetry properties two symmetrical stable fixed points, namely $A_1^{\text{stat}} = \pm\sqrt{\varepsilon/\alpha_1}$, occur which are indicated by the black balls.

where $O(\varepsilon)$ denotes terms $\sim \varepsilon$. Note that the negative sign in (5.30) was skipped in (5.32) because it can easily be taken into account by a suitable choice of θ_1. On the other hand for subcritical coupling strength, i.e. $\varepsilon < 0$, all Fourier modes with non-vanishing wave numbers are damped. Hence, for subcritical coupling strength the random forces dominate the collective dynamics, and n decays to the uniform density $1/(2\pi)$ (cf. (4.30)).

Let us consider the dynamics of the order parameter equations (5.26) and (5.27) in an alternative way. The dynamics of both amplitude and phase can appropriately be investigated in the two-dimensional plane. This can easily be achieved by rotating the symmetric potential (5.28). This means that (A_1, ϕ_1) serve as polar coordinates, and Cartesian coordinates (x, y) are introduced by

$$x = A_1 \cos\phi_1 \ , \quad y = A_1 \sin\phi_1 \ . \tag{5.33}$$

In this way for $\varepsilon < 0$ one obtains a potential which has one minimum at $(0,0)$ (Fig. 5.3). Whenever the system starts outside $(0,0)$ it spirals into the minimum which is a stable focus. As illustrated in Fig. 5.4 for supercritical coupling strength $(\varepsilon > 0)$ the potential has a local maximum at $(0,0)$ and a circular valley with radius A_1^{stat} from (5.30). The local maximum at $(0,0)$

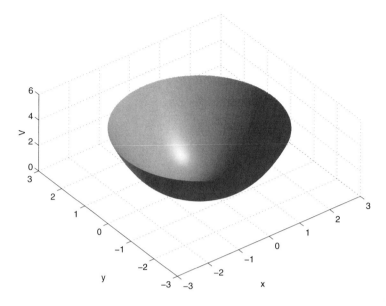

Fig. 5.3. *Subcritical potential landscape:* Plot shows the rotationally symmetric potential belonging to the order parameter equations (5.26) and (5.27) for subcritical coupling strength ($\varepsilon < 0$). Trajectories spiral into the stable focus located in $(0,0)$.

is an unstable focus of the dynamics. Wherever the system starts it spirals into the valley. In other words, when the increasing ε changes its sign from negative to positive, a Hopf bifurcation takes place giving rise to a limit cycle. The latter is explicitely given by (5.30) and (5.31) and corresponds to the bottom of the potential's valley. Hence, on the center manifold the system runs along the bottom of the valley with constant velocity Ω_1 in tangential direction.

Kuramoto (1984) performed a bifurcation analysis of a first-mode instability which occurs when the noise amplitude Q serves as control parameter, while the coupling constants remain fixed. As Q is decreased and falls below its threshold, a pair of Fourier modes becomes unstable, acts as order parameters, and gives rise to travelling waves of the average number density n. The first-mode instability studied by Kuramoto and that studied in this section can be transformed onto each other in the following way:

First, we introduce a relative phase defined by

$$\Phi_j(t) = \psi_j(t) - \Omega t \tag{5.34}$$

and insert it into (5.1) which yields

$$\dot{\Phi}_j = \frac{1}{N} \sum_{k=1}^{N} \sum_{m=1}^{4} \{K_m \sin[m(\Phi_j - \Phi_k)] + C_m \cos[m(\Phi_j - \Phi_k)]\} + F_j(t) . \tag{5.35}$$

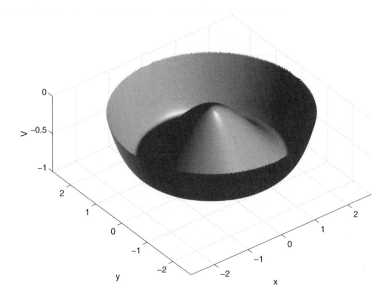

Fig. 5.4. *Supercritical potential landscape:* For supercritical coupling strength $(\varepsilon > 0)$ the potential of the order parameter equations (5.26) and (5.27) exhibits a circular valley corresponding to the bifurcating stable limit cycle.

Next, we introduce the normalized delta-correlated Gaussian white noise

$$G_j(t) = \frac{F_j(t)}{\sqrt{Q}} \tag{5.36}$$

obeying

$$\langle G_j(t)\rangle = 0 \;, \quad \langle G_j(t)G_k(t')\rangle = \delta_{jk}\delta(t - t') \tag{5.37}$$

according to (4.7). Now, we transform the relative phase by setting

$$\theta_j(t) = \frac{\Phi_j(t)}{\sqrt{Q}} \tag{5.38}$$

in this way obtaining the evolution equation

$$\dot{\theta}_j = \frac{1}{N}\sum_{k=1}^{N}\sum_{m=1}^{4}\left\{\tilde{K}_m\sin[m(\theta_j - \theta_k)] + \tilde{C}_m\cos[m(\theta_j - \theta_k)]\right\} + G_j(t) \tag{5.39}$$

with the coupling constants

$$\tilde{K}_m = \frac{K_m}{\sqrt{Q}} \;, \quad \tilde{C}_m = \frac{C_m}{\sqrt{Q}} \;. \tag{5.40}$$

From (5.39) and (5.40) we immediately read off that on the rotating coordinate system given by (5.34) and for the transformed phases (5.38) a control parameter, e.g., $\tilde{K}_1$ exceeds a critical value $\tilde{K}_1^{\mathrm{crit}}$ if either K_1 is increased or $\sqrt{Q}$ is decreased. Hence, in this context increasing the coupling strength and decreasing the noise amplitude are equivalent manipulations.

The evolution equation of the corresponding average number density $n(\theta, t)$ takes a similar form as (5.2) and reads

$$\frac{\partial n(\theta, t)}{\partial t} = \frac{Q}{2} \frac{\partial^2 n(\theta, t)}{\partial \theta^2} - \frac{\partial}{\partial \theta} \left[n(\theta, t) \int_0^{2\pi} d\theta' \, M(\theta - \theta') n(\theta', t) \right] , \quad (5.41)$$

where M is the shortform for the coupling terms (cf. 4.4). The term with Ω is omitted as a consequence of the transformation onto the rotating coordinate system (5.34). The bifurcation analysis of (5.41) is similar to that of (5.2). With a little calculation we straightforwardly see that carrying out a bifurcation analysis of a single-mode instability we come up with a Hopf bifurcation similar to that analyzed in this section. Kuramoto (1984) studied such a bifurcation for supercritical $\tilde{K}_1$.

Unlike *Kuramoto's* analysis our investigation is concerned with the relationship of different types of coupling mechanisms and the corresponding patterns of self-synchronized collective activity. Thus, in this chapter we study noisy cluster states emerging as a consequence of different patterns of supercritical coupling constants. In terms of physiology, for instance, the way neurons interact with each other is modeled by the coupling mechanism. In this sense our investigation is dedicated to the correlation of different types of neuronal interactions and their corresponding patterns of self-synchronized activity.

5.4.3 Second-Mode Instability

In this section K_2 is increased as control parameter which exceeds its critical value $K_2^{\mathrm{crit}} = 2Q$. Therefore we set

$$K_2 = 2Q + \varepsilon , \quad (5.42)$$

where ε is a bifurcation parameter obeying $|\varepsilon| \ll 1$. The other coupling parameters are assumed to be constant. In particular, K_1, K_3, K_4 are assumed to be weak:

$$K_j \ll jQ \quad \text{for} \quad j = 1, 3, 4 . \quad (5.43)$$

Accordingly the Fourier modes with wave numbers $k = \pm 1, \pm 3, \pm 4$ are stable. Moreover as shown in Sect. 5.4.1 higher order Fourier modes ($|k| > 4$) are always stable in this model.

From (5.14) it follows that when ε is increased from negative to positive values, the complex conjugate pair of stable eigenvalues $\lambda_{\pm 2}$ crosses the imaginary axis. Consequently the Fourier modes $\hat{n}(2, t)$ and $\hat{n}(-2, t)$ become unstable, and dominate as order parameters the dynamics of all other stable modes. The analysis of the order parameter dynamics is similar to that in the former section, in this way finally yielding the order parameter equation of the second-mode instability

$$\frac{\partial \hat{n}(2,t)}{\partial t} = [\varepsilon + \mathrm{i}(C_2 - 2\Omega)] \, \hat{n}(2,t)$$
$$- (\alpha_2 + \mathrm{i}\beta_2) \, \hat{n}(2,t) \, |\hat{n}(2,t)|^2 + O(\|x_c\|^4) \tag{5.44}$$

with coefficients α_2 and β_2 given in Appendix C.2. Obviously (5.44) is of the same from as (5.24). Hence, on the analogy of (5.25) one profits from introducing real amplitude A_2 and phase ϕ_2 of the order parameter by setting

$$\hat{n}(2,t) = A_2(t) \exp[\mathrm{i}\phi_2(t)] \,. \tag{5.45}$$

Inserting (5.45) into (5.44) and neglecting terms of fourth and higher order the order parameter equations take the form

$$\dot{A}_2 = \varepsilon A_2 - \alpha_2 A_2^3 \,, \tag{5.46}$$

$$\dot{\phi}_2 = C_2 - 2\Omega - \beta_2 A_2^2 \,. \tag{5.47}$$

Clearly, (5.46) and (5.47) are similar to (5.26) and (5.27). Correspondingly the order parameter dynamics takes place in a two-dimensional potential landscape as illustrated in Figs. 5.3 and 5.4. For subcritical coupling strength ($\varepsilon < 0$) the incoherent state is stable, whereas for supercritical coupling ($\varepsilon > 0$) a stable limit cycle bifurcates which is given by

$$A_2^{\mathrm{stat}} = \sqrt{\frac{\varepsilon}{\alpha_2}} \,, \quad \phi_2(t) = -2\Omega_2 t + \theta_2 \,, \quad \Omega_2 = \Omega - \frac{C_2}{2} + \frac{\varepsilon \beta_2}{2\alpha_2} \,. \tag{5.48}$$

θ_2 is a constant, whereas Ω_2 is the frequency in the synchronized state shifted due to the oscillators' interactions and due to random forces (cf. the formulas for α_2 and β_2 in Appendix C.2).

For subcritical coupling strength ($\varepsilon < 0$) the random forces dominate, so that with increasing time all distributions $n(\psi, 0)$ decay to the uniform density $1/(2\pi)$. When K_2 becomes supercritical ($\varepsilon > 0$) a stable travelling wave appears which can straightforwardly be determined by inserting (4.30) and (5.48) into (4.24):

$$n(\psi,t) = \frac{1}{2\pi} + 2\sqrt{\frac{\varepsilon}{\alpha_2}} \cos[2(\psi - \Omega_2 t) + \theta_2] + O(\varepsilon) \,. \tag{5.49}$$

5.4.4 Third-Mode Instability

This section is devoted to an instability of the third mode occurring when K_3 exceeds the threshold $K_3^{\mathrm{crit}} = 3Q$. Thus, we set

$$K_3 = 3Q + \frac{2}{3}\varepsilon \tag{5.50}$$

with the bifurcation parameter fulfilling $|\varepsilon| \ll 1$. The other coupling modes are assumed to be weak, so that

$$K_j \ll jQ \qquad (5.51)$$

holds for $j = 1, 2, 4$. When K_3 becomes supercritical, i.e. when ε becomes positive, the complex pair of eigenvalues $\lambda_{\pm 3}$ crosses the imaginary axis, and the corresponding Fourier modes $\hat{n}(3, t)$, $\hat{n}(-3, t)$ become unstable. The order parameter equation is determined as in the former sections. With a few calculations one obtains

$$\frac{\partial \hat{n}(3, t)}{\partial t} = \left[\varepsilon + \mathrm{i}3 \left(\frac{C_3}{2} - \Omega \right) \right] \hat{n}(3, t)$$
$$- (\alpha_3 + \mathrm{i}\beta_3)\, \hat{n}(3, t)\, |\hat{n}(3, t)|^2 + O(\|\boldsymbol{x}_\mathrm{c}\|^4) \qquad (5.52)$$

with α_3 and β_3 listed in Appendix C.3. Introducing real amplitude A_3 and phase ϕ_3 according to

$$\hat{n}(3, t) = A_3(t) \exp[\mathrm{i}\phi_3(t)] \qquad (5.53)$$

and neglecting terms of fourth and higher order one ends up with the order parameter equations

$$\dot{A}_3 = \varepsilon A_3 - \alpha_3 A_3^3 , \qquad (5.54)$$

$$\dot{\phi}_3 = \frac{3C_3}{2} - 3\Omega - \beta_3 A_3^2 . \qquad (5.55)$$

These equations are of the same form as those in the former sections. So, the dynamics of (5.54) and (5.55) evolves in a potential landscape plotted in Figs. 5.3 and 5.4. For subcritical K_3, i.e. $\varepsilon < 0$, the uniform density $1/(2\pi)$ is stable, whereas for supercritical coupling strength ($\varepsilon > 0$) a Hopf bifurcation takes place giving rise to a stable limit cycle of the order parameter given by

$$A_3^{\mathrm{stat}} = \sqrt{\frac{\varepsilon}{\alpha_3}} , \quad \phi_3(t) = -3\Omega_3 t + \theta_3 , \quad \Omega_3 = \Omega - \frac{C_3}{2} + \frac{\varepsilon \beta_3}{3\alpha_3} . \qquad (5.56)$$

This limit cycle corresponds to the stable travelling wave

$$n(\psi, t) = \frac{1}{2\pi} + 2\sqrt{\frac{\varepsilon}{\alpha_3}} \cos[3(\psi - \Omega_3 t) + \theta_3] + O(\varepsilon) , \qquad (5.57)$$

where θ_3 is a constant.

5.4.5 Fourth-Mode Instability

To investigate what happens when the fourth mode becomes unstable while all other modes remain stable, it is appropriate to set

$$K_4 = 4Q + \frac{\varepsilon}{2} \tag{5.58}$$

where ε is the bifurcation parameter as used in the former sections. The other couplings are assumed to be weak, i.e.

$$K_j \ll jQ \tag{5.59}$$

holds for $j = 1, 2, 3$. When ε changes its sign from negative to positive the complex pair of eigenvalues $\lambda_{\pm 4}$ belonging to the Fourier modes $k = \pm 4$ crosses the imaginary axis, and correspondingly the system undergoes a Hopf bifurcation. With a few calculations one determines the order parameter equation

$$\frac{\partial \hat{n}(4, t)}{\partial t} = \left[\varepsilon + i4 \left(\frac{C_4}{2} - \Omega \right) \right] \hat{n}(4, t)$$
$$- (\alpha_4 + i\beta_4) \, \hat{n}(4, t) \, |\hat{n}(4, t)|^2 + O(\|\mathbf{x}_c\|^4) \tag{5.60}$$

with α_4 and β_4 given in Appendix C.4. With real amplitude A_4 and phase ϕ_4 according to

$$\hat{n}(4, t) = A_4(t) \exp[i\phi_4(t)] \,, \tag{5.61}$$

and neglecting terms of fourth and higher order the order parameter equations take the form

$$\dot{A}_4 = \varepsilon A_4 - \alpha_4 A_4^3 \,, \tag{5.62}$$

$$\dot{\phi}_4 = 2C_4 - 4\Omega - \beta_4 A_4^2 \,. \tag{5.63}$$

For subcritical coupling strength ($\varepsilon < 0$) the uniform density is stable, and for supercritical coupling ($\varepsilon > 0$) a stable limit cycle emerges which is given by

$$A_4^{\text{stat}} = \sqrt{\frac{\varepsilon}{\alpha_4}} \,, \quad \phi_4(t) = -4\Omega_4 t + \theta_4 \,, \quad \Omega_4 = \Omega - \frac{C_4}{2} + \frac{\varepsilon \beta_4}{4\alpha_4} \,, \tag{5.64}$$

with constant θ_4. Equation (5.64) corresponds to the bifurcating stable travelling wave

$$n(\psi, t) = \frac{1}{2\pi} + 2\sqrt{\frac{\varepsilon}{\alpha_4}} \cos[4(\psi - \Omega_4 t) + \theta_4] + O(\varepsilon) \,. \tag{5.65}$$

5.4.6 Two-Mode Instability

Sections 5.4.2–5.4.5 were devoted to synchronization patterns evolving as a consequence of a single supercritical coupling constant: As K_j passes its critical value K_j^{crit} ($= jQ$) a j-cluster state emerges. The synchronization scenario is different and, in particular, more complex when two or more coupling constants become supercritical. In this section this will be illustrated by analyzing a two-mode instability which occurs when both K_1 and K_2 pass their critical values. Thus, we set

$$K_1 = Q + 2\varepsilon \ , \quad K_2 = 2Q + \mu\varepsilon \quad (|\varepsilon| \ll 1 \, , \ \mu > 0) \qquad (5.66)$$

and

$$K_j \ll jQ \qquad (5.67)$$

for $j = 3, 4$. From (5.14) it follows that when ε becomes positive two complex pairs of eigenvalues, $\lambda_{\pm 1}$ and $\lambda_{\pm 2}$, cross the imaginary axis. Therefore the Fourier modes with wave numbers $k = \pm 1, \pm 2$ become order parameters, whereas the remaining Fourier modes are stable. In (5.66) μ corresponds to the ratio of the supercritical coupling strength of K_1 and K_2.

Inserting (5.66) into (5.4) and taking into account (4.25) yields

$$\frac{\partial \hat{n}(1, t)}{\partial t} = \left[\varepsilon + \mathrm{i} \left(\frac{C_1}{2} - \Omega \right) \right] \hat{n}(1, t)$$
$$+ \pi \left[Q + (\mu - 2)\varepsilon + \mathrm{i}(C_1 + C_2) \right] \hat{n}(1, t)^* \hat{n}(2, t)$$
$$+ \pi \left[K_3 - 2Q + \mathrm{i}(C_2 + C_3) \right] \hat{n}(2, t)^* \hat{n}(3, t) + O(\|\boldsymbol{x}_{\mathrm{C}}\|^4) \qquad (5.68)$$

$$\frac{\partial \hat{n}(2, t)}{\partial t} = [\mu\varepsilon + \mathrm{i}(C_2 - 2\Omega)]\hat{n}(2, t) + 2\pi(Q + 2\varepsilon + \mathrm{i}C_1)\hat{n}(1, t)^2$$
$$+ 2\pi[K_3 - Q + \mathrm{i}(C_1 + C_3)]\hat{n}(1, t)^* \hat{n}(3, t)$$
$$+ 2\pi \left[K_4 - 2Q + \mathrm{i}(C_2 + C_4) \right] \hat{n}(2, t)^* \hat{n}(4, t) + O(\|\boldsymbol{x}_{\mathrm{C}}\|^4) \ , \quad (5.69)$$

where center modes $\boldsymbol{x}_{\mathrm{C}}$ and stable modes $\boldsymbol{x}_{\mathrm{S}}$ are given by

$$\boldsymbol{x}_{\mathrm{C}} = \begin{pmatrix} \hat{n}(2, t) \\ \hat{n}(1, t) \\ \hat{n}^*(1, t) \\ \hat{n}^*(2, t) \end{pmatrix} \ , \quad \boldsymbol{x}_{\mathrm{S}} = \begin{pmatrix} \vdots \\ \hat{n}(4, t) \\ \hat{n}(3, t) \\ \hat{n}^*(3, t) \\ \hat{n}^*(4, t) \\ \vdots \end{pmatrix} . \qquad (5.70)$$

On the center manifold the center modes determine the stable modes according to

$$\hat{n}(j, t) = h_j\big(\hat{n}(2, t), \hat{n}(1, t), \hat{n}^*(1, t), \hat{n}^*(2, t)\big) \ , \quad (j = \pm 3, \pm 4, \pm 5 \dots) \quad (5.71)$$

(cf. Pliss 1964 and Kelley 1967). From (5.68) and (5.69) one reads off that for the derivation of the order parameter equation one has to determine h_3 and h_4. This is achieved by differentiating (5.71) with respect to time for $j = 3, 4$. In this way one obtains h_3 and h_4 which are listed in Appendix D.1. Inserting (D.1) and (D.3) into (5.68) and (5.69) one ends up with the order parameter equations

$$\frac{\partial \hat{n}(1,t)}{\partial t} = \left[\varepsilon + i\left(\frac{C_1}{2} - \Omega\right)\right] \hat{n}(1,t) + a_1 \hat{n}(1,t)^* \hat{n}(2,t)$$
$$+ a_2 \hat{n}(1,t) |\hat{n}(2,t)|^2 + O(\|x_c\|^4) , \tag{5.72}$$

$$\frac{\partial \hat{n}(2,t)}{\partial t} = [\mu\varepsilon + i(C_2 - 2\Omega)]\hat{n}(2,t)$$
$$+ b_1 \hat{n}(1,t)^2 + b_2 \hat{n}(2,t) |\hat{n}(1,t)|^2$$
$$+ b_3 \hat{n}(2,t) |\hat{n}(2,t)|^2 + O(\|x_c\|^4) \tag{5.73}$$

with a_j and b_j from (D.5) to (D.7) in Appendix D.2. For the sake of simplicity let us assume that

$$K_3 = K_4 = C_1 = \ldots = C_4 = 0 \tag{5.74}$$

holds. In this case the coefficients of the order parameter equation are given by

$$a_1 = \pi[Q + (\mu - 2)\varepsilon] , \quad b_1 = 2\pi(Q + 2\varepsilon) , \tag{5.75}$$

$$a_2 = b_2 = b_3 = -4\pi^2 Q . \tag{5.76}$$

From (5.75) and (5.76) it follows that the terms with coefficients a_2, b_2 and b_3 are the only saturating terms in (5.72) and (5.73). The ansatz

$$\hat{n}(j,t) = A_j(t) \exp[i(-j\Omega t + \varphi_j(t))] , \quad (j = 1, 2) \tag{5.77}$$

will be used for the derivation of the evolution equations for amplitudes A_j and phases φ_j. Note that both A_j and φ_j are real quantities. Inserting (5.77) into (5.72) and (5.73) and neglecting terms of fourth and higher order yields

$$\dot{A}_1 = \varepsilon A_1 + \pi[Q + (\mu - 2)\varepsilon]A_1 A_2 \cos(\varphi_2 - 2\varphi_1) - 4\pi^2 Q A_1 A_2^2 , \tag{5.78}$$

$$\dot{A}_2 = \mu\varepsilon A_2 + 2\pi(Q + 2\varepsilon)A_1^2 \cos(2\varphi_1 - \varphi_2) - 4\pi^2 Q A_2 (A_1^2 + A_2^2) , \tag{5.79}$$

$$\dot{\varphi}_1 = \pi[Q + (\mu - 2)\varepsilon]A_2 \sin(\varphi_2 - 2\varphi_1) , \tag{5.80}$$

$$\dot{\varphi}_2 A_2 = 2\pi(Q + 2\varepsilon)A_1^2 \sin(2\varphi_1 - \varphi_2) . \tag{5.81}$$

The stable bifurcating solutions of the order parameter equations have to be determined. Obviously the only stationary state of (5.78) to (5.81) with vanishing A_2 is $(A_1, A_2) = (0,0)$, i.e. the quiescent state corresponding to a uniformly distributed density $n(\psi, t) = 1/(2\pi)$. For positive ε this state is unstable as shown in Sect. 5.4.1. The investigation of bifurcating solutions with non-vanishing A_2 is decisively simplified by introducing the relative phase

$$\phi = 2\varphi_1 - \varphi_2 . \tag{5.82}$$

Inserting (5.82) into (5.78) to (5.81) leads to the order parameter equations

$$\dot{A}_1 = \varepsilon A_1 + \pi[Q + (\mu - 2)\varepsilon]A_1 A_2 \cos\phi - 4\pi^2 Q A_1 A_2^2 , \tag{5.83}$$

$$\dot{A}_2 = \mu\varepsilon A_2 + 2\pi(Q + 2\varepsilon)A_1^2 \cos\phi - 4\pi^2 Q A_2 (A_1^2 + A_2^2) , \tag{5.84}$$

$$\dot{\phi} = -2\pi \left\{ [Q + (\mu - 2)A_2] + (Q + 2\varepsilon)\frac{A_1^2}{A_2} \right\} \sin\phi \,. \tag{5.85}$$

One encounters two types of fixed points:

Type I fixed point ($A_1^{\mathrm{I}} = 0$, whereas $A_2^{\mathrm{I}} \neq 0$): Inserting $A_1^{\mathrm{I}} = 0$ into (5.84) yields

$$A_1^{\mathrm{I}} = 0 \,, \quad A_2^{\mathrm{I}} = \pm\frac{1}{2\pi}\sqrt{\frac{\mu\varepsilon}{Q}} \,. \tag{5.86}$$

The superscript I indicates that the stationary amplitudes belong to the type I fixed point. The investigation of type I fixed points in Appendix D.3 reveals that their linear stability critically depends on the eigenvalue

$$\lambda_1^{\mathrm{I}} = (1 - \mu)\varepsilon + \frac{1}{2}\sqrt{\mu\varepsilon Q} + \varepsilon\frac{\mu - 2}{2}\sqrt{\frac{\mu\varepsilon}{Q}} \,. \tag{5.87}$$

For $\lambda_1^{\mathrm{I}} > 0$ the fixed point is unstable, and for $\lambda_1^{\mathrm{I}} < 0$ it is stable.

Type II fixed point ($A_1^{\mathrm{II}} \neq 0$ and $A_2^{\mathrm{II}} \neq 0$): As shown in Appendix D.4 the relative phase of type II fixed points reads

$$\phi^{\mathrm{II}} = 0 \,, \tag{5.88}$$

and the amplitudes are given by $(A_1^{\mathrm{II}}, A_2^{\mathrm{II}})$ and $(-A_1^{\mathrm{II}}, A_2^{\mathrm{II}})$, where

$$A_2^{\mathrm{II}} = \frac{Q + (\mu - 2)\varepsilon + \sqrt{[Q + (\mu - 2)\varepsilon]^2 + 16Q\varepsilon}}{8\pi Q} \,, \tag{5.89}$$

$$A_1^{\mathrm{II}} = \sqrt{D_1^{\mathrm{II}}} \quad \text{and} \quad D_1^{\mathrm{II}} = \frac{A_2^{\mathrm{II}}\left[\mu\varepsilon - 4\pi^2 Q(A_2^{\mathrm{II}})^2\right]}{2\pi\left[2\pi Q A_2^{\mathrm{II}} - (Q + 2\varepsilon)\right]} \,. \tag{5.90}$$

Both type II fixed points, $(A_1^{\mathrm{II}}, A_2^{\mathrm{II}})$ and $(-A_1^{\mathrm{II}}, A_2^{\mathrm{II}})$, have the same eigenvalues (cf. Appendix D.4). A_1^{II} and, thus, the fixed point as a whole does only exist provided $D_1^{\mathrm{II}} \geq 0$ holds. Hence, the existence of type II fixed points depends on Q, ε and μ. For small values of the bifurcation parameter the denominator of D_1^{II} does not vanish. This is discussed in detail in Appendix D.5.

Let us next focus on how the linear stability of type I and II fixed points is related to each other. Are there ranges of parameters Q, ε and μ where both types of fixed points coexist in a stable way? Or do they occur exclusively? Most of the features concerning the linear stability can easily be read off the formulas of the fixed points' amplitudes and eigenvalues. Figure 5.5 illustrates these stability properties for $Q = 0.5$, $\mu = 4.5$ and $0 \leq \varepsilon \leq 1$. Let us first focus on the type II fixed point: $A_1^{\mathrm{II}} = A_2^{\mathrm{II}} = 1/(4\pi)$ for $\varepsilon = 0$ (for all μ). Comparing the expressions in the square brackets of (5.90) with (5.86) it turns out that whenever $\Delta A_2 = A_2^{\mathrm{I}} - A_2^{\mathrm{II}}$ vanishes A_1^{II} vanishes, too. In Fig. 5.5 ΔA_2 vanishes twice, namely for $\varepsilon_1 \approx 0.111$ and $\varepsilon_2 \approx 0.36$. In between D_1^{II} is negative and, hence, in this range type II fixed points do not exist

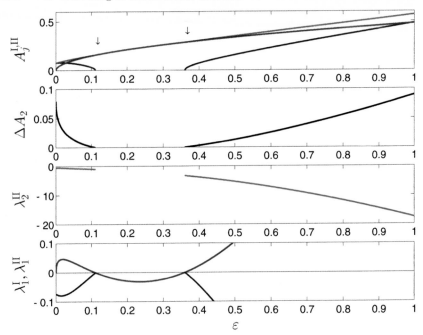

Fig. 5.5. *Linear stability of type I and II fixed points:* Amplitudes and eigenvalues of fixed point I and II are plotted over ε for $Q = 0.5$ and $\mu = 4.5$. Uppermost plot shows amplitudes of fixed point I (A_2^I: *green*) and II (A_1^{II}: *blue*, A_2^{II}: *red*). For $\varepsilon_1 \approx$ 0.111 and $\varepsilon_2 \approx 0.36$ $\Delta A_2 = A_2^I - A_2^{II}$ vanishes as shown in the plot below. ε_1 and ε_2 are indicated by arrows in the uppermost plot. From (5.90) it follows that for vanishing ΔA_2 one obtains $A_1^{II} = 0$. For $\varepsilon_1 < \varepsilon < \varepsilon_2$ fixed point II does not exist because in this range D_1^{II} is negative. λ_2^{II} is negative provided fixed point II exists. Lowest plot shows λ_1^I (*green*) and λ_1^{II} (*blue*). For $\varepsilon = \varepsilon_{1,2}$ the amplitude A_1^{II} vanishes and, thus, λ_1^{II} vanishes, too (cf. (D.15), (D.16) and (D.22)). According to numerical integration λ_1^I vanishes for $\varepsilon = \varepsilon_{1,2}$. Hence, for $\varepsilon_1 < \varepsilon < \varepsilon_2$ fixed point II does not exist, whereas fixed point I is locally stable. For $0 < \varepsilon < \varepsilon_1$ and $\varepsilon > \varepsilon_2$ fixed point II is locally stable, and fixed point I is locally unstable.

according to (5.90). From (D.15), (D.16) and (D.22) it follows that (at least) one of the eigenvalues $\lambda_{1,2}^{II}$ vanishes provided $A_1^{II} = 0$ holds. Combining all these features shows that type II fixed points do not exist for $\varepsilon_1 < \varepsilon < \varepsilon_2$, while they are locally stable for $0 < \varepsilon < \varepsilon_1$ and $\varepsilon > \varepsilon_2$. In between these ranges, i.e. for $\varepsilon = \varepsilon_{1,2}$, one of the eigenvalues (λ_1^{II}) vanishes.

A numerical comparison of λ_1^I and λ_1^{II} reveals the complementary stability properties of both types of fixed points. In summary, when type II fixed points do not exist type I fixed points are locally stable. On the other hand, when type II fixed points are locally stable type I fixed points are locally unstable. In Fig. 5.6 for different values of the noise amplitude Q it is illustrated that there is no stable coexistence of both types of fixed points: In the black area fixed point I is locally stable, whereas the white area corresponds to the local stability of type II fixed points.

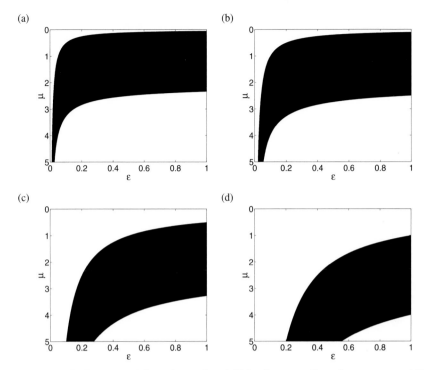

Fig. 5.6a–d. *Linear stability of type I and II fixed points:* Complementary stability properties of both types of fixed points are shown for noise amplitude $Q = 0.05$ (**a**), 0.1 (**b**), 0.5 (**c**), 1 (**d**). The stability analysis was performed as outlined in Fig. 5.5. In the black area of the μ-ε-plane type I fixed points are locally stable, and type II fixed points do not exist. Type II fixed points are locally stable in the white area, whereas type I fixed points are unstable in this area.

As shown in Figs. 5.7–5.9 results of analytical and numerical analysis of the order parameter equations perfectly agree.

The analysis in Appendix D.4 revealed two type II fixed points, namely $(\pm A_1^{\mathrm{II}}, A_2^{\mathrm{II}})$, both showing exactly the same linear stability features. The densities which correspond to these two fixed points are determined in lowest order by expanding n in terms of the leading modes according to

$$n_\pm(\psi, t) = \sum_{k=-2}^{2} \hat{n}(k, t) \exp(ik\psi) + O(A_1^2 + A_2^2) \,, \tag{5.91}$$

where the stable modes contribute to terms of second and higher order denoted by $O(A_1^2 + A_2^2)$. The subscript $+$ $(-)$ refers to the fixed point with positive (negative) A_1. With (5.77), (5.82), (5.88), (5.89), and (5.90) one finally obtains

$$n_\pm(\psi, t) = \frac{1}{2\pi} \pm 2A_1^{\mathrm{II}} \cos(\psi - \Omega t + \phi_0)$$

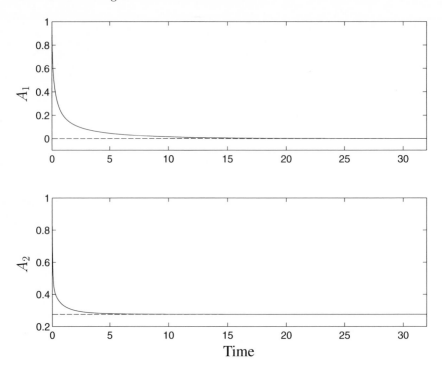

Fig. 5.7. *Amplitude dynamics of type I fixed point:* Order parameter equations (5.72) and (5.73) were integrated numerically for $\varepsilon = 0.6$, $\mu = 2.5$, $Q = 0.5$, $C_j = K_3 = K_4 = 0$ with initial conditions $A_1(0) = A_2(0) = 1$, $\varphi_1(0) = 0$ and $\varphi_2(0) = 0.2$. Plots show the amplitudes of the order parameters (A_1 and A_2, *solid lines*) and the analytically derived amplitudes of the bifurcating stable solutions (A_1^I and A_2^I, *dashed lines*). The results of both the analytical analysis and the numerical integration perfectly agree. The relative phase ϕ in the evolving stable state vanishes (not shown) in accordance with the analytical stability analysis.

$$+\, 2A_2^{II} \cos\left[2(\psi - \Omega t + \phi_0)\right] + O(A_1^2 + A_2^2)\,, \tag{5.92}$$

where ϕ_0 is constant. Taking into account the lowest order terms of (5.92) one obtains the important symmetry relationship

$$n_+(\psi + \pi, t) = n_-(\psi, t) \quad \text{for all } t. \tag{5.93}$$

Hence, except for a phase shift of π the densities n_+ and n_- are identical as illustrated in Fig. 5.10. For the derivation of (5.93) only terms of lowest order were taken into account. However, (5.93) is also fulfilled when higher order terms are considered by numerically integrating the total system, i.e. (5.4) and (5.5) (cf. Sect. 5.6.2).

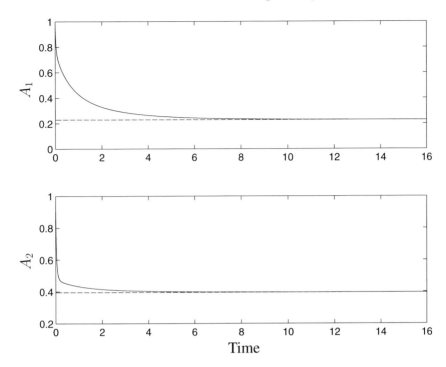

Fig. 5.8. *Amplitude dynamics of type II fixed point:* Numerical integration of the order parameter equations (5.72) and (5.73). Same format and parameter values as in Fig. 5.7 except for $\mu = 4.5$. A_1^{II} and A_2^{II} (*dashed lines*) perfectly agree with numerical results. Normalized relative phase $\phi/(4\pi)$ is shown in Fig. 5.9. A second simulation was performed with negative initial amplitude of the first mode ($A_1(0) = -1$). In this case the curve of A_2 is exactly the same as that of the first simulation, whereas the curve of A_1 is the negative of that of the first simulation. Accordingly in this case the phase dynamics is different as shown in Fig. 5.9.

5.5 Scaling of Noisy Cluster States

In the order parameter dynamics of single-mode instabilities analyzed in Sects. 5.4.2–5.4.5 the order parameters scaled as $\sim \sqrt{\varepsilon}$, where ε is the bifurcation parameter. These results are in accordance with Crawford's (1995) investigation of a more general version of model (4.3), which additionally took into account randomly distributed eigenfrequencies and higher harmonics of arbitrary order in the coupling. The relationship between coupling mechanism and emerging synchronization patterns was not the main concern of *Crawford's* study. Rather from a more fundamental point of view he analyzed scaling properties of single-mode instabilities of the phase-incoherent state. He showed that in the presence of noise ($Q > 0$) there are two different scaling regions: Denoting the number of the critical mode by k and the linear growth rate by γ, he revealed that for $\gamma > k^2 Q$ the order parameter scales as

(a)

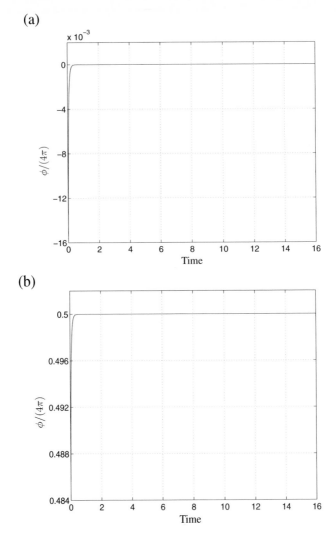

(b)

Fig. 5.9a,b. *Phase dynamics of type II fixed point:* Same simulations as in Fig. 5.8. Normalized relative phase $\phi/(4\pi)$ is plotted versus time. In-phase synchronization and anti-phase synchronization correspond to $\phi/(4\pi) = 0$ (modulo 1) and 0.5 (modulo 1), respectively. Due to the differences of the dynamics of A_1 in both simulations discussed in Fig. 5.8 the phase dynamics of these simulations are different, too. Model parameters and initial conditions were the same in both simulations except for the values of $A_1(0)$. (a) refers to the simulation with $A_1(0) = 1$, and in (b) $A_1(0) = -1$ was chosen. In-phase state as well as anti-phase state perfectly agree with the analytically derived two sorts of type II fixed points, namely (A_1^{II}, A_2^{II}) and $(-A_1^{II}, A_2^{II})$ (cf. (5.93)).

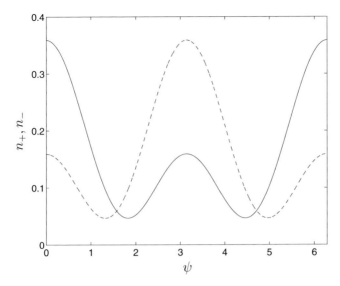

Fig. 5.10. *Densities n_+ and n_- corresponding to both type II fixed points: $n_+(\psi, t)$ and $n_-(\psi, t)$ fulfill the symmetry relationship (5.93). Plot shows a snapshot of n_+ and n_- versus ψ for $t = \phi_0/\Omega$ and $A_1^{II} = A_2^{II} = 0.025$ (cf. (5.92)).*

$\sim \varepsilon$, whereas for $\gamma < k^2 Q$ it scales as $\sim \sqrt{\varepsilon}$. As $\varepsilon \approx \gamma$ near onset, our results agree with his scaling analysis. By the way, in the noise-free limit, i.e. for $Q = 0$, the bifurcating mode scales as $\sim \varepsilon$ (Crawford 1995, cf. Daido 1994).

Unlike Crawford's (1995) study which focusses on a single Hopf bifurcation, the two-modes instability analyzed in Sect. 5.4.6 relies on a double Hopf bifurcation (with 1:2 resonance). The bifurcation analysis of the two-modes instability revealed that depending on the parameters ε, μ and Q type I or type II fixed points are stable (cf. Fig. 5.6). The scaling behavior is given by the amplitudes of the order parameters. According to the formulas of the fixed points' amplitudes $A_j^{I,II}$ the scaling behavior of both types of fixed points is different. Type I fixed points scale as $\sim \sqrt{\varepsilon}$ (cf. (5.86)), whereas type II fixed points exhibit a more complex scaling behavior (cf. (5.89) and (5.90)). The transitions between the different scaling regions are continuous as illustrated in Fig. 5.5. Obviously the scaling behavior of the two-modes instability is more faceted compared to single-mode instabilities.

For a detailed analysis of the onset of linear instability and the subsequent bifurcation for the specific case of a sine coupling ($K_1 \neq 0, K_2 = K_3 = K_4 = C_1 = \ldots = C_4 = 0$ in (4.3)) I refer to Strogatz and Mirollo (1991), Strogatz, Mirollo and Matthews (1992), Crawford (1994).

5.6 The Experimentalist's Inverse Problem

In this section the relationship between the oscillators' collective dynamics and the corresponding experimental data will be discussed. In particular, it will be emphasized that the interpretation of macroscopic signals describing the dynamics of a cluster of oscillators is nothing but an inverse problem lacking a unique solution. Bearing this ambiguity in mind the travelling waves of the density n analyzed in the former sections will be related to macroscopic variables which correspond, for example, to the firing activity of clusters of neurons. This link has to be established as otherwise the modelling could not stimulate the interpretation of experimental data.

5.6.1 Travelling Waves

The emerging synchronization patterns revealed by the bifurcation analysis in Sects. 5.4.2–5.4.6 are travelling waves of the density $n(\psi, t)$. They will be illustrated in this section by numerically integrating the Fourier transformed model equations (5.4) and (5.5). In the numerical simulations Fourier modes with wave numbers $|k| > 200$ were neglected. This approximation turned out to be justified because taking into account a larger number of Fourier modes did not change the simulations' outcome. Simulations were performed by means of a 4th order Runge-Kutta algorithm.

Figures 5.11–5.18 illustrate the mode dynamics and the corresponding travelling waves which emerge as a consequence of single-mode instabilities. These plots indicate the order parameter's prominent role in generating the respective synchronization pattern. The travelling wave given by (5.32) and plotted in Fig. 5.15 can be illustrated in a different as shown in Fig. 5.1: An identification of 0 and 2π of the phase interval $[0, 2\pi]$ yields a circle on which the density n rotates as travelling wave. In the same way the travelling waves determined by (5.49), (5.57), and (5.65) run on such a phase circle.

On the analogy of the deterministic case discussed in Sect. 3.7.1 the configuration of a noisy cluster state refers to the number of oscillators within each cluster, too. A single cluster is delimited from the neighbouring clusters by minima of the density n. Figure 5.19 illustrates the relationship between the shape of the evolving travelling wave and the configuration of the corresponding noisy cluster state. To this end Fig. 5.19 shows n_e, which is the density encountered at the end of the simulations presented in Figs. 5.15–5.18 respectively. Comparing Fig. 5.19 with Fig. 3.2 shows the analogy between noisy cluster states and deterministic cluster states.

Interestingly the stochastic two-, three- and four-cluster states caused by a single supercritical coupling constant K_2, K_3 and K_4 are symmetrical cluster states, i.e. all clusters contain the same number of oscillators. Why does a single supercritical coupling constant most probably give rise to symmetrical cluster states in the presence of noise? In order to answer this question let us consider a deterministic two-cluster state. The analysis in Sect. 3.8.1 showed

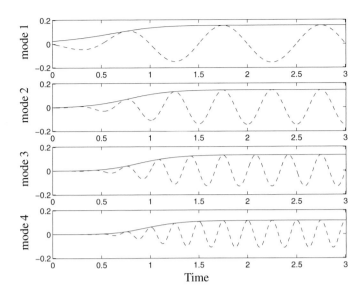

Fig. 5.11. *First-mode instability:* The Fourier transformed model equations (5.4) and (5.5) were integrated numerically with parameters $K_1 = 5$, $K_2 = K_3 = K_4 = C_1 = \ldots = C_4 = 0$, $Q = 0.4$, $\Omega = 2\pi$. The initial state was $n(\psi, 0) = \mathcal{N} [1/(2\pi) + 0.05 \sin \psi]$ where $\mathcal{N}$ is a normalization factor which guarantees that $\int_0^{2\pi} n(\psi, t) \, d\psi = 1$ holds. Plots show the amplitudes $A_1, \ldots, A_4$ (*solid lines*) and real parts $\mathrm{Re}[\hat{n}(1, t)], \ldots, \mathrm{Re}[\hat{n}(4, t)]$ (*dashed lines*) of the first four modes. The first mode acts as order parameter and forces the enslaved modes to increase.

that the number of attractors belonging to a fixed configuration (N_1, N_2) is given by

$$n_{\mathrm{comb}}(N_1, N_2) = \binom{N}{N_1} = \frac{N!}{N_1! \, N_2!} \tag{5.94}$$

(cf. (3.60), Fig. 3.3). Thus, attractors with a symmetrical $(N_1 = N_2)$ and nearly symmetrical $(N_1 \approx N_2)$ configuration predominate. In the derivation of the model equation it was assumed that N is large (cf. Sect. 4.2). Consequently the attractors crowd very tightly in phase space, and noise causes the system to hop among the attractors. For this reason one can expect to encounter the system most probably in the neighbourhood of an attractor which is connected with a (nearly) symmetrical configuration.

Note that this is a qualitatively different phenomenon as compared with attractor crowding revealed by Wiesenfeld and Hadley (1989). In their system splay-phase states scaled as $\sim (N-1)!$, where N was the number of oscillators. In splay-phase states the oscillators' phases are equally spaced in the interval $[0, 2\pi]$. Consequently in the model of Wiesenfeld and Hadley there was only one sort of attractors as far as the configuration is concerned. For this reason the details of the deterministic dynamics faded due to noise induced phase-space diffusion, so that for large enough noise amplitude the synchronization

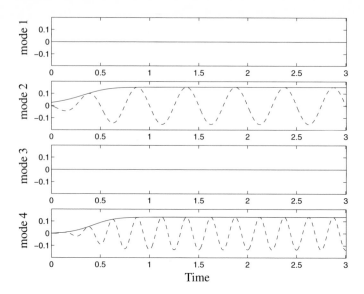

Fig. 5.12. *Second-mode instability:* (5.4) and (5.5) were integrated numerically with parameters $K_2 = 5$, $K_1 = K_3 = K_4 = C_1 = \ldots = C_4 = 0$, $Q = 0.4$, $\Omega = 2\pi$. Initial state was $n(\psi, 0) = \mathcal{N}\left[1/(2\pi) + 0.05\sin(2\psi)\right]$ where $\mathcal{N}$ is the normalization factor. The same format as in Fig. 5.11. As order parameter the second mode drives the fourth mode, whereas due to symmetry properties of the system first and third mode remain equal to zero.

patterns vanished. The situation in our model is different because certain types of attractors predominate numerically, and, thus, as a consequence of noise these predominant attractors show up.

While single-mode instabilities do not give rise to asymmetrical multiple-cluster states in two-modes instabilities one encounters both symmetrical and asymmetrical cluster states. Figure 5.20 illustrates that the configuration crucially depends on the coupling pattern given by K_1 and K_2, the supercritical coupling constants causing a two-modes instability.

5.6.2 Firing Patterns

Let us interpret the synchronization processes of the former section by considering the cluster of oscillators as a model of interacting neurons. As explained in Sect. 3.4.1 the single neuron is approximated by means of a phase oscillator. The jth model neuron fires (or bursts) whenever its phase ψ_j equals $\rho + k2\pi$, where ρ is a fixed parameter, and k is an integer. According to (4.12), the definition of the averaged number density $n(\psi, t)$, the number of model neurons which are most probably firing at time t is given by $n(\rho, t)$. The latter is a macrovariable which corresponds to observables typically measured in experiments. As yet, experimentalists are not able to measure the

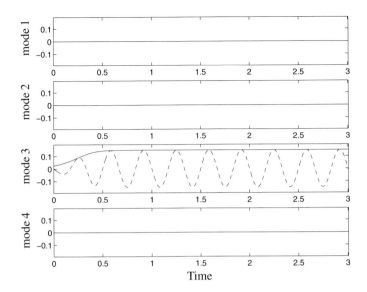

Fig. 5.13. *Third-mode instability:* Numerical integration of (5.4) and (5.5) with parameters $K_3 = 5$, $K_1 = K_2 = K_4 = C_1 = \ldots = C_4 = 0$, $Q = 0.4$, $\Omega = 2\pi$. Initial state was $n(\psi, 0) = \mathcal{N} \left[1/(2\pi) + 0.05 \sin(3\psi) \right]$. The same format as in Fig. 5.11. The third mode is the order parameter. The enslaved modes 1, 2 and 4 remain equal to zero because of the system's symmetry properties.

firing behavior of a large cluster of neurons on a microscopic level. In other words, they cannot measure which particular neurons of such a cluster of, say, 100000 neurons fire at time t. Rather in large clusters they can only assess the amount of firing neurons, i.e., how many neurons fire at time t. For this reason the interpretation of measured collective synchronized activity becomes involved.

The ambiguity of macroscopic signals can easily be illustrated by interpreting Fig. 5.19 in the context of the experimentalist's situation. One-, two-, three- and four-cluster states rotate with the oscillators' eigenfreqency Ω. As if he was looking through a small peephole located at $\psi = \rho$, the experimentalist can only estimate the collective *firing density*

$$p(t) = n(\rho, t) \tag{5.95}$$

(cf. Sec. 2.8). For this reason he cannot distinguish between one-cluster states running at eigenfrequency $n\Omega$ and n-cluster states running at Ω as illustrated in Fig. 5.21. Thus, macroscopic variables corresponding to the firing density $p = n(\rho, t)$ have to be interpreted very carefully. This example clearly shows that qualitatively quite different collective dynamical states may be related to similar macroscopic variables. In order to distinguish these states the experimentalist should strive for additionally measuring the activity of as many single oscillators (i.e. neurons) as possible.

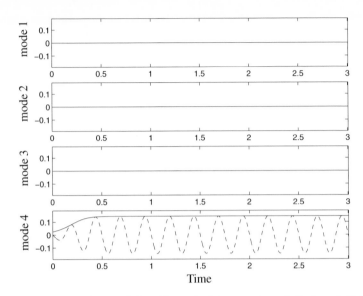

Fig. 5.14. *Fourth-mode instability:* Numerical integration of (5.4) and (5.5) with parameters $K_3 = 5$, $K_1 = K_2 = K_4 = C_1 = \ldots = C_4 = 0$, $Q = 0.4$, $\Omega = 2\pi$. Initial state was $n(\psi, 0) = \mathcal{N}\left[1/(2\pi) + 0.05\sin(4\psi)\right]$. The same format as in Fig. 5.11. The order parameter is the fourth mode. As a consequence of the system's symmetry properties the enslaved modes 1, 2 and 3 remain equal to zero.

Note that (5.95) corresponds to the definition of the firing density (2.62). For $\rho \neq 0$ one can introduce the shifted phase ψ' according to (3.4). As a result of this transformation the model neuron fires (or bursts) whenever its shifted phase ψ' equals zero (modulo 2π). The transformation (3.4) does not affect the model equation (5.1) as discussed in Sect. 3.5. Consequently, in this chapter we assume that ρ is equal to zero.

Firing patterns of one-mode instabilities turned out to be complex and ambiguous. However, the situation becomes even more sophisticated if one considers firing patterns of two-modes instabilities. Two additional dynamical features have to be taken into account:

1. *Asymmetrical two-cluster states:* Besides one-cluster states and symmetrical two-cluster states two-modes instabilities may additionally cause asymmetrical two-cluster states (cf. Fig. 5.20).

2. *Bistability of antiphase firing patterns:* As shown in Sect. 5.4.6 for a certain range of parameters two stable solutions bifurcate. The latter correspond to type II fixed points, which occur pairwise and fulfill the symmetry relationship (5.93). Thus, the firing patterns related to both type II fixed points are of the same shape and exhibit an antiphase relationship. According to the center manifold theorem and numerical simulations both fixed points are attractors. For this reason one encounters bistability, i.e. depending on the initial conditions the system may approach either one fixed point

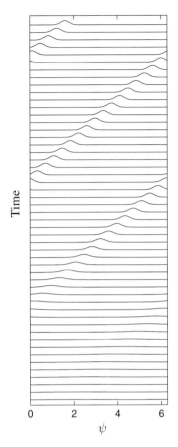

Fig. 5.15. *Travelling wave of the first-mode instability:* Same simulation as in Fig. 5.11. Succesive traces show n as a function of ψ. Earliest time is at the bottom of the figure. Simulation illustrates the travelling wave from (5.32).

or the other. Figures 5.22 and 5.23 show the firing patterns corresponding to a bistable one-cluster state and a bistable asymmetrical two-cluster state, respectively.

5.7 Neural Coding Revisited

Let us come back to the question as to which role the synchronization plays in the context of neural coding (cf. Sect. 3.9). Considering a neural network as a cluster of phase oscillators we became aware of the huge variety of different synchronized states which may emerge in a noise-free setting (cf. Chap. 3). However, noise is inevitable in biological systems, and, thus, has to be taken into account in the discussion of the neuronal code, too. Due to noise

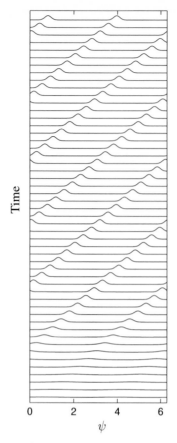

Fig. 5.16. *Travelling wave of the second-mode instability:* Same simulation as in Fig. 5.12. Simulation illustrates the travelling wave (5.49). The same format as in Fig. 5.15.

synchronization patterns may fade away. On the other hand out of a large number of different synchronized states noise may select particular patterns. In the former section the noisy two-cluster state served as an example for this noise induced selection process.

As shown in Sects. 5.4.2–5.4.6 modifying the coupling pattern of the oscillatory network may change the outcome of this selection decisively. So, for given noise amplitude and eigenfrequencies it is the pattern of the oscillators' interactions which determines the emerging synchronized state. In this way encoding facilities are opened up since a single object as a whole may correspond to a stochastic n-cluster state while additional attributes of the object may be encoded by means of the configuration. For instance, depending on whether or not an additional particular visual feature is present a

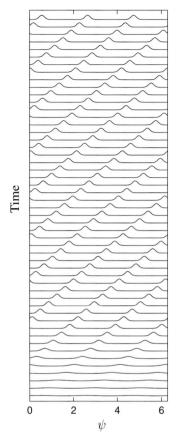

Fig. 5.17. *Travelling wave of the third-mode instability:* Same simulation as in Fig. 5.13. Emerging pattern of the density n corresponds to the travelling wave (5.57). The same format as in Fig. 5.15.

certain object may be encoded by a noisy one-cluster or two-cluster state, respectively.

5.8 Summary and Discussion

This chapter was concerned with spontaneously emerging patterns of synchronization of a cluster of continuously interacting phase oscillators subjected to additive Gaussian white noise. For the analysis of model equation (5.1) the cluster is described by the average number density $n(\psi, t)$. According to its definition (4.12) it is a macrovariable giving the number of oscillators with phase ψ at time t. When a coupling constant K_j becomes supercriti-

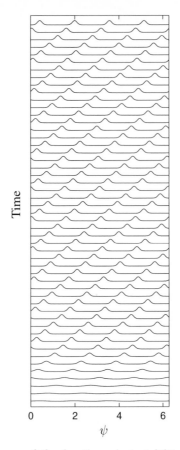

Fig. 5.18. *Travelling wave of the fourth-mode instability:* Same simulation as in Fig. 5.14. Simulation shows the travelling wave (5.65). The same format as in Fig. 5.15.

cal $(K_j > K_j^{\text{crit}} = jQ$, cf. (5.15)) the corresponding pair of Fourier modes $\hat{n}(\pm j, t)$ becomes unstable and acts as order parameters.

The bifurcating solutions of the averaged number density were investigated by deriving the order parameter equations of the one-mode instabilities and of a two-modes instability. According to this analytical analysis and additionally confirmed by numerical investigation the bifurcating solutions turned out to be travelling waves of $n(\psi, t)$.

In order to profit from the insights into the model's dynamics several synchronization processes were considered from the experimentalist's point of view. To this end the cluster served as a model for interacting neurons, and the time course of the firing density $p(t)$ was related to the different sorts of bifurcating travelling waves. For the interpretation of the firing density (5.95) it is necessary to keep in mind that the average number density $n(\psi, t)$

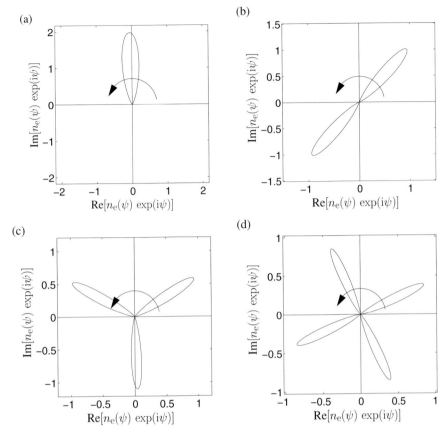

Fig. 5.19a–d. *Noisy n-cluster states of single-mode instabilities:* For the instability of the first (**a**), second (**b**), third (**c**) and fourth (**d**) mode, $n_e(\psi)\exp(i\psi)$ is plotted in the Gaussian plane, where n_e is the final density, i.e. the trace on top of Figs. 5.15 (**a**), 5.16 (**b**), 5.17 (**c**), and 5.18 (**d**), respectively. Due to the vanishing coefficients $C_1,\ldots,C_4$ the synchronization frequency in all simulations equals Ω. This is indicated by arrows having the same length. Note that for non-vanishing coefficients $C_1,\ldots,C_4$ the synchronization frequency generically differs from Ω (cf. (5.31), (5.48), (5.56), (5.64)). The stochastic one- (**a**), two- (**b**), three- (**c**) and four-cluster states (**d**) correspond to the n-cluster states of the deterministic model shown in Fig. 3.2. b

is a macrovariable which reflects the net collective firing activity. In this sense it is comparable to macrosignals such as the electric and magnetic field of the brain registered by means of electroencephalography (EEG) and magnetoencephalography (MEG) (see, e.g., Niedermeyer and Lopes da Silva 1987, Hämäläinen et al. 1993, Nunez 1995).

Let me summarize some results of the analysis of the firing patterns: Macrosignals, like the firing density $p(t)$, have to be interpreted very carefully

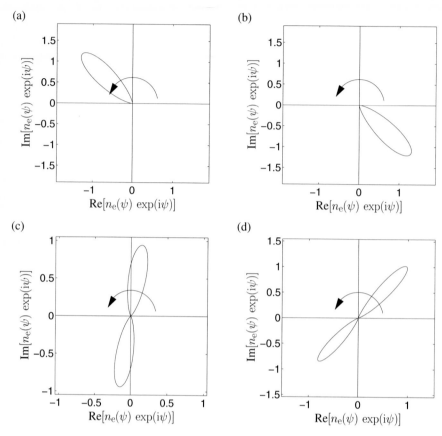

Fig. 5.20a–d. *Noisy n-cluster states of two-modes instabilities:* Apart from one-cluster states ((**a**) and (**b**)) two-modes instabilities may be associated with symmetrical (**c**) as well as asymmetrical (**d**) cluster states. This is illustrated by plotting $n_e(\psi)\exp(i\psi)$ in the Gaussian plane (cf. Fig. 5.19). Cluster states correspond to type I (**c**) and type II ((**a**),(**b**),(**d**)) fixed points, respectively. Type II fixed points occur as bistable pairs fulfilling the symmetry relationship (5.93) ((**a**) and (**b**)). The Fourier transformed model equations (5.4) and (5.5) were numerically integrated with different parameters and initial conditions: (**a**): $\varepsilon = 0.9$, $\mu = 0.5$, $Q = 0.5$, initial mode amplitudes $A_1(0) = A_2(0) = 0.02$. (**b**): $A_1(0) = -0.02$, all other parameters (including simulation duration) as in (**a**). (**c**): $\varepsilon = 0.9$, $\mu = 2.5$, $Q = 0.5$, $A_1(0) = 0$, $A_2(0) = 0.05$. (**d**): $\varepsilon = 0.9$, $\mu = 5$, $Q = 0.5$, $A_1(0) = A_2(0) = 0.02$, where $n(\psi,0) = \mathcal{N}[1/(2\pi) + A_1(0)\sin\psi + A_2(0)\sin(2\psi)]$ with normalization factor $\mathcal{N}$. The coupling pattern (K_1, K_2) is given by ε and μ according to (5.66).

because qualitatively different collective dynamical states may give rise to similar or even the same macroscopic firing patterns. Additionally recording microsignals in parallel may serve as a way out of the ambiguity of macrosignals. Appropriate microsignals are, for instance, single unit activity (SUA)

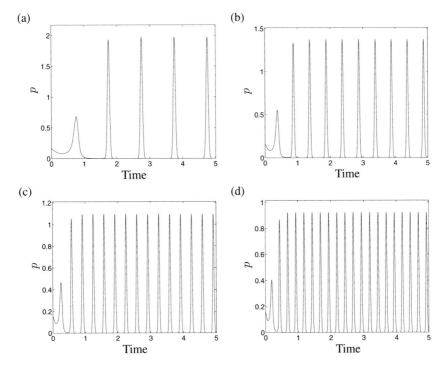

Fig. 5.21a–d. *Firing density of symmetrical cluster states:* Plot displays collective firing probabilities belonging to the simulations of Figs. 5.15 (**a**), 5.16 (**b**), 5.17 (**c**), and 5.18 (**d**). The firing density $p(t) = n(0, t)$ was plotted versus time for the first-mode (**a**), second-mode (**b**), third-mode (**c**) and fourth-mode instability (**d**). In all cases $\Omega = 2\pi$ is the oscillators' eigenfrequency which corresponds to a period $T = 1$. Anyhow, compared to the one-cluster state (**a**) the collective firing behavior in the two- (**b**), three- (**c**) and four-cluster state (**d**) oscillates with 2Ω, 3Ω and 4Ω, respectively.

and multiple unit activity (MUA). The latter refer to the spiking activity of a single neuron or a small group of neurons, respectively.

The oscillators coupling mechanism models the neurons' mutual interactions. Our results point out that different coupling mechanisms may give rise to the same synchronization pattern. For example, a symmetrical two-cluster state may evolve as a result of a single-mode instability (cf. Sect. 5.4.3) or a two-modes instability (cf. Sect. 5.4.6). Nevertheless, as will be explained below, the system's reaction to external stimulation may be quite different in both cases. Put otherwise: Two clusters of interacting neurons may exhibit the same spontaneous firing behavior. Anyhow, due to the different types of neuronal interactions both clusters may react to stimulation in an entirely different way.

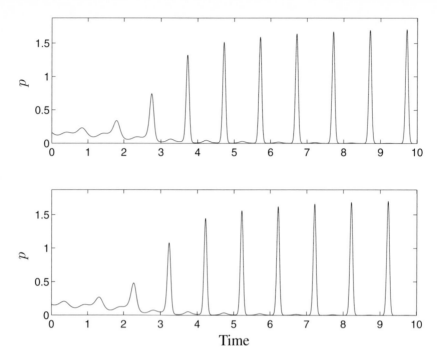

Fig. 5.22. *Firing patterns of bistable one-cluster states:* The bistable pair of type II fixed points corresponds to travelling waves n_+ and n_- exhibiting a phase shift of π (cf. (5.92) and (5.93)). Consequently one encounters an antiphase relationship of the corresponding firing pattern. The firing density p is plotted over time in the upper and lower plot for different initial conditions, respectively. In both simulations the two different stable firing patterns occur which correspond to the pair of type II fixed points. The initial conditions determine which firing pattern emerges. Parameters for the numerical simulation: $\varepsilon = 0.9$, $\mu = 0.5$, $Q = 0.5$. Initial mode amplitudes: $A_1(0) = 0.02$ *(upper plot)*, $A_1(0) = -0.02$ *(lower plot)*, $A_2(0) = 0.02$ *(for both simulations)*, where the initial density is given by $n(\psi, 0) = \mathcal{N}[1/(2\pi) + A_1(0)\sin\psi + A_2(0)\sin(2\psi)]$. Same simulations as in Figs. 5.20a,b.

The investigation of a two-modes instability revealed that a pair of stable antiphase firing patterns may occur. Hence, stimulation might cause a switching between both stable states, which might affect the phase resetting behavior decisively. This issue will be addressed in subsequent chapters.

Finally the relationship between clustering and neural coding was discussed. As explained in this chapter clustering may open up additional encoding facilities. However, it can only be decided experimentally whether or not clustering and, in particular, the cluster states' configuration are relevant in the context of neural coding. The data analysis tool for estimating the configuration of cluster states in real data will be described in Chap. 8.

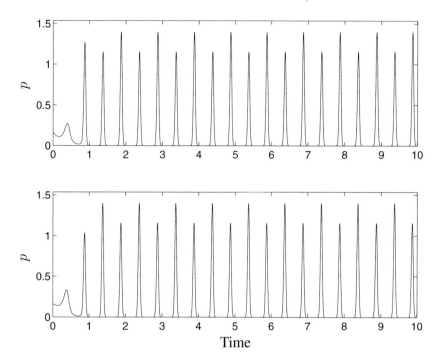

Fig. 5.23. *Firing patterns of bistable two-cluster states:* Apart from bistable one-cluster states (cf. Fig. 5.22) a two-modes instability may be related to bistable asymmetrical two-cluster states, too. Same format as in Fig. 5.22. As a consequence of the different initial conditions the system is attracted by the two different type II fixed points which correspond to antiphase firing patterns. Parameters for the numerical simulation: $\varepsilon = 0.9$, $\mu = 5$, $Q = 0.5$. Initial mode amplitudes: $A_1(0) = 0.02$ (*upper plot*), $A_1(0) = -0.02$ (*lower plot*), $A_2(0) = 0.02$ (*for both simulations*), and initial density $n(\psi, 0) = \mathcal{N}[1/(2\pi) + A_1(0)\sin\psi + A_2(0)\sin(2\psi)]$.

6. Single Pulse Stimulation

6.1 Introductory Remarks

This chapter is dedicated to the impact of pulsatile stimuli on noisy cluster states. In order to avoid misunderstandings it should be mentioned that in physics the duration of pulses is typically small compared to the time scale of the reference processes under consideration. On the contrary, in the context of phase resetting in biology (cf. Winfree 1980, Glass Mackey 1988) pulses can but need not be short compared to the period of an oscillation. The duration of the pulse can even be of the same order as a period of an oscillation.

As explained in Sects. 2.2.1 and 4.2 the stimulus is modeled by

$$S(\psi) = \sum_{m=1}^{4} I_m \cos(m\psi + \gamma_m) \,, \tag{6.1}$$

where ψ is the phase of an oscillator, and $\{I_m\}$ and $\{\gamma_m\}$ are parameters determining the stimulation mechanism. We shall analyze how the stimulus' effect depends on these parameters $\{I_m\}$ and $\{\gamma_m\}$. To this end we shall study a variety of numerical simulations where different types of stimuli are applied to the different noisy cluster states which we investigated in Chap. 5. Typically all of these simulations have the same structure which consists of three parts:

1. In the first part of a simulation there is no stimulus ($S = 0$). We shall observe spontaneously emerging noisy cluster states which we already know from Chap. 5. We there performed an analytical analysis which enables us to identify order parameters and stable modes and to comprehend their contributions to the spontaneous formation of the synchronized states and the corresponding firing patterns.

2. During the second part of a simulation the pulsatile stimulus is administered ($S \neq 0$). We shall study how different types of stimuli act on different types of noisy cluster states. For instance, we shall probe how the noisy one-cluster state from Sect. 5.4.2 is affected by two different stimuli, one with parameters $I_1 \neq 0$, $I_2 = I_3 = I_4 = \gamma_1 = \ldots \gamma_4 = 0$, the other one with $I_1 \neq 0, I_4 \neq 0, I_2 = I_3 = \gamma_1 = \ldots \gamma_4 = 0$. In this way we shall encounter several and qualitatively different ways of how a cluster reacts to a pulsatile stimulus. Again we shall profit from our analytical analysis of Chap. 5: As

a result of it we know the modes which act as order parameters and stable modes during synchronization processes, respectively. We shall see that both types of modes react on pulsatile stimuli in a characteristically different way. This will turn out to be particularly important for the understanding of the stimulation induced firing pattern. Concerning the stimulus' intensity we have to distinguish between supercritical short pulses and subcritical long pulses investigated in Sects. 6.2–6.5 and 6.7, respectively. The cluster's dynamics during stimulation is dominated by the stimulus if the stimulation intensity is supercritical. For a subcritical intensity the competition between couplings and stimulus leads to qualitatively different stimulation induced phenomena.

3. At the end of the second part of each stimulation the stimulus is turned off, and during the third part there is no stimulation any more ($S = 0$). We, thus, study how the cluster relaxes towards a stable state, for example, how it resynchronizes and reestablishes the synchronized state observed during the first part of the stimulation. In terms of mathematics the third part of the stimulation can be considered as an initial-value problem: The stimulus removes the cluster from a stable state and determines the cluster's initial state for its relaxation process.

Let us recall the evolution equation for the average number density $n(\psi, t)$ derived in Sect. 4.2:

$$\frac{\partial n(\psi, t)}{\partial t} = \underbrace{-\frac{\partial}{\partial \psi} \left[n(\psi, t) \int_0^{2\pi} d\psi' \, M(\psi - \psi')n(\psi', t) \right]}_{\text{II}} \tag{6.2}$$

$$\underbrace{-\frac{\partial}{\partial \psi} n(\psi, t) \, S(\psi)}_{\text{III}} \quad \underbrace{-\Omega \, \frac{\partial}{\partial \psi} n(\psi, t)}_{\text{I}} \quad \underbrace{+\frac{Q}{2} \frac{\partial^2 n(\psi, t)}{\partial \psi^2}}_{\text{IV}} \tag{6.3}$$

(cf. 4.21). We are already familiar with the meaning of the different terms: As shown in Chap. 5 noisy cluster states emerge if the mutual interactions (term II) are strong enough compared to the random forces (term IV). Moreover we saw that the synchronization patterns belonging to the noisy cluster states move as travelling waves (modulo 2π) as a consequence of the drift term (I).

What still remains to be analyzed is the impact of term III, that means the impact of the stimulus S modeled by (6.1). To this end we shall study the above mentioned numerous simulations having a three-part structure: During the first part we shall ecounter the spontaneous formation of order, i.e. the self-synchronization and correspondingly in (6.3) we set $S = 0$ since there is no stimulation. During the second part the stimulus is turned on ($S \neq 0$) and, hence, in (6.3) we have to take it into account according to (6.1). In the third part of each simulation we probe the cluster's relaxation and resynchronization occurring in the absence of any stimulation. Accordingly, during the third part we put $S = 0$ in (6.3).

We can illustrate the three-part scenario of the simulations from the neurophysiological standpoint. Imagine a patient who is suffering from a movement disorder where in the brain there is a cluster of synchronously firing neurons which feed their pathological activity into several other neuronal clusters in this way causing a pathological brain dynamics associated with severe symptoms. Let us assume that the patient does not respond well to drug therapy and, thus, a deep brain electrode is stereotactically implanted in the brain so that it is placed in that particular pathologically active neuronal cluster. In Chap. 10 movement disorders and sterotactic stimulation techniques of this kind will be explained in detail. For the time being it is important to simply imagine that somewhere in the brain there is a cluster displaying pathological synchronized activity which has to be desynchronized in order to give relief to the patient. This is one possible setting within which we can interpret the simulations:

The physician, a neurologist or a neurosurgeon, can use the chronically implanted deep brain electrode in two ways. On the one hand he can assess the particular cluster's strength of pathological synchronization, e.g., by registering the cluster's local field potential (LFP) with this electrode. On the other hand he can also administer pulsatile stimuli via this electrode. Hence, by measuring the LFP the physician can observe how the cluster becomes synchronized (part 1 of the simulation). Via the chronically implanted deep brain electrode he applies a well-timed pulsatile stimulus (part 2) and induces a desynchronization. Again measuring the cluster's LFP the physician observes how the cluster resynchronizes, that means how the amplitude of the LFP increases (part 3). As soon as a threshold amplitude of the LFP is exceeded (part 1 of the next simulation with the same model parameters) the physician again applies a pulsatile stimulus (part 2) and after that observes the resynchronization again (part 3). In this way a repetitive pulsatile deep brain stimulation with feedback control could be performed. Of course, this procedure should be automated in order to be convenient and practicable. We shall come back to this issue in Chap. 10. First, we have to learn more about desynchronization processes caused by stimulation. To utilize our stochastic approach to phase resetting for applications in neuroscience all stimulation induced effects will be visualized in terms of the firing pattern of a group of interacting neurons.

6.2 How Stimulation Affects Order Parameters

This section is devoted to stimulation induced desynchronization and spontaneous resynchronization which occurs after the stimulation is finished. To this end it will be explained how stimulation affects the order parameter of a single-mode instability and how these effects can appropriately be described.

6.2.1 Cluster Variables and Order Parameters

In the deterministic model the dynamics of n-cluster states was investigated with the cluster variables (3.71). In order to use this concept also for the stochastic model the *cluster variable* Z_m has to be introduced by setting

$$Z_m(t) = R_m(t) \exp[i\varphi_m(t)] = \int_0^{2\pi} n(\psi, t) \exp(im\psi) \, d\psi , \qquad (6.4)$$

where $R_m(t)$ and $\varphi_m(t)$ are the (real) cluster amplitude and the (real) cluster phase, respectively. Note that Z_1 corresponds to the ensemble variable Z defined by (2.33). As illustrated in Fig. 2.3 $Z_1(t)$ gives the center of mass of the density $n(\psi, t)$.

Inserting the Fourier expansion (4.24) of $n(\psi, t)$ into (6.4) one immediately obtains the relationship between cluster variables and Fourier modes:

$$Z_m(t) = 2\pi\hat{n}(-m, t) . \qquad (6.5)$$

Denoting (real) amplitude and (real) phase of the mth Fourier mode by $A_m(t)$ and $\phi_m(t)$, the mth mode takes the form

$$\hat{n}(m, t) = A_m(t) \exp[i\phi_m(t)] . \qquad (6.6)$$

Inserting (6.4) and (6.6) into (6.5) and taking into account that $\hat{n}(-m, t)$ is the complex conjugate of $\hat{n}(m, t)$ one ends up with

$$R_m(t) = 2\pi A_m(t) , \quad \varphi_m(t) = -\phi_m(t) = \phi_{-m}(t) . \qquad (6.7)$$

Equations (6.5) and (6.7) substantiate the in a way phenomenological definition of the cluster variables (3.71) and (6.4) for the deterministic and stochastic model. Via (6.7) cluster amplitudes and phases are directly related to the amplitude and phase of the corresponding Fourier modes of the average number density $n(\psi, t)$. The bifurcation analysis in Sect. 5.4 revealed that the Fourier modes $\hat{n}(m, t)$ and $\hat{n}(-m, t)$ become order parameters when the coupling K_m becomes supercritical. In this way cluster amplitude and phase reflect the order parameters' amplitude and phase.

The extent of synchronization of a cluster state can conveniently be estimated with the cluster variables Z_m because due to the normalization condition (4.23) their amplitudes fulfill

$$|Z_m(t)| = R_m(t) \le 1 \qquad (6.8)$$

for all times t. Fully desynchronized states exhibiting a uniform density $n(\psi, t) = 1/(2\pi)$ are associated with vanishing cluster amplitudes R_m ($m = 1, 2, 3, \ldots$).

6.2.2 Uniform and Partial Desynchronization

In order to analyze the desynchronizing effect of a stimulus we have to describe the extent of desynchronization appropriately. Certainly we should not simply identify a desynchronized state with a uniform distribution of the average number density $n(\psi, t)$. Apart from such a strong type of desynchronization one should take into account weaker forms, too. Let us, e.g., assume that a stimulus induces a transition from a strongly synchronized one-cluster state to a weakly synchronized two-cluster state. In such a case the stimulus obviously acts in a desynchronizing way although the desynchronized state is not connected with a uniform distribution of $n(\psi, t)$. For this reason two definitions are introduced:

1. *Uniform desynchronization:* A cluster is uniformly desynchronized at time t provided its average number density is close to a uniform distribution. In this case all non-vanishing Fourier modes are close to zero, i.e. $|\hat{n}(k, t)| \ll 1$ for $k \neq 0$. The extent of uniform desynchronization will be estimated by means of

$$\delta(t) = \left\| n(\psi, t) - \frac{1}{2\pi} \right\| , \tag{6.9}$$

where the norm is defined by $\|g\| = \sqrt{\int_0^{2\pi} g^2(\xi)\, d\xi}$. Writing the Fourier modes in terms of real amplitudes $A_k(t)$ and phases $\phi_k(t)$ according to $\hat{n}(k, t) = A_k(t) \exp[i\phi_k(t)]$ and using the Fourier expansion (4.24) one obtains

$$\delta(t) = \sqrt{2\pi \sum_{k \in \mathbb{Z},\, k \neq 0} A_k^2(t)} = \sqrt{\frac{1}{2\pi} \sum_{k \in \mathbb{Z},\, k \neq 0} R_k^2(t)} , \tag{6.10}$$

where (6.7) was taken into account. As explained in Appendix A only a finite number of Fourier modes is taken into account for the numerical simulations, namely those with wave numbers $k = \pm 1, \pm 2, \ldots, \pm 200$. According to extensive numerical tests we are allowed to neglect contributions of Fourier modes of higher order ($|k| > 200$), which is why in all subsequent numerical investigations the sum in (6.10) will run only over wave numbers $k = \pm 1, \pm 2, \ldots, \pm 200$. The smaller $\delta(t)$ the larger is the extent of the uniform desynchronization at time t. Complete uniform desynchronization at time t is associated with a uniform average number density, i.e. $n(\psi, t) = 1/(2\pi)$.

2. *Partial desynchronization:* A cluster which is partially desynchronized with respect to the jth mode at time t has a pair of Fourier modes with wave numbers $\pm j$ which are close to zero, i.e. $|\hat{n}(\pm j, t)| \ll 1$. As the amplitude of Fourier modes typically decreases with increasing wave number j, the term partial desynchronization refers to modes of low order, e.g., $j = 1, \ldots, 4$. One can directly assess the markedness of partial desynchronization of the jth mode by means of the amplitude of the jth Fourier mode A_j and the jth cluster amplitude R_j according to

$$A_j(t) = \frac{R_j}{2\pi} = |\hat{n}(\pm j, t)| \tag{6.11}$$

(cf. (6.7)). The smaller $A_j(t)$ and $R_j(t)$ the more the jth mode is partially desynchronized at time t. Complete partial desynchronization of the jth mode at time t corresponds to $A_j(t) = R_j(t) = 0$. Note that this definition is independent of whether or not the other non-vanishing Fourier modes equal zero.

Additionally one should recall that $\int_0^\infty n(\psi, t)\, d\psi$ is a conserved quantity according to (4.23). For this reason $\hat{n}(0, t)$ equals $1/(2\pi)$ for all times t (cf. (4.30)).

6.2.3 Stimulating a One-Cluster State

This section is concerned with the stimulus' impact on a synchronized one-cluster state. To this end let us recall the first-mode instability which was analyzed in Sect. 5.4.2. When the coupling parameter K_1 exceeds its critical value K_1^{crit} a Hopf bifurcation occurs, where the Fourier modes $\hat{n}(1, t)$ and $\hat{n}(-1, t)$ become order parameters. According to the order parameter equations (5.26) and (5.27) the system behaves as a ball rotating in the circular potential valley (Fig. 5.4). Fig. 5.11 shows how the first mode (given by $\hat{n}(1, t)$ and $\hat{n}(-1, t)$) enslaves the stable modes. The corresponding travelling wave and the firing pattern are displayed in Figs. 5.15 and 5.21a.

The dynamics of the averaged number density is governed by (4.21). Before and after the stimulation $S(\psi)$ vanishes, whereas during the stimulation $S(\psi)$ models the stimulus' impact on the cluster. The stimulation begins and ends at times t_B and t_E, respectively. The stimulation duration will be denoted by $T = t_\mathrm{E} - t_\mathrm{B}$. The jth cluster phase at the beginning and at the end of the stimulation will be denoted as $\varphi_{j,\mathrm{B}}$ and $\varphi_{j,\mathrm{E}}$, respectively:

$$\varphi_{j,\mathrm{B}} = \varphi_j(t_\mathrm{B})\,, \quad \varphi_{j,\mathrm{E}} = \varphi_j(t_\mathrm{E})\,. \tag{6.12}$$

The effect of stimulation on the cluster of oscillators can appropriately be illustrated by considering the potential of the order parameter equations. A stimulation giving rise to a vanishing first mode ($A_1 = 0$) has to force the system out of the potential's valley in order to place it on top of the potential's maximum located in the origin (Fig. 6.1).

From now on let us describe the stimulation induced dynamics by means of cluster amplitudes R_m and cluster phases φ_m defined by (6.4). R_m and φ_m are directly related to amplitude and phase of the mth Fourier mode by (6.7). This way of description will enable us to profit from normalization property (6.8): The larger R_1 the more synchronized is a one-cluster state, where R_1 equals 1 provided all phases coincide (modulo 2π).

Performing transformation (6.7) one can still use the order parameters' potential to visualize the effect of stimulation. One simply has to take into account that due to (6.7) the radius of the potential valley has to be multiplied by 2π, and the system runs in opposite direction within the valley.

In this section let us consider a stimulus which is of first order, that is of the same order as the order parameters $\hat{n}(\pm 1, t)$. In other words, we choose

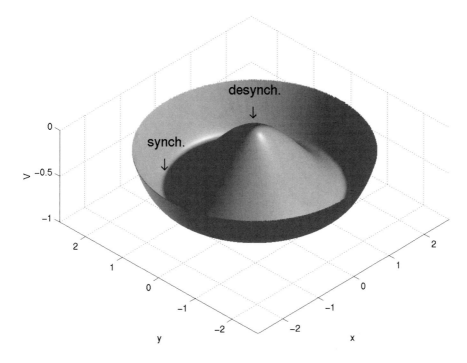

Fig. 6.1. *Principle of the stimulation induced partial desynchronization:* The limit cycle which emerges when the coupling K_1 becomes supercritical can be visualized as a ball rotating in the potential's circular valley (denoted by 'synch.', cf. Sect. 5.4.2). A stimulation which causes a complete partial desynchronization of the first mode, that means, which causes the amplitude of the first mode to vanish ($A_1 = 0$), has to place the ball on top of the potential's hill (denoted by 'desynch.').

$I_1 > 0, I_2 = I_3 = I_4 = 0$ (cf. (6.1)). Fig. 6.2 shows how the cluster variable Z_1 of a one-cluster state vanishes due to a well-timed stimulus of suitable intensity I_1. This figure can be interpreted as if one looks at the potential from above. Before the stimulation starts Z_1 runs in the potential's valley in counterclockwise direction. The stimulus lifts Z_1 out of the valley and places it on top of the potential's hill (Fig. 6.2a). After the stimulation Z_1 spirals back into the circular valley, so that finally it rotates within the valley again as it was doing before the stimulation (Fig. 6.2b).

The time course of the stimulation induced partial desynchronization of the first mode and its spontaneous recovery is illustrated in Fig. 6.3. At the end of the stimulation the partial desynchronization is complete, that means R_1 vanishes. After the stimulation the first mode recovers, and finally R_1 reaches its stationary value again.

To illustrate that the complete partial desynchronization of the first mode causes a phase singularity Fig. 6.4 displays the time course of the cluster phase φ_1 during stimulation. After a critical stimulation duration T_{crit} the

(a)

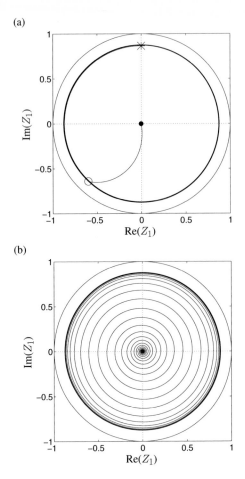

(b)

Fig. 6.2a,b. *Effect of stimulation on Z_1:* The dynamics of cluster variable Z_1 during and after stimulation was analyzed by numerically integrating (4.21). The trajectory of Z_1 is plotted in the Gaussian plane. Stimulation induced partial desynchronization (**a**): Starting at '∗' Z_1 runs several times in counterclockwise direction along its limit cycle. Stimulation starts at 'o' and ends in '•'. At the end of the stimulation R_1 vanishes. Spontaneous resynchronization after the stimulation (**b**): The origin of the Gaussian plane is an unstable focus corresponding to the potential's hill (cf. Fig. 6.1). Therefore Z_1 spirals back towards its stable limit cycle. Coupling and stimulation parameters: $K_1 = 1$, $K_2 = K_3 = K_4 = C_1 = \cdots = C_4 = 0$, $Q = 0.4$, $I_2 = I_3 = I_4 = \gamma_1 = \cdots = \gamma_4 = 0$, $I_1 = 7$ (during stimulation), $I_1 = 0$ (before and after stimulation).

first cluster variable Z_1 is located on top of the potential's hill. Consequently the amplitude R_1 vanishes, and the phase φ_1 is not defined, i.e. φ_1 exhibits a singularity. If the stimulation duration exceeds T_{crit}, that means if one does not switch off the stimulator after the critical stimulation duration, Z_1 rolls

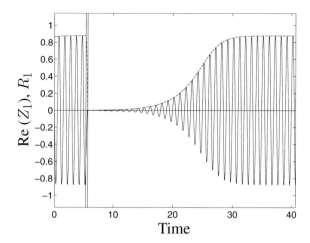

Fig. 6.3. *Time course of Z_1:* For the same simulation as in Fig. 6.2 Re(Z_1) (*solid line*) and R_1 (*dashed line*) are plotted over time. The vertical lines indicate begin and end of the stimulation. R_1 equals zero at the end of the stimulation, and recovers after stimulation. During one period of the oscillation of Re(Z_1) the cluster variable Z_1 once rotates around the origin of the Gaussian plane in Fig. 6.2.

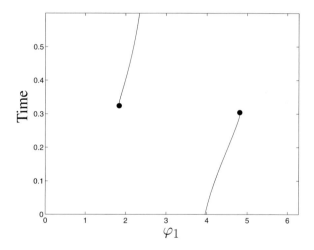

Fig. 6.4. *Partial desynchronization and phase singularity:* The time course of the cluster phase φ_1 during the stimulation is shown for the same initial conditions and model parameters as in Fig. 6.2. Time zero corresponds to the begin of the stimulation in Fig. 6.2. The stimulus lifts Z_1 out of the potential's valley and places it on top of the potential's hill (*indicated by the right black dot*). In contrast to the simulation in Fig. 6.2 the stimulus is not turned off. Consequently the stimulus forces Z_1 to roll down on the opposite side of the hill. For this reason a jump of φ_1 of widh π occurs (*indicated by the left black dot*).

down on the opposite side of the hill. This gives rise to a jump of the cluster phase φ_1 of width π (Fig. 6.4).

Our analysis of the dynamics of the first cluster variable Z_1 revealed that a partial desynchronization of the first mode occurs provided the stimulus is administered at a critical initial cluster phase $\varphi_{1,B}$ for a critical stimulation duration T_{crit}. Hereby the stimulus has to be sufficiently strong which will be discussed below in detail. After having considered how the stimulus affects the first mode, the question suggests itself as to how the stable modes are affected by the stimulus. With this aim in view Fig. 6.5 shows the stimulation induced dynamics of the Fourier modes up to fourth order.

The stimulation starts at $t_B \approx 5.38$ and ends at $t_E \approx 5.69$ (indicated by the left and right vertical line, respectively). At the beginning of the stimulation the cluster is in the stable synchronized state analyzed in Sect. 5.4.2. At the end of the stimulation the amplitude of the first cluster variable vanishes as discussed above. In contrast, the second, third and fourth cluster amplitude do not vanish at time t_E. It is important to stress that apart from the partial desynchronization of the first mode the stimulus damps the amplitudes of the other modes, so that

$$R_j(t_B) > R_j(t_E) \quad \text{for} \quad j = 1, \ldots, 4 \tag{6.13}$$

holds.

After the stimulation the cluster spontaneously resynchronizes. This process of recovery is governed by Haken's (1977, 1983) *slaving principle*. Indeed, the latter is illustrated in Fig. 6.5 par excellence. Due to the stimulation at time t_E the amplitude of the order parameters $\hat{n}(1, t)$ and $\hat{n}(-1, t)$ vanishes ($R_1 = 0$). According to the slaving principle the order parameters determine the enslaved, i.e. stable, modes so that

$$R_u \propto R_s^l \tag{6.14}$$

is fulfilled, where R_u and R_s denote the amplitudes of the order parameters and the enslaved modes, respectively (cf. (3.17)), and l typically equals 2 or 3. Correspondingly the second mode, for instance, fulfills

$$\hat{n}(2, t) \propto \hat{n}(1, t)^2 \tag{6.15}$$

(cf. (5.23)). The order parameters act on a slow time scale, whereas enslaved modes act on a fast time scale. For this reason the relaxation of the enslaved modes is much quicker than the increase of the order parameters' amplitude. This is the very reason why the enslaved modes quickly adjust to the order parameters as determined by (3.17) and (5.23). Hence, vanishing order parameters cause the enslaved modes to relax to zero. From the mathematician's viewpoint the relaxation process of the higher order modes occurring directly after stimulation is due to the fact that the system is attracted by the center manifold.

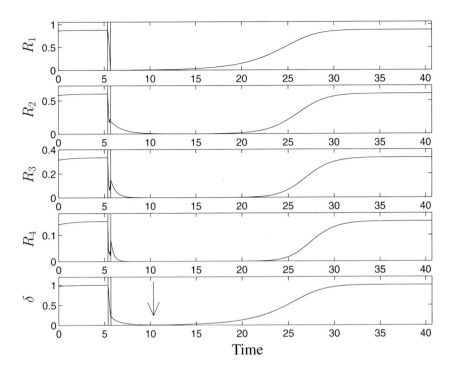

Fig. 6.5. *Resynchronization and slaving principle:* The time course of cluster amplitudes $R_1, \ldots, R_4$ is plotted for the same simulation as in Figs. 6.2 and 6.3. Begin and end of the stimulation are indicated by the two vertical lines. Resynchronization is governed by the slaving principle as explained in the text. The extent of uniform desynchronization is estimated with $\delta(t)$ from (6.10). The vertical arrow marks the minimum of $\delta(t)$. Note that the timing points of strongest partial desynchronization of the first mode (at the end of the stimulation) and strongest uniform desynchronization do not coincide.

As a consequence of the relaxational damping of the enslaved modes directly after stimulation one encounters a delay between complete partial desynchronization of the first mode (i.e. $R_1 = 0$) and the maximal extent of uniform desynchronization (i.e. $\delta(t) = $ min., indicated by the arrow in Fig. 6.5). The former occurs at time $t_E \approx 5.69$, whereas the latter occurs later, namely at time $t_\delta \approx 10.33$.

Figure 6.2b illustrates how the cluster variable Z_1 spirals down the potential's hill until it finally reaches the circular valley again. While the first mode recovers in this way, its amplitude R_1 increases, and again the slaving principle comes into effect: The enslaved modes are governed by the order parameters, that means by Z_1, according to (3.17) and (6.15). Thus, as R_1 increases the amplitudes of the enslaved modes are forced to increase, too, and finally the system attains the stable synchronized state as before the stimulation (Fig. 6.5).

Both the stable modes' relaxational damping as well as the resynchronization are reflected by the firing pattern. This is illustrated in Fig. 6.6
which shows the time course of the firing density $p(t) = n(\rho, t)$ (cf. (5.95)).
In this chapter let us choose $\rho = 0$. For non-vanishing ρ one can straightforwardly introduce the shifted phase ψ' according to (3.4), so that the model
neuron fires (or bursts) whenever its shifted phase ψ' equals zero (modulo
2π). Sections 2.8 and 5.6.2 revealed that as a consequence of transformation
(3.4) one additionally has to introduce transformed stimulation parameters
γ'_m according to (2.64).

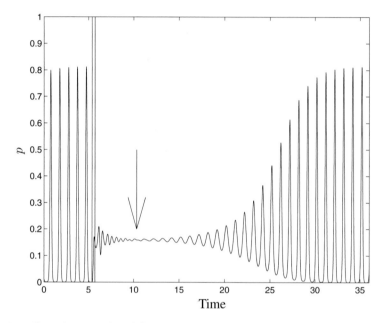

Fig. 6.6. *Stimulation induced firing pattern:* The time course of the firing density
$p = n(0, t)$ is plotted for the same simulation as in Fig. 6.5. The two vertical
lines indicate begin and end of the stimulation. The vertical arrow indicates the
maximal uniform desynchronization occurring at time $t_\delta \approx 10.33$. Between t_E and
t_δ the firing patterns reflects the enslaved modes relaxational damping, whereas
after t_δ it is dominated by the order parameters.

Between the end of the stimulation and the occurence of the maximal uniform desynchronization (at time $t_\delta \approx 10.33$) the firing pattern is dominated
by the stable modes which are of higher order. The resynchronization occurring after t_δ is governed by the increase of the order parameters' amplitude.
We are already familiar with the firing pattern during resynchronization from
Fig. 5.21a.

To clarify the relationship between the cluster variables and the firing
pattern let us recall the Fourier expansion of the average number density

$$n(\psi, t) = \frac{1}{2\pi} + \sum_{k \neq 0} \hat{n}(k, t)\, e^{ik\psi} \tag{6.16}$$

(cf. (4.24) and (4.30)). Using (6.6) and (6.7) the firing density p can be expanded in terms of the amplitudes and phases of Fourier modes or cluster variables, respectively:

$$p(t) = n(0, t) = \frac{1}{2\pi} + 2 \sum_{k > 0} A_k(t) \cos[\phi_k(t)] \tag{6.17}$$

$$= \frac{1}{2\pi} + \pi^{-1} \sum_{k > 0} R_k(t) \cos[\varphi_k(t)] \ . \tag{6.18}$$

Complete uniform desynchronization means that all Fourier modes with non-vanishing wave number k vanish, so that $p(t)$ equals $1/(2\pi) \approx 0.159$ according to (6.18). That is why the firing density of a uniformly desynchronized cluster, in particular, does not vanish. In terms of neurophysiology this means that in a state of complete uniform desynchronization a population of neurons fires incoherently. In particular, the neurons' firing does not cease during periods of complete uniform desynchronization.

To study the impact of the cluster phase $\varphi_{1,B}$ on the outcome of the stimulation let us consider a series of simulations with equally spaced $\varphi_{1,B}$, where all other parameters are the same (Fig. 6.7). This means that a stable stationary one-cluster state is stimulated for different initial values of $\varphi_{1,B}$. We analyze the resetting behavior by means of the winding number which was defined in Sect. 2.7.1. In the first three plots of Fig. 6.7 the winding number equals $+1$, whereas in plots 4 to 6 it is equal to zero. The transition from type 1 resetting to type 0 resetting occurs when the trajectory starting at the filled circle reaches the origin of the Gaussian plane, i.e. after the critical stimulation duration T_{crit}. Thus, the transition of the type of resetting is intimately related with the occurence of stimulation induced complete partial desynchronization of the first mode.

It is important to note that according to a numerical analysis the underlying dynamics of the average number density $n(\psi, t)$ has only one attractor. This was checked by running the simulations for a large number of different initial conditions $\{n(\psi, 0)\}$. In all cases with increasing time $n(\psi, t)$ tended to the same density $n(\psi, \infty)$. Accordingly all trajectories of Z_1 tend to the same attractor of Z_1.

The transition from type 1 to type 0 resetting can additionally be visualized with phase resetting curves which were defined in Sect. 2.7.1. To this end in Fig. 6.8 $\varphi_{1,E}$ is plotted over $\varphi_{1,B}$ for each stimulation. Between the third and fourth plot the transition of the type of resetting occurs.

Figure 6.9 shows the time course of the cluster phase $\varphi_1(t)$. The transition of the type of resetting occurring at time T_{crit} is associated with a singularity of $\varphi_1(t)$.

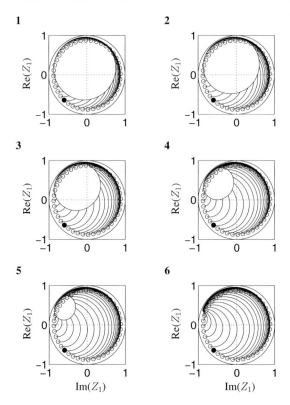

Fig. 6.7. *Winding numbers:* 40 simulations of (4.21) were performed with $\varphi_{1,B}$ equally spaced in $[0, 2\pi]$. From plot 1 to plot 6 the stimulation duration T increases. The trajectories of Z_1 are shown in the Gaussian plane, where the beginning of each trajectory is marked by a circle. The filled circle indicates the trajectory running through the origin (cf. Fig. 6.2). Coupling and stimulation parameters as in Fig. 6.5: $K_1 = 1$, $K_2 = K_3 = K_4 = C_1 = \cdots = C_4 = 0$, $Q = 0.4$, $I_1 = 7$, $I_2 = I_3 = I_4 = \gamma_1 = \cdots = \gamma_4 = 0$. In each plot a cubic spline connects the ends of the trajectories in this way illustrating the corresponding winding number. In the first three plots the winding number is equal to $+1$, and in the other plots it equals zero. Same format as in Fig. 2.10.

Finally the results concerning the stimulation induced dynamics of the first-mode instability are summarized:

1. A stimulus causes a complete partial desynchronization of the first mode provided it is administered at a critical initial cluster phase $\varphi_{1,B}^{\text{crit}}$ for a critical duration T_{crit}. Obviously the stimulation intensity has to be large enough. This issue will be addressed in Sect. 6.7 in detail.
2. A transition from type 1 resetting to type 0 resetting occurs when the stimulation duration exceeds T_{crit}.

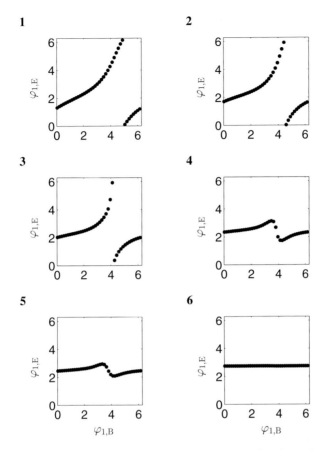

Fig. 6.8. *Phase resetting curves:* $\varphi_{1,E}$ is plotted over $\varphi_{1,B}$ for the series of simulations of Fig. 6.7. Stimulation durations and arrangement of the plots as in Fig. 6.7. The transition from type 1 to type 0 resetting takes place between plots 3 and 4. Same format as in Fig. 2.11.

3. During the period of recovery one observes two dynamical phenomena: the relaxational damping of the stable modes which is followed by the resynchronization driven by the order parameters (Figs. 6.5 and 6.6).

6.2.4 Stimulating a Two-Cluster State

In this section let us focus on how a stimulus affects a synchronized two-cluster state. With this aim in view we briefly recall the bifurcation scenario of the second-mode instability (cf. Sect. 5.4.3): When the coupling K_2 exceeds its critical value K_2^{crit} a Hopf bifurcation occurs giving rise to a stable limit cycle. The Fourier modes $\hat{n}(2,1)$ and $\hat{n}(-2,1)$ act as order parameters, and their dynamics is governed by the order parameter equations (5.46) and

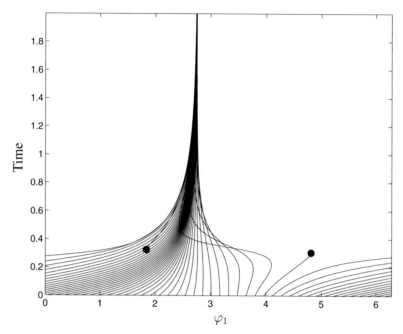

Fig. 6.9. *Cluster phase $\varphi_1(t)$:* The time course of $\varphi_1(t)$ of the series of simulations of Fig. 6.7 is shown. At the critical stimulation duration $T_{\mathrm{crit}} = 0.304$ one observes a phase singularity and a jump of width π as indicated by the two black dots (cf. Fig. 6.4). Same format as in Fig. 2.12.

(5.47). The latter have a potential which is of the form shown in Fig. 6.1. As discussed in the former section a ball moving in the potential landscape visualizes the motion of both the order parameters and the cluster variables.

Let us choose a stimulus which is of second order ($I_2 > 0, I_1 = I_3 = I_4 = 0$, cf. (6.1)) and, thus, of the same order as the order parameters $\hat{n}(\pm 2, t)$. A complete partial desynchronization occurs provided the stimulus lifts the ball out of the potential's valley and places it on top of the hill. On the analogy of the former section one may expect that a desynchronization of this kind is realized by a stimulus administered at a critical initial cluster phase $\varphi_{2,\mathrm{B}}^{\mathrm{crit}}$ (modulo 2π) for a critical duration T_{crit}. During one period $\tilde{T} = 2\pi/\Omega_2$ of the symmetrical double-peak travelling wave of the density $n(\psi, t)$ (cf. Fig. 5.16) the ball runs twice through the circular valley of the potential, where it rotates at frequency $2\Omega_2$ (cf. (5.48) and (5.49)). Thus, one may furthermore expect that within one period $\tilde{T}$ of the travelling wave there are two timing points, $t_{\mathrm{B}}^{\mathrm{crit}}$ and $t_{\mathrm{B}}^{\mathrm{crit}} + \tilde{T}/2$, at which one can start a stimulation which induces a partial desynchronization. Accordingly at these timing points the cluster phase $\varphi_2(t)$ would read

$$\varphi_2(t_{\mathrm{B}}^{\mathrm{crit}}) = \varphi_{2,\mathrm{B}}^{\mathrm{crit}} \ , \quad \varphi_2(t_{\mathrm{B}}^{\mathrm{crit}} + \tilde{T}/2) = \varphi_{2,\mathrm{B}}^{\mathrm{crit}} + 2\pi \ . \qquad (6.19)$$

These suggestions were confirmed by a numerical analysis similar to that one of the former section. Here a series of 40 simulations is presented where t_B, i.e. the timing point when the stimulation starts, was equally spaced within a time interval of length $\tilde{T}$. Correspondingly the intial phases $\varphi_{2,B}$ were equally spaced within the interval $[0, 4\pi]$. It turned out that the second half of the simulations entirely agrees with the first half because a stimulation starting at time t' totally corresponds to a stimulation starting at $t' + \tilde{T}/2$. For this reason in Figs. 6.10–6.12 only half of the simulation series is shown, where the initial phases $\varphi_{2,B}$ are equally spaced in $[0, 2\pi]$. Actually for the second half of the series of simulations (with $\varphi_{2,B}$ equally spaced in $[2\pi, 4\pi]$) one obtains exactly the same plots.

An additional numerical investigation revealed that the dynamics of the average number density $n(\psi, t)$ has only one attractor no matter which initial state $n(\psi, 0)$ was chosen. That is why all of the trajectories of Z_2 in Fig. 6.10 tend to the same attractor of Z_2. In each cycle of φ_2 there is a critical value $\varphi_{2,B}^{\text{crit}}$ as shown in Fig. 6.10. When the stimulation starts at this critical initial phase and lasts for a critical duration T_{crit} the amplitude of the second mode vanishes, i.e. a complete partial desynchronization of the second mode occurs (Figs. 6.10 and 6.12). When the stimulation duration exceeds T_{crit} one observes a transition from type 1 resetting to type 0 resetting (Figs. 6.10 and 6.11).

Figure 6.13 shows the dynamics of the amplitudes $R_1, \ldots, R_4$ during and after a stimulation which causes a complete partial desynchronization of the second mode. In the stable synchronized two-cluster state R_1 and R_3 vanish (cf. Fig. 5.12). Moreover, first and third mode are not affected by the stimulus and, thus, vanish throughout the whole simulation. The stimulus desynchronizes the second mode completely ($R_2(t_E) = 0$) and damps the amplitude of the fourth mode ($R_4(t_E) < R_4(t_B)$). Due to the slaving principle directly after the stimulation one observes a relaxational damping of the stable modes, in particular, of the fourth mode. Therefore the minimum of $\delta(t)$ lags behind the end of the stimulation. According to the analysis of Sect. 5.4.3 the cluster's resynchronization is governed by the dynamics of the second mode.

The corresponding firing pattern is displayed in Fig. 6.14. The relaxational damping occurring between t_E and t_δ mainly reflects the contribution of the fourth mode. After that the firing pattern is dominated by the recovering second mode.

Comparing the results of this and the former section points out the similarities of the stimulation induced dynamics of synchronization patterns which are due to single-mode instabilities. These similarities were additionally confirmed by extensive numerical investigations of the effects of stimulations on the three- and four-cluster states described in Sects. 5.4.4 and 5.4.5. Also in these cases one encounters the prominent role of the order parameters giving rise to relaxational damping of stable modes and resynchronization.

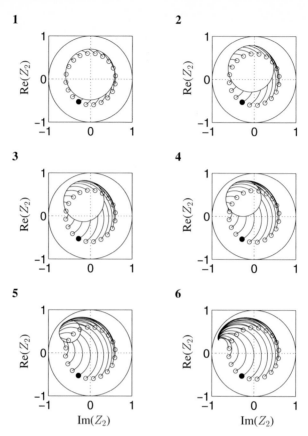

Fig. 6.10. *Winding numbers:* 20 simulations of (4.21) with $\varphi_{2,B}$ equally spaced in $[0, 2\pi]$ are displayed. Stimulation duration T increases from from plot 1 to plot 6. Plots show the trajectories of Z_2 in the Gaussian plane. Same format as in Fig. 6.7. Coupling and stimulation parameters: $K_2 = 1$, $K_1 = K_3 = K_4 = C_1 = \cdots = C_4 = 0$, $Q = 0.4$, $I_2 = 7$, $I_1 = I_3 = I_4 = \gamma_1 = \cdots = \gamma_4 = 0$. The winding number in the first three plots equals $+1$, whereas in the other plots it vanishes.

Finally, let us take up the relationship between the frequency of the order parameters and that of the related travelling wave again. Sections 5.4.3–5.4.5 revealed that the order parameters of a two-, three- and four-modes instability rotate in their potential's valley at a frequency which is two, three and four times as large as the frequency of the corresponding travelling wave of $n(\psi, t)$. Thus, per period of the travelling wave there are two, three and four equally spaced critical timing points t_B^{crit} for starting a desynchronizing stimulation (cf. (6.19)). In a nutshell, each burst of the cluster's collective firing activity has its own critical timing point t_B^{crit} (Fig. 6.15).

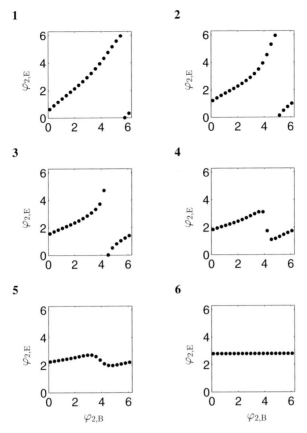

Fig. 6.11. *Phase resetting curves:* $\varphi_{2,E}$ is plotted over $\varphi_{2,B}$ for the series of simulations of Fig. 6.10. The same stimulation durations and arrangement of the plots as in Fig. 6.10. Between plots 3 and 4 one observes a transition from type 1 resetting to type 0 resetting.

6.3 Transient Mode Excitation and Early Response

In the former sections we focussed on how order parameters are affected by stimulation, how they are perturbed and how they recover. To understand stimulation induced firing patterns, additionally the stimulus' influence on the stable modes has to be clarified in more detail. With this aim in view let us again recall the stimulus

$$S(\psi) = \sum_{m=1}^{4} I_m \cos(m\psi + \gamma_m) \tag{6.20}$$

(cf. 4.5 and (6.1)). Depending on the parameters $I_1, \ldots, I_4$ the stimulus $S(\psi)$ contains terms which are of the same or of a different order as the order

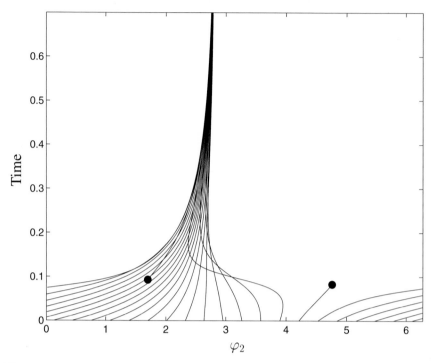

Fig. 6.12. *Cluster phase $\varphi_2(t)$:* The time course of $\varphi_2(t)$ of the series of simulations of Fig. 6.10 is shown. At the critical stimulation duration $T_{\text{crit}} = 0.088$ one observes a phase singularity and a jump of width π.

parameters. To come to the most important point first, stimulation terms which are not of the same order as the order parameters come into effect, in particular, when the order parameters vanish at the end of the stimulation. For instance, when the stimulus induces a complete partial desynchronization of the order parameters a *transient excitation of stable modes* is caused by stimulation terms which are of a different order as the order parameters. This type of excitation will be denoted as *excitation of higher order* provided the excited modes are of higher order compared to the order parameters. Likewise an excitation of stable modes which are of lower order will be called *excitation of lower order*. With increasing noise amplitude the strength of the excitation is damped. Thus, for the comparison of the different stimulation mechanisms the noise amplitude will be kept constant.

In this section we shall dwell on the excitation of higher and lower order occurring in the first- and second-mode instability studied in the former section. Stimulation terms giving rise to a transient excitation of stable modes in both cases do not change the resetting behavior qualitatively. This means that subjected to the stimulus the cluster behaves in the following way:

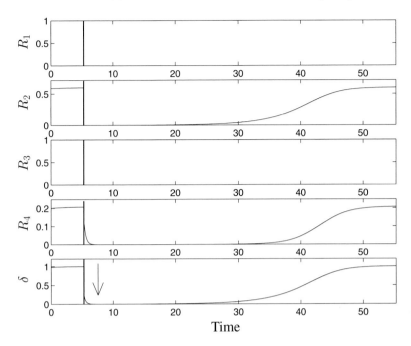

Fig. 6.13. *Amplitude dynamics during stimulation and recovery:* The time course of cluster amplitudes $R_1, \ldots, R_4$ is plotted for a simulation which induces a complete partial desynchronization of the second mode (corresponding to the trajectory starting at the filled circle in Fig. 6.10). Begin and end of the stimulation are indicated by two vertical lines which melt due to the scaling of the time axis. The vertical arrow indicates the maximal uniform desynchronization, i.e. the minimum of $\delta(t)$, occurring at time $t_\delta \approx 7.53$. Note that R_1 and R_3 equal zero during the whole simulation.

1. A complete partial desynchronization of the order parameters is caused provided the stimulus is administered at a critical initial cluster phase $\varphi_{j,\mathrm{B}}^{\mathrm{crit}}$ for a critical duration T_{crit}.
2. When the stimulation duration exceeds T_{crit} a transition from type 1 resetting to type 0 resetting occurs.

6.3.1 Excitation of Higher Order

Excitation of higher order is typically caused by stimulation terms which are of higher order compared to the order parameters. To study this effect let us come back to the one-cluster state analyzed in Sect. 6.2.3. The dynamics of this state is governed by the Fourier modes with wave numbers $+1$ and -1, i.e. the order parameters are of first order. The impact of stimulation terms which are of higher order can exemplarily be illustrated by comparing the outcome of two stimuli which are different with respect to these terms. In both cases a complete partial desynchronization of the first mode is induced.

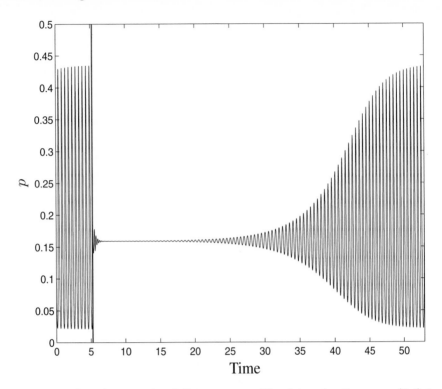

Fig. 6.14. *Stimulation induced firing pattern:* The firing density $p = n(0, t)$ is plotted over time for the simulation from Fig. 6.13. Begin and end of the stimulation are indicated by two (melting) vertical lines.

However, the stable modes' reactions on the stimuli differ remarkably. The stimulation parameters read:

1. $I_1 = 7$, $I_2 = I_3 = I_4 = \gamma_1 = \cdots = \gamma_4 = 0$: The stimulation term is of the same order as the order parameters. The corresponding stimulation induced dynamics was analyzed in Sect. 6.2.3.

2. $I_1 = I_4 = 7$, $I_2 = I_3 = \gamma_1 = \cdots = \gamma_4 = 0$: This stimulus additionally contains a term of fourth order ($I_4 = 7$).

In this section both stimulation mechanisms will be referred to as first and second stimulation mechanism, respectively. The very difference between both mechanisms can easily be illustrated by comparing suitably timed stimuli giving rise to a complete partial desynchronization of the first mode. It should be emphasized that the critical initial phase and the critical stimulation duration necessary to induce a desynchronization are different for both mechanisms. The corresponding cluster amplitudes' dynamics and firing patterns are displayed in Figs. 6.5, 6.6, 6.16 and 6.17. Comparing Figs. 6.5 and 6.16 clearly shows that the higher order term of the second stimulation mechanism excites

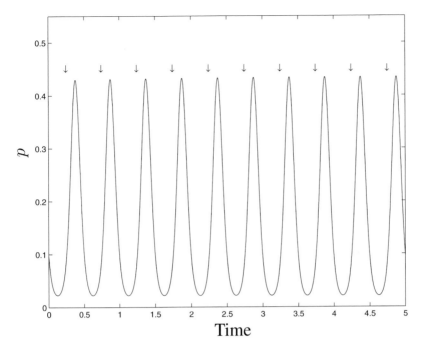

Fig. 6.15. *Critical timing points of a two-cluster state:* When a stimulus with intensity parameters as in Fig. 6.14 is administered at the critical timing points $t_{\mathrm{B}}^{\mathrm{crit}}$ (*indicated by small arrows*) for the critical duration T_{crit} the second mode of $n(\psi, t)$ vanishes. Same coupling parameters as in Fig. 6.14. Hence, the period of the travelling wave equals 1.

stable modes, in particular, the second and fourth mode. Accordingly

$$R_j(t_{\mathrm{B}}) < R_j(t_{\mathrm{E}}) \ \text{ for } \ j = 2, 3, 4 \tag{6.21}$$

holds in contrast to the first stimulation mechanism which damps stable modes (cf. (6.13)).

Due to the slaving principle a relaxational damping of the stable modes takes place after the stimulation. By virtue of the excitation of higher order the maximal uniform desynchronization occurs delayed compared to the first stimulation mechanism. A further consequence of this excitation is the delayed occurence of the resynchronization following the relaxational damping.

During the relaxational damping (i.e. for $t_{\mathrm{E}} \leq t \leq t_\delta$) the amplitude of the first mode vanishes or is close to zero, and, hence, the stable modes dominate the firing pattern (Fig. 6.17). To emphasize the contribution of the stable modes to the firing density p directly subsequent to the stimulation this firing pattern will, in general, be called *early firing response* or, briefly, *early response*. As a consequence of the excitation of higher order the second

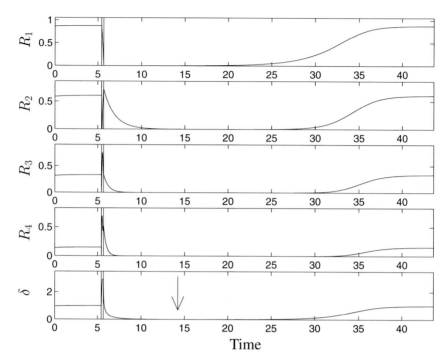

Fig. 6.16. *Impact of the stimulus' higher order term:* Cluster amplitudes $R_1, \ldots, R_4$ are plotted over time. The two vertical lines mark begin and end of the stimulation which gives rise to a complete partial desynchronization of the first mode. As compared with Fig. 6.5 the stimulation term of fourth order excites the stable modes, in particular, R_2 and R_4. Coupling and stimulation parameters: $K_1 = 1$, $K_2 = K_3 = K_4 = C_1 = \cdots = C_4 = 0$, $Q = 0.4$, $I_2 = I_3 = \gamma_1 = \cdots = \gamma_4 = 0$, $I_1 = I_4 = 7$ (during stimulation), $I_1 = I_4 = 0$ (before and after stimulation). Maximal uniform desynchronization occurs at time $t_\delta \approx 14.23$ and, thus, later as compared with a stimulation mechanism of first order (Fig. 6.5).

stimulation mechanism gives rise to a *pronounced early response*. By comparing the latter with the early response of the first stimulation mechanism (Fig. 6.6) one immediately realizes the important role of the stable modes during the early period after the stimulation. In accordance with Sect. 5.6.2 the firing pattern during the cluster's resynchronization reflects the recovery of the order parameters.

The phase resetting curves of the second stimulation mechanism are similar to those of the first one. Unlike that a big difference between both stimulation mechanisms turns out if we consider the dynamics of the average number density $n(\psi, t)$. To this end Fig. 6.18 shows the time course of $n(\psi, t)$ before and during stimuation, where in contrast to the simulations shown in Figs. 6.6 and 6.17 the stimulus is not turned off. The cluster's synchronized firing activity before the stimulation corresponds to a stable travelling

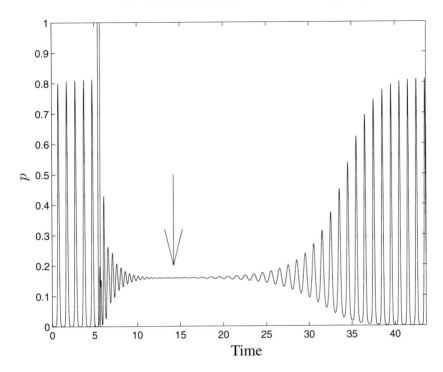

Fig. 6.17. *Early response caused by excitation of higher order:* The time course of the firing density $p(t) = n(0, t)$ belonging to the simulation from Fig. 6.16. The two vertical lines mark begin and end of the stimulation. Excitation of higher order induces a pronounced early response.

wave of $n(\psi, t)$ which we already encountered in Sect. 5.6.1 (cf. Fig. 5.15). The first stimulation mechanism forces the travelling wave to split into an advancing peak and a delaying peak (Fig. 6.18a). Put otherwise: the first stimulation mechanism causes a straddling process similar to that described in the context of the ensemble model (cf. Sect. 2.7.1). Unlike that, due to the second stimulation mechanism the travelling wave splits into separate peaks in a way similar to the splitting mechanism observed in the ensemble model (Fig. 6.18b, cf. Sect. 2.7.3). Hence, in terms of noisy cluster states excitation of higher order means that a synchronized cluster splits into several subclusters. This is additionally illustrated by snapshots showing the density $n(\psi, t)$ which corresponds to a complete partial desynchronization of the first mode (Fig. 6.19).

6.3.2 Excitation of Lower Order

Excitation of lower order can be caused by a stimulus (6.20) containing terms which are of lower or higher order compared to the order parameters. This

(a)   (b)

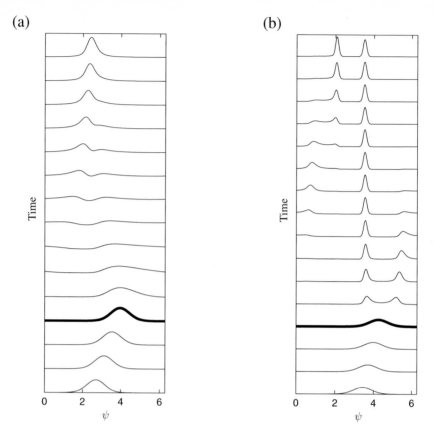

Fig. 6.18a,b. *Straddling and splitting processes of the density:* Same simulations as in Figs. 6.6 (**a**) and 6.17 (**b**) except for the fact that the stimulus is not turned off. Succesive traces show n as a function of ψ, where earliest time is at the bottom of the figure. The bold curve indicates the density at the beginning of the stimulation, i.e. $n(\psi, t_B)$. Stimulation of first order gives rise to a splitting process of the density, whereas stimulation terms of higher order cause a splitting process. Note that the scaling in (**a**) and (**b**) is different. The four lowest curves in (**a**) and (**b**) refer to the same density, respectively.

will be illustrated by considering the impact of three different stimulation mechanisms on the two-cluster state analyzed in Sect. 5.4.3. The corresponding order parameters are the Fourier modes with wave numbers $+2$ and -2, and, accordingly, they are of second order. Although the three well-timed stimuli give rise to a complete partial desynchronization of the second mode, the early response is completely different in the three cases. The parameters of the stimulation mechanisms are:

(a)   **(b)**

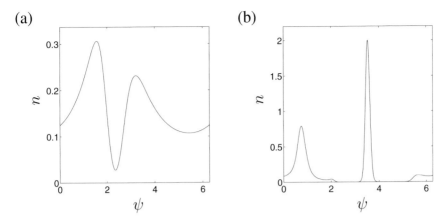

Fig. 6.19a,b. *Excitation of higher order:* $n(\psi, t_B + T_{\mathrm{crit}})$, the density after the critical stimulation duration, is displayed for the same simulations as in Figs. 6.6 (**a**) and 6.17 (**b**). The first stimulation mechanism is associated with a double-peak distribution (**a**) which was already observed in the ensemble model. Excitation of higher order provokes a distribution with several bumps (**b**). Note that both densities correspond to a complete partial desynchronization of the first mode.

1. $I_2 = 7$, $I_1 = I_3 = I_4 = \gamma_1 = \cdots = \gamma_4 = 0$: The stimulation term is of the same order as the order parameters. The corresponding stimulation induced dynamics was studied in Sect. 6.2.4.
2. $I_1 = 7$, $I_2 = I_3 = I_4 = \gamma_1 = \cdots = \gamma_4 = 0$: This stimulus is of first order $(I_1 = 7)$.
3. $I_3 = 7$, $I_1 = I_2 = I_4 = \gamma_1 = \cdots = \gamma_4 = 0$: The stimulus is of higher order compared to the order parameters.

In this section these three stimulation mechanisms will be referred to as first, second and third stimulation mechanism. The first mechanism induces a weak early response as shown in Fig. 6.14.

Apart from an excitation of stable modes of higher order the second stimulation mechanism causes a strong excitation of the first mode (Fig. 6.20). Moreover, the higher the order of the stable modes the quicker the relaxational damping takes place. For this reason the pronounced early response is dominated by the slowly relaxing first mode (Fig. 6.21), so that the firing patterns belonging to the relaxational damping and the resynchronization are no longer separated in time.

The third stimulation mechanism excites, in particular, the third and first mode so that the pronounced early response consists of two parts as far as the prominent frequency component is concerned (Fig. 6.22). The first part is dominated by the quickly relaxing third mode, and the second part reflects the dynamics of the slowly relaxing first mode (Fig. 6.23).

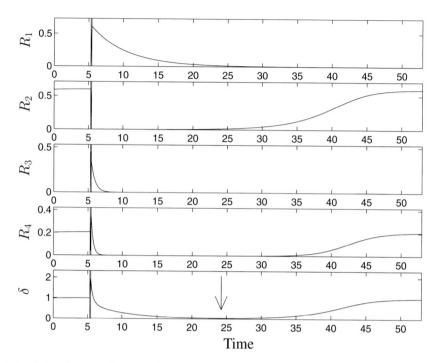

Fig. 6.20. *Impact of a stimulation term of lower order:* The time course of cluster amplitudes $R_1, \ldots, R_4$ is shown with the same format as in Fig. 6.13. The stimulation causes a complete partial desynchronization of the second mode. In comparison to the second order stimulation mechanism (Fig. 6.13) the stimulation term of first order excites the stable modes, in particular, the first mode. The strongly excited first mode relaxes slowly compared to the stable modes of higher order. Correspondingly the maximal uniform desynchronization occurs rather late, namely at time $t_\delta \approx 24.2$ (cf. Fig. 6.13). Coupling and stimulation parameters: $K_2 = 1$, $K_1 = K_3 = K_4 = C_1 = \cdots = C_4 = 0$, $Q = 0.4$, $I_2 = I_3 = I_4 = \gamma_1 = \cdots = \gamma_4 = 0$, $I_1 = 7$ (during stimulation), $I_1 = 0$ (before and after stimulation).

6.4 Couplings Determine Reaction to Stimulation

The examples of the former section illustrate the influence of the stimulation mechanism on the cluster's recovery and, in particular, on its early response. However, these considerations should not give the impression that the cluster's reaction to stimulation is exclusively determined by the stimulation mechanism. Rather it has to be emphasized that the way the cluster reacts to stimulation depends on both the stimulus' influence and the oscillators' mutual interactions. In particular, the latter are very important for the cluster's behavior during the period of recovery. Aware of the huge variety of a cluster's reactions on a particular stimulus two aspects will be pointed out:

1. The couplings determine how fast the resynchronization takes place.

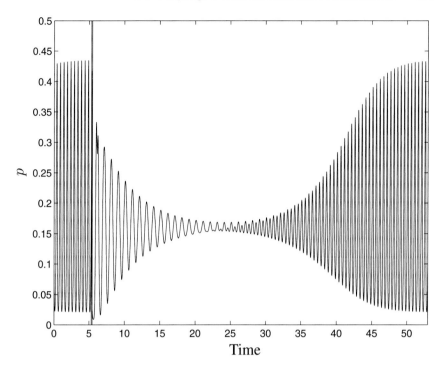

Fig. 6.21. *Early response caused by an excitation of lower order:* The firing density $p(t) = n(0, t)$ belonging to the simulation from Fig. 6.20 is plotted over time. The excited and slowly relaxing first mode dominates the pronounced early response. Same format as in Fig. 6.14.

2. During recovery transient states may occur which are qualitatively different from the cluster's behavior before the stimulation.

These dynamical features will be elucidated with examples concerning rapid recovery and harmonic early response.

6.4.1 Rapid Recovery

A well-timed stimulus with parameters $I_1 = I_3 = 7$, $I_2 = I_4 = \gamma_1 = \cdots \gamma_4 = 0$ is administered to two one-cluster states which are both due to a first-mode instability. Both one-cluster states have similar couplings: $K_1 = 1$, $K_2 = K_3 = K_4 = C_1 = \cdots = C_4 = 0$ are the parameters of the first state, whereas the second has the same parameters except for $C_3 = 1$. Accordingly both synchronized states do not differ remarkably. In particular, they have practically the same synchronization frequency because in lowest order C_3 does not contribute to Ω_1 according to (5.31) and (C.3). Nevertheless, especially after a stimulation induced partial desynchronization of the first mode

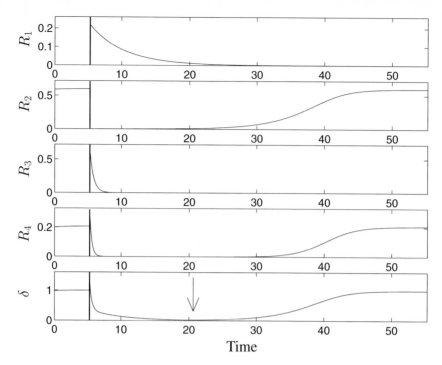

Fig. 6.22. *Impact of a stimulation term of higher order:* Cluster amplitudes $R_1, \ldots, R_4$ are plotted over time with the same format as in Fig. 6.20. A complete partial desynchronization of the second mode is induced by the stimulus. The stimulation term of third order excites the stable modes, in particular, the third and first mode. Coupling and stimulation parameters: $K_2 = 1$, $K_1 = K_3 = K_4 = C_1 = \cdots = C_4 = 0$, $Q = 0.4$, $I_1 = I_2 = I_4 = \gamma_1 = \cdots = \gamma_4 = 0$, $I_3 = 7$ (during stimulation), $I_3 = 0$ (before and after stimulation). The strongly excited third mode relaxes much quicker compared to the excited first mode. In comparison with the stimulation shown in Fig. 6.20 the excitation of the first mode is weaker, and, thus, the maximal uniform desynchronization occurs earlier ($t_\delta \approx 20.62$).

the coupling parameter C_3 has an effect: For vanishing C_3 we observe an excitation of higher order (Fig. 6.24) which causes a pronounced early resonse (Fig. 6.25). In contrast, for $C_3 = 1$ the pronounced early response is followed by a rapid recovery (Figs. 6.26 and 6.27). According to detailed numerical investigations this strong tendency to a rapid recovery also occurs provided the initial cluster phase $\varphi_{1,\mathrm{B}}$ deviates from its critical value $\varphi_{1,\mathrm{B}}^{\mathrm{crit}}$.

6.4.2 Harmonic Early Response

All simulations studied above referred to single-mode instabilities. In all of these cases a well-timed stimulus induced a complete partial desynchronization of the order parameters $\hat{n}(k,t)$ and $\hat{n}(-k,t)$. Depending on the stimulation terms additionally an excitation of higher and/or lower order occured.

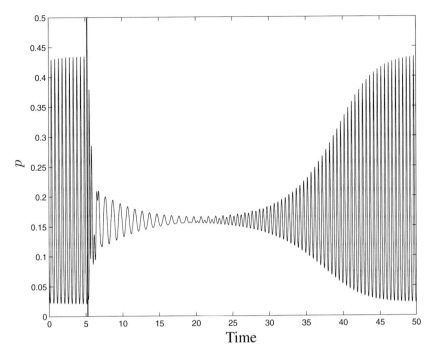

Fig. 6.23. *Early response reflecting excitation of lower and higher order:* The time course of the firing density $p(t) = n(0, t)$ is plotted for the simulation from Fig. 6.22. The first part of the pronounced early response (between t_E and $t \approx 7$) is dominated by the quickly relaxing third mode, whereas the second part is dominated by the slowly relaxing first mode. Same format as in Fig. 6.21.

As a consequence of the slaving principle directly after the stimulation a relaxational damping of the stable modes took place, no matter whether or not the stable modes were excited (cf. Figs. 6.5 and 6.16).

A different situation occurs if the dynamics of the synchronized state is governed by two or more pairs of Fourier modes acting as order parameters. Let us, for instance, consider the two-modes instability analyzed in Sect. 5.4.6. In particular, we focus on the type II fixed points which are associated with one-cluster states (cf. Fig. 5.22). In this case the first mode (i.e. $\hat{n}(\pm 1, t)$) and the second mode (i.e. $\hat{n}(\pm 2, t)$) act as order parameters. According to our numerical analysis no relaxational damping is observed when a stimulus induces a complete partial desynchronization of the first mode (i.e. $\hat{n}(\pm 1, t_E) = 0$). In contrast, a pair of non-vanishing order parameters (i.e. $\hat{n}(\pm 2, t)$) remains, which dominates the cluster's response to stimulation (Fig. 6.28). Correspondingly, a harmonic early response occurs which reflects the transient dynamics of the second mode (Fig. 6.29).

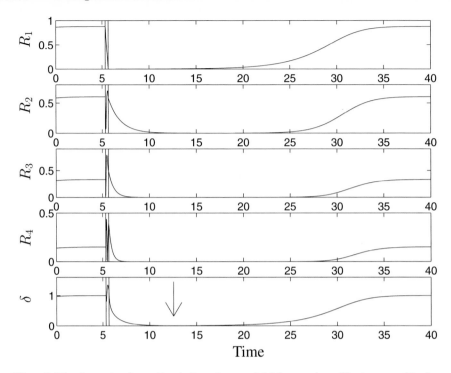

Fig. 6.24. *Impact of a stimulation term of higher order:* Cluster amplitudes $R_1, \ldots, R_4$ are plotted over time with the same format as in Fig. 6.22. The stimulus causes a complete partial desynchronization of the first mode. The stimulation term of third order excites the stable modes, in particular, the third mode. Coupling and stimulation parameters: $K_1 = 1$, $K_2 = K_3 = K_4 = C_1 = \cdots = C_4 = 0$, $Q = 0.4$, $I_2 = I_4 = \gamma_1 = \cdots = \gamma_4 = 0$, $I_1 = I_3 = 7$ (during stimulation), $I_1 = I_3 = 0$ (before and after stimulation).

6.5 Vulnerability and Recovery

In the former sections we mainly focussed on what happens when the stimulus induces a partial desynchronization of a certain mode. For this reason the stimuli under consideration had to be administered at a critical initial cluster phase $\varphi_{j,\mathrm{B}}^{\mathrm{crit}}$ for a critical duration T_{crit}. Obviously it is no less important to understand the impact of a stimulus which is not suitably timed. Hence, in this section we shall study how stimulation induced effects depend on the initial cluster phase $\varphi_{j,\mathrm{B}}$ and the stimulation duration T. In this way we will realize the important protective effect of couplings, in particular, in comparison with the behavior of the ensemble model. Moreover, we shall see how closely partial desynchronization of the order parameters is related to transient phenomena such as excitation of higher or lower order and harmonic early response.

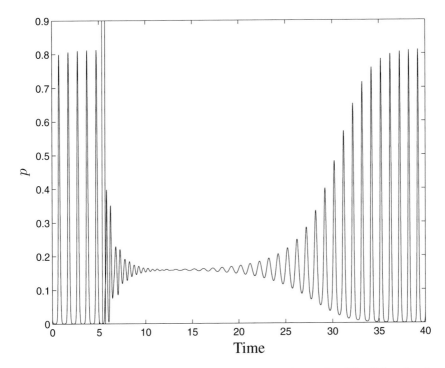

Fig. 6.25. *Early response reflecting excitation of higher order:* The firing density $p(t) = n(0, t)$ for the simulation of Fig. 6.24 is plotted over time. Same format as in Fig. 6.23. The early response reflects the excitation of the third mode.

6.5.1 Phase Errors Versus Duration Errors

Let us start with a simple example, where the stimulus is of the same order as the order parameters. To this end it is appropriate to consider the stimulation of a one-cluster state analyzed in detail in Sect. 6.2.3. In this case the order parameters as well as the stimulus are of first order. Accordingly we introduce the *normalized phase error* of the first cluster variable

$$\Delta \psi = \frac{\varphi_{1,B} - \varphi_{1,B}^{\text{crit}}}{2\pi} \qquad (6.22)$$

and the *normalized duration ratio*

$$\Delta T = \frac{T}{T_{\text{crit}}} \ . \qquad (6.23)$$

$\varphi_{1,B}$ and T denote the actual initial cluster phase of the first mode and the actual stimulation duration, whereas $\varphi_{1,B}^{\text{crit}}$ and T_{crit} denote the critical initial cluster phase and the critical stimulation duration of a well-timed stimulus giving rise to a partial desynchronization of the first mode.

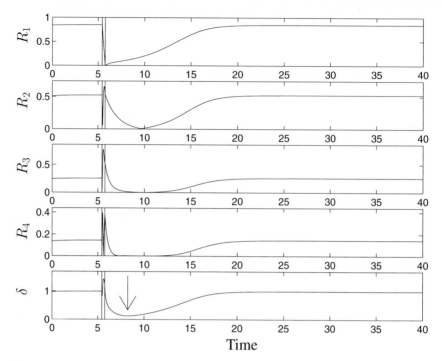

Fig. 6.26. *Impact of a stimulation term of higher order:* Cluster amplitudes $R_1, \ldots, R_4$ are plotted over time with the same format as in Fig. 6.22. A complete partial desynchronization of the first mode is induced by the stimulation. Coupling and stimulation parameters: $K_1 = 1$, $C_3 = 1$, $K_2 = K_3 = K_4 = C_1 = C_2 = C_4 = 0$, $Q = 0.4$, $I_2 = I_4 = \gamma_1 = \cdots = \gamma_4 = 0$, $I_1 = I_3 = 7$ (during stimulation), $I_1 = I_3 = 0$ (before and after stimulation). As a consequence of the different coupling pattern ($C_3 = 1$) the cluster rapidly resynchronizes. Critical initial cluster phase $\varphi_{1,\mathrm{B}}^{\mathrm{crit}}$ and stimulation duration T_{crit} are different compared to those of the simulation of Fig. 6.25.

The stimulus' impact on the jth mode can be estimated by comparing the jth cluster amplitude before and after the stimulation. To this end

$$r_j = \frac{R_j(t_\mathrm{E})}{R_j(t_\mathrm{B})} \tag{6.24}$$

is introduced, provided $R_j(t_\mathrm{B})$ does not vanish. The stimulus damps or excites the jth mode if $r_j < 1$ or $r_j > 1$ holds, respectively. A complete partial desynchronization of the jth mode corresponds to $r_j = 0$.

For the simulation shown in Fig. 6.5 initial cluster phase and stimulation duration were varied by performing simulations for $-0.1 \le \Delta\psi \le 0.1$ and $0.9 \le \Delta T \le 1.1$. The determination of $r_1 \ldots, r_4$ visualizes the stimulus' influence on the amplitudes of the first four modes (Fig. 6.30). A complete partial desynchronization of the first mode can only be achieved for vanishing

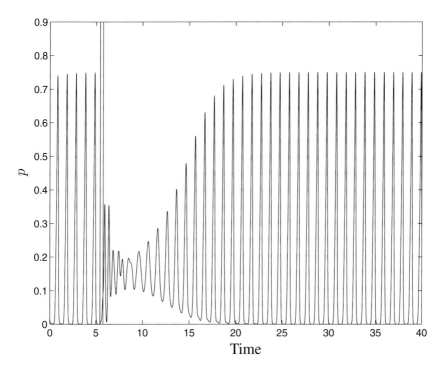

Fig. 6.27. *Rapid resynchronization:* The time course of the firing density $p(t) = n(0, t)$ for the simulation of Fig. 6.26. The early response is followed by a rapid resynchronization due to the coupling pattern ($C_3 = 1$). Same format as in Fig. 6.25.

phase error $\Delta\psi$ and accurate duration ratio $\Delta T = 1$. Phase errors reduce the desynchronizing effect of the stimulation in a much more pronounced way compared to errors of the stimulation duration. For instance, r_1 exceeds 0.1 if the phase error fulfills $|\Delta\psi| > 0.013$. Thus, the partial desynchronization of the first mode crucially depends on the proper choice of the initial cluster phase $\varphi_{1,\mathrm{B}}$. Within the whole range of parameters $\Delta\psi$ and ΔT under consideration the second mode is damped. Depending on the parameters $\Delta\psi$ and ΔT the amplitudes of the third and the fourth mode are damped, excited, or remain unchanged.

The recovery of the cluster amplitudes $R_1, \ldots, R_4$ after the stimulation for a range of phase errors $\Delta\psi$ is shown in Fig. 6.31. The smaller $|\Delta\psi|$ the more the first mode is desynchronized by the stimulation, i.e. the smaller is $R_1(t_{\mathrm{E}})$. Moreover, the smaller $|\Delta\psi|$ the longer it takes the first mode to recover, and the more pronounced is the relaxational damping of the stable modes (R_2, R_3, and R_4). Thus, in order to achieve a partial desynchronization with delayed recovery the phase error $|\Delta\psi|$ has to be as small as possible.

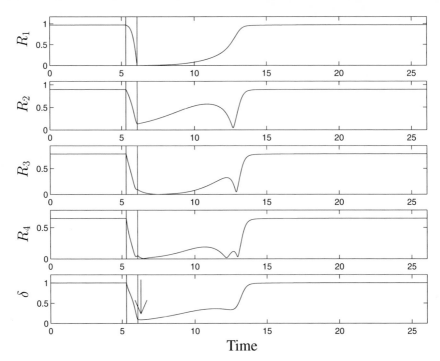

Fig. 6.28. *Stimulating a one-cluster state of a two-modes instability:* Cluster amplitudes $R_1, \ldots, R_4$ are plotted over time with the same format as in Fig. 6.26. A complete partial desynchronization of the first mode is induced by the stimulation. The stimulation is followed by a transient epoch which is dominated by the second mode. Coupling and stimulation parameters: $K_1 = 2.3$, $K_2 = 1.45$, $K_3 = K_4 = C_1 = \cdots = C_4 = 0$, $Q = 0.4$, $I_2 = I_3 = I_4 = \gamma_1 = \cdots = \gamma_4 = 0$, $I_1 = 7$ (during stimulation), $I_1 = 0$ (before and after stimulation).

Figures 6.30 and 6.31 refer to the stimulus' effect on the cluster amplitudes. Especially from the experimentalist's point of view it should additionally be demonstrated how the firing pattern depends on the phase error. Considering the firing density $p(t)$ of a series of simulations with equally spaced initial cluster phase $\varphi_{1,B}$, the desynchronizing effect turns out to show up only if the initial cluster phase is close to the critical value $\varphi_{1,B}^{\text{crit}}$ (Fig. 6.32). Otherwise, apart from phase shifts stimulation does not change the firing pattern remarkably (Fig. 6.33).

Figures 6.30–6.33 stress that partial desynchronization only occurs if the stimulus is administered when the initial cluster phase $\varphi_{1,B}$ is close to the critical value $\varphi_{1,B}^{\text{crit}}$ (modulo 2π). In the next two sections we will see that, accordingly, also dynamical phenomena which are associated with partial desynchronization only appear provided the phase error $|\Delta\psi|$ is small. Phenomena of this kind are excitation of higher or lower order and harmonic early response.

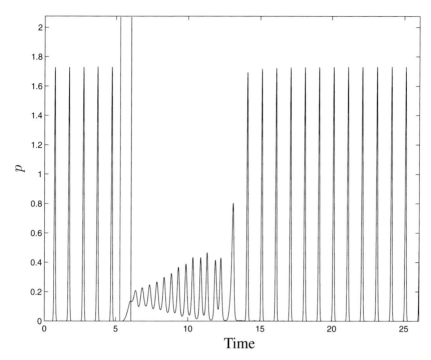

Fig. 6.29. *Harmonic early response:* The time course of the firing density $p(t) = n(0, t)$ for the simulation of Fig. 6.28. The first mode vanishes at the end of the stimulation. In contrast, the second mode does not vanish and provokes a harmonic early response. Same format as in Fig. 6.27.

6.5.2 Protective Effect of Couplings

In this section we shall study how excitation of higher order depends on the phase error (6.22) and the duration error (6.23). Excitation of higher order is a typical side effect of stimulation terms which are of higher order compared to the order parameters. Therefore we turn to the stimulation mechanism of Sect. 6.3.1 with parameters $I_1 = I_4 = 7$, $I_2 = I_3 = \gamma_1 = \cdots = \gamma_4 = 0$.

On the analogy of Fig. 6.30 simulations were performed for $-0.1 \leq \Delta\psi \leq 0.1$ and $0.9 \leq \Delta T \leq 1.1$ (Fig. 6.34). The stimulus' effect on the four leading modes was estimated by computing $r_1, \ldots, r_4$ defined by (6.24). A complete partial desynchronization of the first mode only occurs for $\Delta\psi = 0$ and $\Delta T = 1$. As already observed in the former section, the effect of the phase error on reducing the desynchronization of the first mode is much stronger compared to the effect of the duration error: Here r_1 exceeds 0.1 for $|\Delta\psi| > 0.01$. Accordingly, partial desynchronization of the first mode can only be achieved provided the initial cluster phase is close to $\varphi_{1,B}^{\text{crit}}$. The stimulation term of fourth order excites the second and, in particular, the fourth mode

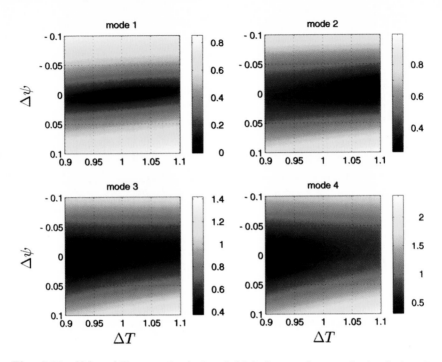

Fig. 6.30. *Vulnerability to stimulation:* Initial cluster phase and stimulation duration were varied for the simulation of Fig. 6.5 by running simulations for $-0.1 \leq \Delta\psi \leq 0.1$ and $0.9 \leq \Delta T \leq 1.1$. For each simulation the stimulus' impact on the amplitude of the first four modes was estimated by calculating $r_1, \ldots, r_4$ from (6.24). The latter are plotted in a colour coded way. Coupling and stimulation parameters: $K_1 = 1$, $K_2 = K_3 = K_4 = C_1 = \cdots = C_4 = 0$, $Q = 0.4$, $I_2 = I_3 = I_4 = \gamma_1 = \cdots = \gamma_4 = 0$, $I_1 = 7$ (during stimulation), $I_1 = 0$ (before stimulation).

for all $\Delta\psi$ and ΔT under consideration. For certain values of $\Delta\psi$ and ΔT the amplitude of the third mode is damped, excited, or remains unchanged.

Figure 6.35 shows how the recovery of the cluster amplitudes $R_1, \ldots, R_4$ depends on the phase error $\Delta\psi$. The smaller $|\Delta\psi|$ the more pronounced is the desynchronization of the first mode, and the longer it takes the cluster to recover. In the same way, the smaller $|\Delta\psi|$ the more the second and, in particular, the fourth mode are excited initially. The excitation of the second mode lasts longer because the fourth mode relaxes much quicker than the second mode (cf. Fig. 6.16).

Figures 6.36 and 6.37 show the firing density $p(t)$ during series of simulations with equally spaced initial cluster phase $\varphi_{1,\mathrm{B}}$. Both desynchronization of the first mode and pronounced early response only occur provided the initial cluster phase is close to the critical value $\varphi_{1,\mathrm{B}}^{\mathrm{crit}}$ (Fig. 6.36). For larger values of the phase error $|\Delta\psi|$ apart from phase shifts the stimulation does

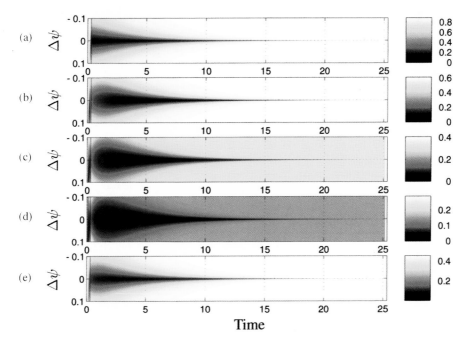

Fig. 6.31a–e. *Recovery of mode amplitudes:* The simulation from Fig. 6.5 was performed where the initial cluster phase $\varphi_{1,\mathrm{B}}$ was varied. To this end simulations were carried out for $-0.1 \le \Delta\psi \le 0.1$. Plots show the time course of the cluster amplitudes R_1 (**a**), R_2 (**b**), R_3 (**c**), and R_4 (**d**). Uniform desynchronization was estimated with $\delta(t)$ (**e**) from (6.9) and (6.10). $R_1, \ldots, R_4$ and δ are displayed in a colour coded fashion. Stimulation starts at time $t_{\mathrm{B}} = 0$ and ends at time $t_{\mathrm{E}} = 0.314$ (indicated by the vertical blue line). Coupling parameters as in Fig. 6.30. Stimulation parameters: $I_2 = I_3 = I_4 = \gamma_1 = \cdots = \gamma_4 = 0$, $I_1 = 7$ (during stimulation), $I_1 = 0$ (after stimulation).

not provoke pronounced changes of the firing pattern (Fig. 6.37). In Sect. 2.8 it was shown that in the ensemble model stimulation terms of higher order give rise to burst splitting. Remarkably the latter also occurs for large values of the phase error $|\Delta\psi|$. Comparing the ensemble's burst splitting (Fig. 2.26) with the cluster's pronounced early response (Figs. 6.36 and 6.37) it, thus, turns out that the oscillators' synchronizing interactions prevent the cluster from burst splitting provided $|\Delta\psi|$ is not small. In this way the couplings have a protective effect on the cluster's state of synchronization.

6.5.3 Partial Desynchronization and Transient Phenomena

Finally the close relationship between the partial desynchronization of the order parameters and the harmonic early response will be demonstrated. To this end the example of Sect. 6.4.2 concerning the stimulation of a two-mode instability is chosen. Numerical investigations similar to those shown in Figs.

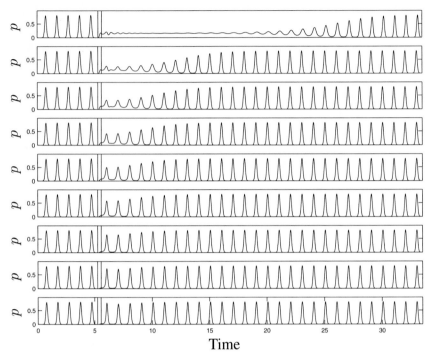

Fig. 6.32. *Desynchronization and firing patterns:* The firing density $p(t) = n(0,t)$ is plotted over time for equally spaced initial cluster phase $\varphi_{1,\mathrm{B}}$. In this series of simulations $\varphi_{1,\mathrm{B}}$ is close to the critical value $\varphi_{1,\mathrm{B}}^{\mathrm{crit}}$: In the uppermost plot the stimulus induces a complete partial desynchronization of the first mode (same simulation as in Fig. 6.6), whereas in the lowest plot $\varphi_{1,\mathrm{B}}$ equals $\varphi_{1,\mathrm{B}}^{\mathrm{crit}} + \pi/4$. From the uppermost plot to the lowest plot $\varphi_{1,\mathrm{B}}$ is increased by steps of $\pi/32$. Begin and end of the stimulation are indicated by a pair of vertical lines. Coupling parameters as in Fig. 6.30. Stimulation parameters: $I_2 = I_3 = I_4 = \gamma_1 = \cdots = \gamma_4 = 0$, $I_1 = 7$ (during stimulation), $I_1 = 0$ (before and after stimulation).

6.34 and 6.35 clearly revealed that both partial desynchronization of the first mode and harmonic early response vanish if $|\Delta\psi|$ is not small. Indeed, the range of $\Delta\psi$ associated with a partial desynchronization is even smaller as compared with the examples studied in Sects. 6.5.1 and 6.5.2. Figure 6.38 visualizes that pronounced changes of the firing pattern only occur if $|\Delta\psi|$ is small. In this case due to the desynchronization of the first mode the second mode dominates the dynamics and gives rise to a harmonic early response. For larger values of $|\Delta\psi|$ the firing pattern practically remains unchanged except for phase shifts.

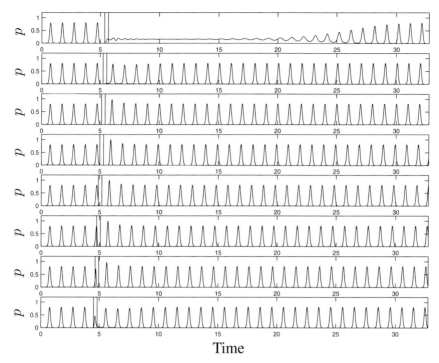

Fig. 6.33. *The stimulus' impact on the firing pattern:* The firing density $p(t) = n(0, t)$ is plotted over time for a series of simulations with initial cluster phase $\varphi_{1,\mathrm{B}}$ equally spaced within a cycle. The stimulus in the uppermost plot induces a complete partial desynchronization of the first mode (i.e. $\varphi_{1,\mathrm{B}} = \varphi_{1,\mathrm{B}}^{\mathrm{crit}}$). $\varphi_{1,\mathrm{B}}$ is increased by steps of $\pi/4$ from the uppermost plot to the lowest plot, so that the second uppermost plot of this figure and the lowest plot of Fig. 6.32 are identical. All other stimulation and coupling parameters as in Fig. 6.32.

6.6 Black Hole and Recovery

From the physiological point of view it is of crucial importance whether or not synchronized collective activity may vanish due to stimulation. To point out how the oscillators' interactions govern the cluster's behavior after stimulation let us dwell on the stimulation induced dynamics of the single-mode instabilities investigated above. The very reason for the recovery of the corresponding synchronized states is that the dynamics of the order parameters' amplitude is governed by a potential illustrated in the upper plot of Fig. 6.39. The amplitude of the order parameters behaves as a ball moving within the potential landscape. When the system is synchronized in a stable way the ball is located in the potential's minimum (indicated by the black ball). A suitable stimulus induces a partial desynchronization of the order parameters. This means that the stimulus lifts the ball out of the minimum placing it on top of the potential's hill. After the stimulation the ball rolls down in

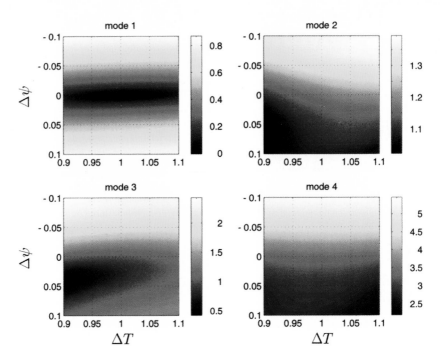

Fig. 6.34. *Vulnerability to stimulation:* The simulation of Fig. 6.16 was carried out with varying initial cluster phase ($-0.1 \leq \Delta\psi \leq 0.1$) and stimulation duration ($0.9 \leq \Delta T \leq 1.1$). For each simulation $r_1, \ldots, r_4$ were determined. Same format as in Fig. 6.30. Coupling and stimulation parameters: $K_1 = 1$, $K_2 = K_3 = K_4 = C_1 = \cdots = C_4 = 0$, $Q = 0.4$, $I_2 = I_3 = \gamma_1 = \cdots = \gamma_4 = 0$, $I_1 = I_4 = 7$ (during stimulation), $I_1 = I_4 = 0$ (before stimulation).

the left or right minimum. Correspondingly, the cluster resynchronizes after the stimulus is turned off.

One encounters a different situation provided the potential's shape is of the form displayed in the lower plot of Fig. 6.39. In this case as a consequence of a stimulation giving rise to a complete partial desynchronization the ball gets trapped in the potential's minimum located in $R = 0$, where R denotes the amplitude of the order parameters. In other words, after the stimulation the order parameters remain equal to zero. According to the slaving principle the stable modes will relax to zero (cf. (6.14)). From (6.18) it follows that the firing density $p(t)$ will, thus, tend to $1/(2\pi)$, so that the cluster of neurons fires incoherently. A stable state where the collective rhythmic activity ceases is called a *black hole* (Winfree 1980). Obvioulsy the minimum at $R = 0$ of the lower potential in Fig. 6.39 gives rise to a black hole which is associated with incoherent firing.

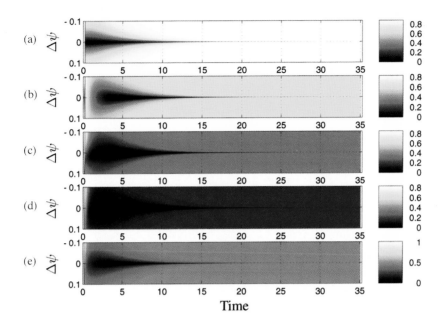

Fig. 6.35a–e. *Recovery of mode amplitudes:* The simulation of Fig. 6.16 was performed with varying initial cluster phase $\varphi_{1,\mathrm{B}}$ ($-0.1 \leq \Delta\psi \leq 0.1$). The cluster amplitudes R_1 (**a**), R_2 (**b**), R_3 (**c**), and R_4 (**d**) are plotted over time. $\delta(t)$ (**e**) serves for estimating the extent of the uniform desynchronization. Stimulation starts at time $t_\mathrm{B} = 0$ and ends at time $t_\mathrm{E} = 0.248$ (indicated by the vertical blue line). Coupling parameters as in Fig. 6.34. Stimulation parameters: $I_2 = I_3 = \gamma_1 = \cdots = \gamma_4 = 0$, $I_1 = I_4 = 7$ (during stimulation), $I_1 = I_4 = 0$ (after stimulation).

Figure 6.39 points out that the reaction of a cluster to partial desynchronization depends on the oscillators' mutual interactions. The latter determine whether or not the desynchronized state (i.e. $R = 0$) is stable. A cluster can only resynchronize provided the desynchronized state is not stable. Otherwise the cluster remains trapped in the black hole, and, correspondingly, the stimulation induced incoherency persists. All single-mode instabilities analyzed above have a potential similar to that one in the upper plot of Fig. 6.39. Consequently in all these cases collective synchronized activity recovers after stimulation induced partial desynchronization. Also in the two-modes instability investigated in Sect. 6.4.2 no black hole behavior was observed because the desynchronized state was unstable.

6.7 Subcritical Long Pulses

In the former sections we analyzed the impact of stimuli with high stimulation intensity. For this kind of stimuli stable fixed points are the attractors of the

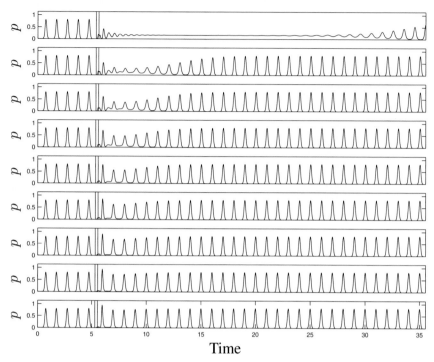

Fig. 6.36. *Desynchronization and firing patterns:* The firing density $p(t) = n(0, t)$ is plotted over time for equally spaced initial cluster phase $\varphi_{1,B}$ which is close to the critical value $\varphi_{1,B}^{\mathrm{crit}}$. The uppermost plot shows a stimulation induced complete partial desynchronization of the first mode (same simulation as in Fig. 6.17), whereas in the lowest plot $\varphi_{1,B}$ is equal to $\varphi_{1,B}^{\mathrm{crit}} + \pi/4$. Same format as in Fig. 6.32. Coupling parameters as in Fig. 6.34. Stimulation parameters: $I_2 = I_3 = \gamma_1 = \cdots = \gamma_4 = 0$, $I_1 = I_4 = 7$ (during stimulation), $I_1 = I_4 = 0$ (before and after stimulation).

dynamics of the density $n(\psi, t)$, and, thus, of the cluster variables' dynamics, too. Due to the high stimulation intensity the stimulus dominates the cluster's dynamics and, accordingly, the cluster variables rapidly and directly tend towards the attractor in a way shown, for example, in Fig. 6.7. For this reason it was sufficient to administer the stimulus as a single short pulse in order to provoke a complete partial desynchronization of a certain mode.

By decreasing the stimulation intensity the oscillators' mutual interactions gain increasing influence, and, correspondingly, the cluster's dynamics is governed by the interplay of stimulation and the tendency to self-synchronization. Consequently other types of attractors may occur which are, in particular, approached in a qualitatively different way. Extensive numerical investigations show that the cluster variable may, for instance, spiral towards a stable fixed point or a periodic orbit. As a consequence of the cluster variable's spiraling the desynchronization inhibiting effect of the phase

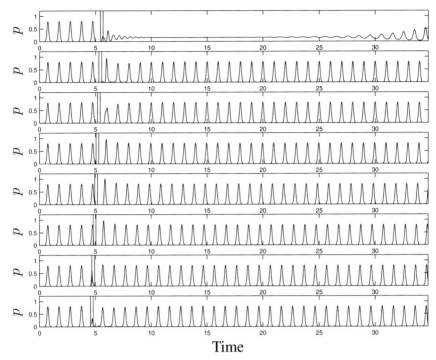

Fig. 6.37. *The stimulus' impact on the firing pattern:* The time course of the firing density $p(t) = n(0, t)$ is plotted for a series of simulations with initial cluster phase $\varphi_{1,\mathrm{B}}$ equally spaced within a cycle. The stimulus in the uppermost plot induces a complete partial desynchronization of the first mode (i.e. $\varphi_{1,\mathrm{B}} = \varphi_{1,\mathrm{B}}^{\mathrm{crit}}$). $\varphi_{1,\mathrm{B}}$ is increased by steps of $\pi/4$ from the uppermost plot to the lowest plot, so that the second uppermost plot of this figure and the lowest plot of Fig. 6.36 are identical. All other stimulation and coupling parameters as in Fig. 6.36.

error $\Delta\psi$ is reduced. The spiraling takes time, and, hence, a stimulus of this kind has to be administered as a single long pulse. Desynchronization related phenomena such as, for example, excitation of higher order may also be induced by the long pulses.

6.7.1 Spiraling Towards the Desynchronized State

One can illustrate the very effect of spiraling cluster variables, e.g., by administering a stimulus with subcritical intensity I_1 to the one-cluster state of the first-mode instability analyzed in Sect. 5.4.2. As a result of the subcritical stimulation intensity the cluster variables Z_j are not attracted by stable fixed points in a straight way as, for example, shown in Fig. 6.7. Rather stable periodic orbits of $Z_1, Z_2, Z_3, \ldots$ emerge, and according to a thorough numerical investigation the cluster variables spiral towards these periodic orbits,

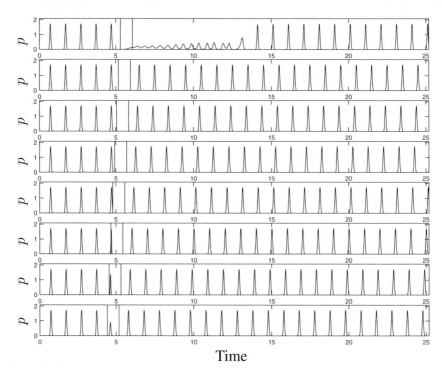

Time

Fig. 6.38. *Phase error $\Delta\psi$ and harmonic early response:* The firing density $p(t) = n(0, t)$ is plotted over time for equally spaced initial cluster phase $\varphi_{1,\mathrm{B}}$. The uppermost plot shows the stimulation from Fig. 6.29, where the stimulus causes a complete partial desynchronization of the first mode. From the uppermost plot to the lowest plot $\varphi_{1,\mathrm{B}}$ is increased by steps of $\pi/4$. Same format as in Fig. 6.37. Coupling parameters as in Fig. 6.28. Stimulation parameters: $I_2 = I_3 = I_4 = \gamma_1 = \cdots = \gamma_4 = 0$, $I_1 = 7$ (during stimulation), $I_1 = 0$ (before and after stimulation).

irrespective of where they initially start from. A stimulation induced spiral trajectory of Z_1 is shown in Fig. 6.40a.

To cause a complete partial desynchronization of the first mode an appropriate initial cluster phase $\varphi_{1,\mathrm{B}}^{\mathrm{crit}}$ has to be chosen, and the stimulus has to be administered during a critical duration T_{crit}, so that the spiral trajectory reaches the origin of the Gaussian plane where Z_1 vanishes (Fig. 6.40b)

The time courses of the cluster amplitudes and of the corresponding firing density $p(t)$ are plotted in Figs. 6.41 and 6.42. During the stimulation the density $n(\psi, t)$ approaches its limit cycle. Correspondingly one observes a rhythmic motion of the cluster amplitudes and of $p(t)$. The firing pattern during the stimulation will be denoted as *rhythmic stimulation driven response*. Note that the main frequency component of the rhythmic stimulation driven response typically differs from that of the stable synchronized

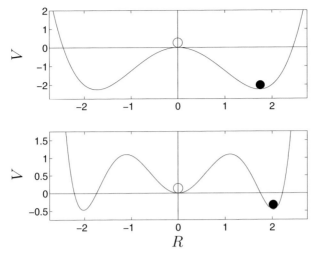

Fig. 6.39. *Recovery and black hole:* Plot shows two different types of potentials V of the order parameters' amplitude R. The shape of V determines the cluster's reaction to stimulation induced partial desynchronization (i.e. $R = 0$, *indicated by the white ball*). In the upper plot the cluster recovers, whereas in the lower plot the system is trapped in the black hole, so that the cluster remains in a state where it fires incoherently. The black dots correspond to the stable synchronized states before the stimulation, whereas the white balls indicate the state directly after the stimulation.

state before the stimulation. The recovery after a long subcritical pulse (Fig. 6.41) corresponds to the recovery after a short pulse (Fig. 6.5).

To study the influence of the phase error $\Delta\psi$ from (6.22) on the stimulation induced dynamics of the cluster amplitudes, the simulation of Fig. 6.41 was performed for $-0.1 < \Delta\psi < 0.1$ (Fig. 6.43). Within the whole range of $\Delta\psi$ the stimulus causes a rhythmic motion of the cluster amplitudes. A complete partial desynchronization of R_1 is only induced for vanishing $\Delta\psi$. With increasing $|\Delta\psi|$ the desynchronizing effect is reduced. However, as the trajectory of Z_1 densely spirals in the neighbourhood of the origin of the Gaussian plane (Fig. 6.40), the desynchronizing effect of the phase error $\Delta\psi$ is diminished compared to that of a supercritical short pulse, e.g., shown in Fig. 6.31.

6.7.2 Excitation of Higher Order

In this section two points will be emphasized:

1. If the stimulus $S(\psi) = \sum_{m=1}^{4} I_m \cos(m\psi + \gamma_m)$ contains terms of higher order, long pulses typically give rise to an excitation of higher order.

2. The attractor which is associated with spiral trajectories of the cluster variables needs not to be a limit cycle. It may also be a stable fixed point.

(a)

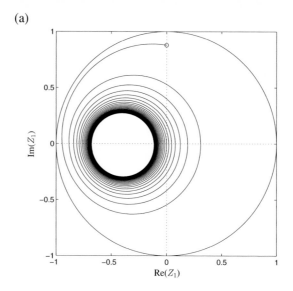

(b)

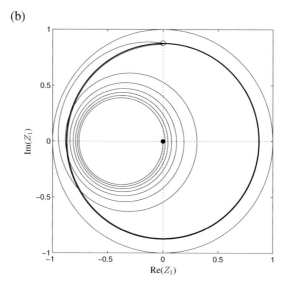

Fig. 6.40a,b. *Spiraling cluster variable Z_1:* The trajectory of the first cluster variable Z_1 is plotted in the Gaussian plane. (**a**) shows the trajectory of Z_1 during an infinitely long stimulation. The beginning of the stimulation is indicated by the small circle. As a consequence of the subcritical stimulation intensity I_1 a periodic orbit of Z_1 emerges. $I_1 = 4.5$ (subcritical compared to, e.g., $I_1 = 7$ in Fig. 6.7), $I_2 = I_3 = I_4 = \gamma_1 = \cdots = \gamma_4 = 0$. (**b**) shows the trajectory of Z_1 before and during the stimulation. The circular part of Z_1's trajectory corresponds to the synchronized state before stimulation. Begin and end of the stimulation are indicated by the small circle and the dot. The desynchronizing stimulation ends when Z_1 reaches the origin of the Gaussian plane. $I_1 = 0$ (before stimulation), $I_1 = 4.5$ (during stimulation), all other stimulation parameters as in (**a**). Coupling parameters in (**a**) and (**b**): $K_1 = 1$, $K_2 = K_3 = K_4 = C_1 = \cdots = C_4 = 0$, $Q = 0.4$.

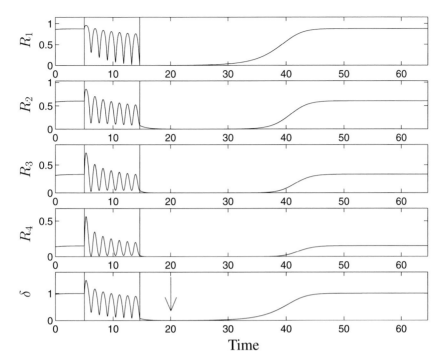

Fig. 6.41. *Dynamics of cluster amplitudes during a long subcritical pulse:* The time courses of the cluster amplitudes $R_1, \ldots, R_4$ for the stimulation from Fig. 6.40b are shown. The two vertical lines mark the stimulation's begin and end. Uniform desynchronization is estimated with $\delta(t)$ defined by (6.10). The vertical arrow indicates the minimum of $\delta(t)$. Coupling parameters as in Fig. 6.40. Stimulation parameters: $I_1 = 0$ (before and after stimulation), $I_1 = 4.5$ (during stimulation), $I_2 = I_3 = I_4 = \gamma_1 = \cdots = \gamma_4 = 0$.

Both dynamical features will be illustrated with one example: The one-cluster state already considered in the former section will be exposed to a subcritical stimulus with parameters $I_1 = I_2 = 4.585$, $I_3 = I_4 = \gamma_1 = \cdots = \gamma_4 = 0$. Due to the subcritical stimulation the cluster variables Z_j do not approach their attractors along short paths, rather they spiral towards stable fixed points as illustrated in Fig. 6.44a. If the appropriate initial cluster phase $\varphi_{1,\mathrm{B}}^{\mathrm{crit}}$ is chosen, the stimulation has to be stopped when Z_1 reaches the origin of the Gaussian plane (Fig. 6.44b). In this way a complete partial desynchronization of the first mode is attained.

The time course of the cluster amplitudes is shown in Fig. 6.45. Due to the eccentric location of the stable fixed points which attract the cluster variables Z_j (Fig. 6.44), double peaks of the cluster amplitudes R_j occur during the stimulation. The stimulation term of second order ($I_2 = 4.585$) excites, in particular, the second mode. As a consequence of this excitation of higher order a pronounced early response of the firing density $p(t)$ is observed (Fig.

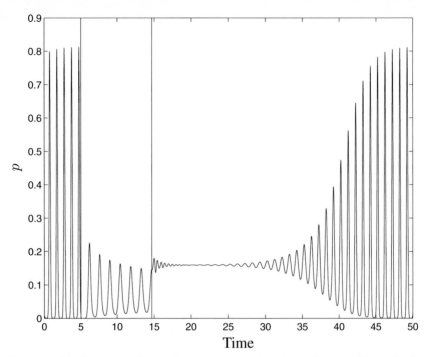

Fig. 6.42. *Rhythmic stimulation driven response:* The firing density $p(t) = n(0, t)$ is plotted over time for the simulation of Fig. 6.41. During the subcritical stimulation we observe a rhythmic stimulation driven response.

6.46). It is important to distinguish between the rhythmic stimulation driven response and the pronounced early response. The former occurs during the stimulation and results from the spiral trajectories of the cluster variables Z_j. In contrast, the latter follows the stimulation, and is due to stimulation terms which are of a different order compared to the order parameters. The cluster's recovery is similar to that after a short pulse (cf. Figs. 6.16 and 6.17).

Figure 6.47 visualizes the role the phase error $\Delta\psi$ defined by (6.22) plays. For $-0.1 < \Delta\psi < 0.1$ a rhythmic motion of the cluster amplitudes $R_1, \ldots, R_4$ is induced by the stimulus. Only for vanishing $\Delta\psi$ a complete partial desynchronization of R_1 is achieved. An increasing error $|\Delta\psi|$ reduces the desynchronizing effect. Because of the spiral trajectory of Z_1 the desynchronization reducing effect of the phase error $\Delta\psi$ is less pronounced than that of a supercritical short pulse, e.g., shown in Fig. 6.35. Z_1's trajectory of the simulation shown in Fig. 6.40 spirals more densely in the neighbourhood of the origin of the Gaussian plane compared to the dynamics of Z_1 analyzed in this section (Fig. 6.44). Correspondingly, the desynchronizing effect of $\Delta\psi$ illustrated in Fig. 6.43 is weaker compared to that in Fig. 6.47.

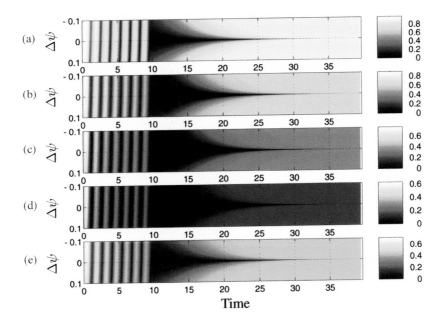

Fig. 6.43a–e. *Recovery of mode amplitudes:* The simulation of Fig. 6.41 was performed with varying initial cluster phase $\varphi_{1,\mathrm{B}}$, i.e. for $-0.1 \leq \Delta\psi \leq 0.1$. Cluster amplitudes R_1 (**a**), R_2 (**b**), R_3 (**c**), and R_4 (**d**) are plotted over time. $\delta(t)$ (**e**) from (6.10) serves for estimating the uniform desynchronization. Stimulation starts at time $t_\mathrm{B} = 0$ and ends at time $t_\mathrm{E} = 9.668$ (indicated by the vertical blue line). Coupling parameters as in Fig. 6.40. Stimulation parameters: $I_2 = I_3 = I_4 = \gamma_1 = \cdots = \gamma_4 = 0$, $I_1 = 4.5$ (during stimulation), $I_1 = 0$ (after stimulation).

6.7.3 Excitation of Lower Order

An excitation of lower order following a long pulse is observed, in particular, if the stimulus $S(\psi)$ contains terms of lower order compared to the order parameters. Figure 6.48 visualizes the firing pattern corresponding to an excitation of lower order. During the stimulation one observes the rhythmic stimulation driven response. Subsequent to the stimulation a pronounced early response occurs which results from the excitation due to the stimulation term of lower order ($I_1 > 0$). The cluster's behavior during recovery is similar to that after a short pulse (Fig. 6.21).

6.8 Summary and Discussion

This chapter was devoted to single pulse stimulation. We encountered a variety of stimulation induced dynamical phenomena occurring both during and

(a)

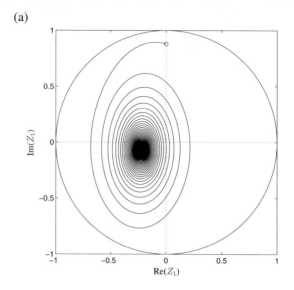

(b)

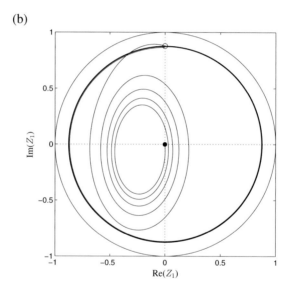

Fig. 6.44a,b. *Spiraling cluster variable Z_1:* The trajectory of the first cluster variable Z_1 in the Gaussian plane is shown during an infinitely long stimulation (**a**) and during a stimulation which provokes a complete partial desynchronization of the first mode (**b**). Same format as in Fig. 6.40. Stimulation parameters: $I_1 = I_2 = 0$ (before stimulation), $I_1 = I_2 = 4.585$ (during stimulation), $I_3 = I_4 = \gamma_1 = \cdots = \gamma_4 = 0$. Coupling parameters as in Fig. 6.40: $K_1 = 1$, $K_2 = K_3 = K_4 = C_1 = \cdots = C_4 = 0$, $Q = 0.4$.

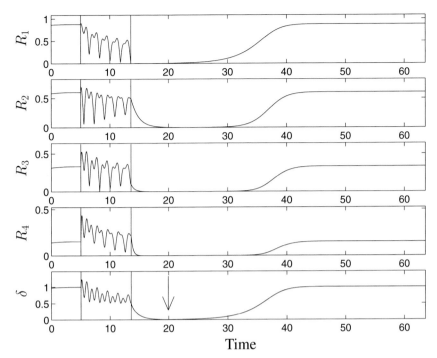

Fig. 6.45. *Dynamics of cluster amplitudes during a long subcritical pulse:* Cluster amplitudes $R_1, \ldots, R_4$ are plotted over time for the stimulation from Fig. 6.44b. The stimulation's begin and end are marked by the two vertical lines. Uniform desynchronization is estimated with $\delta(t)$ from (6.10). The vertical arrow indicates the minimum of $\delta(t)$. Coupling parameters as in Fig. 6.44. Stimulation parameters: $I_1 = I_2 = 0$ (before and after stimulation), $I_1 = I_2 = 4.585$ (during stimulation), $I_3 = I_4 = \gamma_1 = \cdots = \gamma_4 = 0$.

after stimulation. However, the order parameter concept serves as a framework within which this diversity of different reactions on a stimulus becomes easy to digest.

Let the summary start with the *short pulses* of high intensity. In Sect. 6.2 we learned how a stimulus acts on order parameters. In particular, in the case of a one-mode instability the system behaves as a ball moving in a potential landscape. Applying a stimulation pulse means shifting the ball from one place in the potential landscape to another one. Accordingly, after the stimulus is turned off, the ball relaxes towards the nearest minimum of the potential. Our approach clearly revealed that the oscillators' mutual interactions determine the potential of the order parameters and, thus, they determine whether or not a cluster resynchronizes after a desynchronizing stimulation (Sect. 6.6).

In all systems analyzed in Sect. 6.2 the desynchronized states were not stable, and, hence, in all cases the cluster resynchronizes after the stimulation.

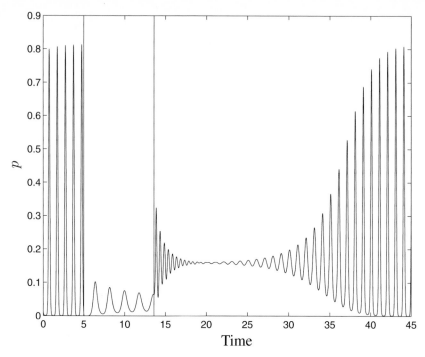

Fig. 6.46. *Rhythmic stimulation driven response and early pronounced response:* The firing density $p(t) = n(0, t)$ from (6.18) is plotted over time for the simulation of Fig. 6.45. During stimulation a rhythmic stimulation driven response occurs, whereas the early pronounced response comes after the stimulation.

Nevertheless, a stimulus of a critical duration applied at a critical initial phase gives rise to a complete partial desynchronization of the order parameters, i.e. the order parameters' amplitude vanishes at the end of the stimulation. In this way a suitably timed stimulus provokes a transient partial desynchronization which is followed by the cluster's resynchronization. A transition from type 1 resetting to type 0 resetting occurs when the stimulation duration exceeds T_{crit}. The transient desynchronization may be less or more pronounced; it may even be obscured by the enslaved modes' reaction to the stimulus: The stimulus may excite enslaved modes of higher or lower order compared to the order parameters. In terms of the firing pattern of a cluster of neurons this *transient excitation of higher or lower order* corresponds to a *pronounced early response* occurring directly after the end of the stimulation (Sect. 6.3).

In order to achieve a pronounced partial desynchronization for a given stimulation intensity, the stimulation duration and the intial cluster phase have to be as close to the critical stimulation duration and the critical intial cluster phase (modulo 2π) as possible. As shown in Sect. 6.5 errors of the initial cluster phase turned out to be more serious compared to errors of the stimulation duration.

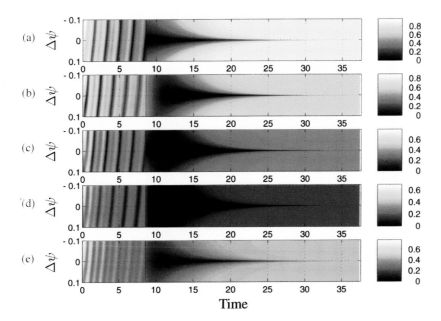

Fig. 6.47a–e. *Recovery of mode amplitudes:* The simulation of Fig. 6.45 was performed with varying initial cluster phase $\varphi_{1,\mathrm{B}}$, that means for $-0.1 \leq \Delta\psi \leq 0.1$. The plots show the time courses of the cluster amplitudes R_1 (**a**), R_2 (**b**), R_3 (**c**), and R_4 (**d**). $\delta(t)$ (**e**) from (6.10) serves for estimating the uniform desynchronization. Stimulation starts at time $t_\mathrm{B} = 0$ and ends at time $t_\mathrm{E} = 8.6$ (indicated by the vertical blue line). Coupling parameters as in Fig. 6.44. Stimulation parameters: $I_3 = I_4 = \gamma_1 = \cdots = \gamma_4 = 0$, $I_1 = I_2 = 4.585$ (during stimulation), $I_1 = I_2 = 0$ (after stimulation).

We observed that for a given coupling pattern the resynchronization typically takes the longer the more pronounced the partial desynchronization at the end of the stimulation is (Sect. 6.5). However, certain coupling patterns are associated with a *rapid recovery*, even in the case of a complete partial desynchronization (Sect. 6.4.1). The decisive influence of the couplings on the reaction to stimulation also became evident when we considered a two-modes instability (Sect. 6.4.2). In this case, for instance, one pair of order parameters vanishes at the end of the stimulation, while the other pair of order parameters undergoes a transient, in this way dominating the early period of recovery. The example we dwelled on was a *harmonic early response*.

The cluster's reaction during the stimulation is qualitatively different provided the stimulation intensity is decreased (Sect. 6.7). In this case the cluster's dynamics during stimulation is essentially governed by the interplay of both couplings and stimulus, so that new types of attractors as well as transients occur: The stimulus causes a spiraling of the order parameters which corresponds to a *rhythmic stimulation driven response* of the firing pattern

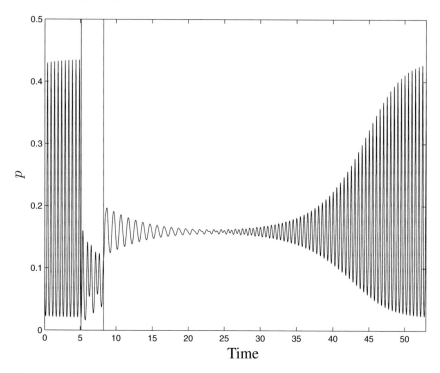

Fig. 6.48. *Rhythmic stimulation driven response and early pronounced response:* A two-cluster state of a second-mode instability is exposed to a stimulus with terms of first and second order. The stimulus causes a complete partial desynchronization of the second mode. The firing density $p(t) = n(0,t)$ is plotted over time. The rhythmic stimulation driven response occurs during the stimulation, whereas the early pronounced response follows the stimulation. Coupling and stimulation parameters: $K_2 = 1$, $K_1, K_3, K_4 = C_1 = \cdots = C_4 = 0$ (cf. Sect. 5.4.3), $I_1 = I_2 = 0$ (before and after stimulation), $I_1 = I_2 = 2.4$ (during stimulation), $I_3 = I_4 = \gamma_1 = \cdots = \gamma_4 = 0$.

(Sect. 6.7). Compared to the supercritical short pulses analyzed in Sects. 6.2–6.5 stimulation pulses with decreased intensity have to be applied for a longer time in order to induce a complete partial desynchronization of the order parameters. For this reason we denoted the latter type of pulses as *subcritical long pulses*. Interestingly errors of the initial cluster phase are less disturbing for the desynchronizing outcome of the stimulation in the case of the subcritical long pulses. For, e.g., medical applications subcritical long pulse are important for two reasons: On the one hand they may be more gentle by preventing from tissue damage due to the low stimulation intensity. On the other hand this kind of stimulation may be very effective because of the diminished influence of the phase error.

To realize the weighty influence of the couplings on the cluster's behavior during and after stimulation, let us compare the ensemble model presented in

Chap. 2 with the cluster model investigated in this chapter. In the ensemble model there are no synchronizing order parameters. Hence, after a desynchronizing stimulation the ensemble is not able to resynchronize. Stimulation terms of higher order excite higher order Fourier modes of the ensemble's distribution $f(\psi, t)$, in this way inducing burst splitting which is not restricted to small values of the phase error (Sect. 2.8). In contrast, excitation of higher or lower order in the cluster model is restricted to small phase errors. Phenomena induced by subcritical long pulses, such as a rhythmic stimulation driven response, indispensably rely on the interplay of couplings and stimulus. For this reason they cannot be observed in the ensemble model.

7. Periodic Stimulation

7.1 Introductory Remarks

The effects of periodic stimuli on rhythmic biological activity were experimentally studied in detail in a variety of physiological paradigms. A number of studies were dedicated, for instance, to the entrainment of the respiration to a mechanical ventilator (cf. Fallert and Mühlemann 1971, Vibert, Caille, Segundo 1981, Baconnier et al. 1983, Petrillo, Glass, Trippenbach 1983, Petrillo and Glass 1984) and to periodic electrical stimulation of cardiac pacemaker tissue (cf. Reid 1969, Levy, Iano, Zieske 1972, Van der Tweel, Meijler, Van Capelle 1973, Jalife and Moe 1976, 1979, Ypey, Van Meerwijk, DeHaan 1982, Guevara, Glass, Shrier 1981, Glass et al. 1983, Glass et al. 1984, Jalife and Michaels 1985, Guevara, Shrier, Glass 1988). Other authors investigated the periodic forcing of circadian oscillators (Pittendrigh 1965, Winfree 1980), of neurons (Perkel et al. 1964, Pinsker 1977, Guttman, Feldman, Jakobsson 1980) and of tremor activity (Elble and Koller 1990, Elble, Higgins, Hughes 1992).

A vast literature addresses the dynamics of periodically forced nonlinear oscillators from the mathematical point of view (cf. Haken 1977, 1983, Arnold 1983, Guckenheimer and Holmes 1990). Nevertheless, our understanding of this issue is still far from complete. Accordingly, many modelling approaches were mainly dedicated to the stimulation of single oscillators. This lead to numerous fundamental results, e.g., in the context of periodic forcing of integrate and fire models (cf. Keener, Hoppensteadt, Rinzel 1981, Hoppensteadt 1986, Glass and Mackey 1988). However, we have to choose a different approach if we want to analyze the impact of a periodic stimulus on the synchronization pattern of a network of oscillators. Effects of this kind are important, e.g., in the context of deep brain stimulation (cf. Siegfried and Hood 1985, Volkmann and Sturm 1998, Volkmann, Sturm, Freund 1998, cf. Sect. 10). Therefore in this chapter it will be outlined how to investigate a cluster of oscillators subjected to periodic forcing and noise.

It is important to note that the derivation of the evolution equation (4.21) for the average number density $n(\psi, t)$ also holds if the stimulus is time dependent. That is why we merely have to replace $S(\psi)$ by $S(\psi, t)$ in (4.21). For this reason our stochastic approach is valid in the case of time dependent stimuli, too.

This chapter is not intended to be a comprehensive or even concluding investigation. Rather let us restrict ourselves to focus on dynamical features which are relevant in physiology, namely on $n : m$ phase-locking between the stimulus and the cluster's activity and on stimulation induced annihilation of collective rhythms. For the sake of comparability we will consider how different types of periodic stimuli act on the one-cluster state analyzed in Sect. 5.4.2.

Along the lines of our approach we will describe stimulation induced processes in terms of an $n : m$ phase locking between the stimulus and the cluster variables. Additionally we will relate the cluster's dynamics to the corresponding firing patterns.

7.2 Smooth Periodic Stimulation

This section is dedicated to the effects arising due to a persisting smooth periodic stimulation. Accordingly, the stimulation coefficients X_m and Y_m in the formula of the stimulus acting on the jth oscillator

$$S(\psi_j, t) = \sum_{m=1}^{4} [X_m(t) \sin(m\psi) + Y_m(t) \cos(m\psi)] \tag{7.1}$$

(cf. (4.5)) are now time dependent smooth periodic functions: $X_m(t) = X_m(t + P)$ and $Y_m(t) = Y_m(t + P)$, where P is the period. To illustrate $n : m$ phase locking between stimulus and cluster let us choose an example where we can profit from our analysis of the former chapter. With this aim in view we consider a stimulation mechanism given by

$$X_m(t) = I_m \sin(m\omega t - \gamma_m) \; , \; Y_m(t) = I_m \cos(m\omega t - \gamma_m) \; , \tag{7.2}$$

where I_m and γ_m are constant parameters. Inserting (7.2) into (7.1) we obtain

$$S(\psi_j, t) = \sum_{m} I_m \cos[m(\psi_j - \omega t) + \gamma_m] \; . \tag{7.3}$$

Turning to a *rotating coordinate system* by introducing relative phases

$$\theta_j(t) = \psi_j(t) - \omega t \; , \tag{7.4}$$

the model equation (4.3) reads

$$\dot{\theta}_j = \tilde{\Omega} + \frac{1}{N} \sum_{k=1}^{N} M(\theta_j - \theta_k) + \tilde{S}(\theta_j) + F_j(t) \; , \tag{7.5}$$

where

$$\tilde{S}(\theta_j) = \sum_m I_m \cos(m\theta_j + \gamma_m) , \quad \tilde{\Omega} = \Omega - \omega . \tag{7.6}$$

Equation (7.5) is of the same form as (4.3). In particular, in the rotating coordinate system the stimulation is no longer time dependent. Thus, based on the derivation in Sect. 4.2 we obtain an evolution equation for the average number density $n(\theta, t)$ for the relative phase θ by merely replacing $n(\psi, t)$, Ω and $S(\psi)$ by $n(\theta, t)$, $\tilde{\Omega}$ and $\tilde{S}(\theta)$ in (4.21).

To study the cluster's behavior during the stimulation and the cluster's reorganization after the stimulation in this chapter several simulations will be presented, where the stimulus is turned on at time t_B and switched off at time t_E. As in the former chapter $T = t_E - t_B$ denotes the stimulation duration. Before and after the stimulation $\tilde{S}(\theta)$ vanishes, whereas during the stimulation $\tilde{S}(\theta)$ is given by (7.6). On the analogy of Sect. 6.2.3 we investigate the impact of periodic stimuli on the one-cluster state analyzed in Sect. 5.4.2. The latter is due to a first-mode instability , where for supercritical coupling, $K_1 > K_1^{\text{crit}}$, the Fourier modes $\hat{n}(1, t)$ and $\hat{n}(-1, t)$ become order parameters, and the system behaves as a ball rotating in the circular potential valley shown in Fig. 5.4. We are already familiar with the corresponding firing pattern (Fig. 5.21a).

7.2.1 1:1 Phase Locking

Let us first dwell on an example, where a one-cluster state with the spontaneous synchronization frequency Ω is entrained by a strong periodic stimulus with frequency $\omega \neq \Omega$. Both stimulus and order parameters are of first order. To this end we choose

$$I_1 = 4 , \quad I_2 = I_3 = \ldots = 0 , \quad \gamma_1 = \gamma_2 = \ldots = 0 \tag{7.7}$$

as stimulation parameters, and noise amplitude $Q = 0.4$. By switching to the rotating coordinate system (7.4) we can directly make use of our results obtained in Sect. 6.2.3. There we have shown that the cluster variables are strongly attracted by stable fixed points as illustrated in Fig. 6.7, provided the stimulation intensity I_1 is large compared to the noise amplitude Q. In the case under consideration the latter requirement is fulfilled, which is why all cluster variables and, thus, also the average number density $n(\theta, t)$ is attracted by a stationary stable solution: $n(\theta, t) \rightarrow n_{\text{stat}}(\theta)$ for $t \rightarrow \infty$ (cf. Fig. 6.18a). Taking into account that the rotation frequency of the coordinate system (7.4) and the frequency of the stimulation are the same, we immediately see that the stationary solution $n_{\text{stat}}(\theta)$ corresponds to a 1:1 phase locking between stimulus and first cluster variable.

Let us illustrate the entrainment in detail. To this end we first compare the dynamics of the leading modes with the one of the stimulus given by

$$X_1(t) = \sin[\phi_{\text{stim}}(t)] , \quad Y_1(t) = \cos[\phi_{\text{stim}}(t)] , \tag{7.8}$$

where

$$\phi_{\text{stim}}(t) = \omega t \tag{7.9}$$

is the stimulus' phase (cf. (7.2) and (7.7)). Before the stimulation the cluster's synchronization frequency is $\Omega = 2\pi$. The stimulus quickly entrains the cluster variables (Fig. 7.1). In the entrained state the cluster adopts the stimulus' frequency $\omega = 3\pi$, and the cluster amplitudes, $R_1, \ldots, R_4$, are increased. After the stimulation the cluster relaxes towards the initial synchronized state with frequency Ω.

We clearly see the 1:1 phase locking by plotting $\Phi_{1,1}(t)$, the 1:1 phase difference between the stimulus and the first cluster variable, defined by

$$\Phi_{1,1}(t) = \phi_{\text{stim}}(t) - \varphi_1(t) = \omega t - \varphi_1(t) . \tag{7.10}$$

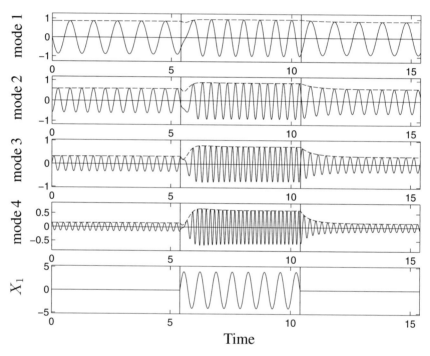

Fig. 7.1. *Entrainment of the cluster variables:* Time course of amplitudes, R_j (*dashed lines*) and real parts $\text{Re}(Z_j)$ (*solid lines*) of the first four cluster variables ($j = 1, \ldots, 4$) obtained by numerically integrating (4.21) with periodic stimulus. The lowest plot shows the time course of the stimulus X_1 from (7.8). Begin and end of the periodic stimulation are indicated by vertical lines. Coupling parameters: $K_1 = 1$, $K_2 = K_3 = K_4 = C_1 = \ldots = C_4 = 0$. Noise amplitude $Q = 0.4$. Spontaneous synchronization frequency of the cluster $\Omega = 2\pi$, frequency of the stimulus $\omega = 3\pi$. Stimulation parameters: $I_1 = 4$, $I_2 = I_3 = \ldots = 0$, $\gamma_1 = \gamma_2 = \ldots = 0$.

As a result of the stimulation $\Phi_{1,1}$ quickly tends towards a constant value, so that the phases of stimulus and first cluster variable are tightly locked (Fig. 7.2).

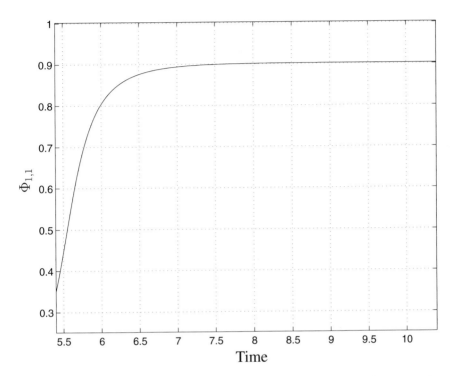

Fig. 7.2. *1:1 phase locking:* Time course of $\Phi_{1,1}(t)$ from (7.10) during the periodic stimulation (cf. Fig. 7.1).

Figure 7.3 displays the time course of the corresponding firing density $p(t) = n(0,t)$. In addition to the frequency entrainment we observe that during the stimulation the peak amplitude of $p(t)$ increases, which means that the cluster becomes more synchronized. This stimulation induced increase of synchronization is additionally reflected by the increase of the cluster amplitudes (Fig. 7.1). Figures 7.1 and 7.3 clearly show that the stimulus does not provoke a qualitative change of the cluster's synchroniztion pattern: Before, during and after the stimulation we observe a one-cluster state.

7.2.2 1:2 Phase Locking

To demonstrate 1:2 entrainment let us consider the one-cluster state of the former section subject to a strong periodic stimulus of second order. Accordingly, in (7.3) we put

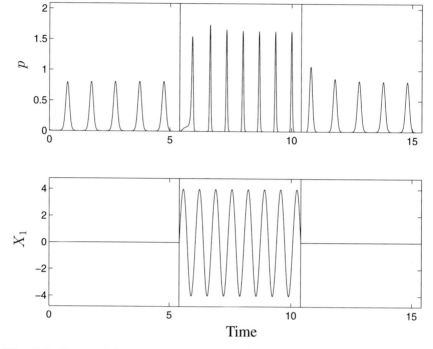

Fig. 7.3. *Entrained firing pattern:* Time course of the firing density $p(t) = n(0,t)$ from (6.18) and of the stimulus $X_1(t)$ from (7.8). The vertical lines indicate begin and end of the periodic stimulation. Same simulation as in Fig. 7.1.

$$I_2 = 6 , \quad I_1 = I_3 = I_4 = \ldots = 0 , \quad \gamma_1 = \gamma_2 = \ldots = 0 . \qquad (7.11)$$

The other parameters are chosen as in the former section: The cluster's synchronization frequency is $\Omega = 2\pi$, the noise amplitude is $Q = 0.4$, and $\omega = 3\pi$. Inserting (7.11) into (7.2) the time dependent coefficients of the stimulation mechanism read

$$X_2(t) = \sin\left[\phi_{\text{stim}}(t)\right] , \quad Y_2(t) = \cos\left[\phi_{\text{stim}}(t)\right] , \qquad (7.12)$$

with the stimulus' phase given by

$$\phi_{\text{stim}}(t) = 2\omega t . \qquad (7.13)$$

In particular, the stimulus' frequency is 2ω, and, thus, twice as large as in the former section. Figure 7.4 illustrates the entrainment and, especially, the 1:2 locking between the stimulus and the first cluster variable. Before the stimulus is turned on, $\Omega = 2\pi$ is the cluster's synchronization frequency which is given by the frequency of the first cluster variable. During the stimulation the cluster variables are quickly entrained, and the cluster amplitudes are

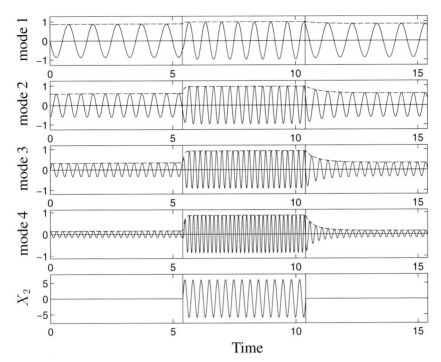

Fig. 7.4. *Entrainment of the cluster variables:* Amplitudes R_j (*dashed lines*) and real parts $\mathrm{Re}(Z_j)$ (*solid lines*) of the first four cluster variables ($j = 1, \ldots, 4$) are plotted versus time. The time course of the stimulus X_2 from (7.12) is shown in the lowest plot. Vertical lines indicate begin and end of the periodic stimulation. Noise amplitude and coupling parameters as in the simulation of Fig. 7.1. The spontaneous synchronization frequency of the cluster is $\Omega = 2\pi$, whereas the frequency of the stimulus reads $2\omega = 6\pi$. Stimulation parameters: $I_2 = 6$, $I_1 = I_3 = I_4 = \ldots = 0$, $\gamma_1 = \gamma_2 = \ldots = 0$.

increased. In the phase locked state the frequency of the first cluster variable is $\omega = 3\pi$. After the stimulation the cluster's initial state occurs again.

To study the 1:2 phase locking let us consider $\Phi_{1,2}(t)$, the 1:2 phase difference between the stimulus and the first cluster variable, introduced by

$$\Phi_{1,2}(t) = \phi_{\mathrm{stim}}(t) - 2\varphi_1(t) = 2\omega t - 2\varphi_1(t) . \tag{7.14}$$

During the stimulation $\Phi_{1,2}$ tends towards a constant value, in this way reflecting the strong 1:2 locking (Fig. 7.5).

The 1:2 entrained firing density $p(t) = n(0, t)$ is illustrated in Fig. 7.6. As in the former section the stimulation induced increase of the peak amplitude of $p(t)$ indicates an increase of the cluster's strength of synchronization. According to Figs. 7.4 and 7.6 there is no stimulation induced qualitative change of the cluster's synchronization pattern: Throughout the whole simulation we observe a one-cluster state.

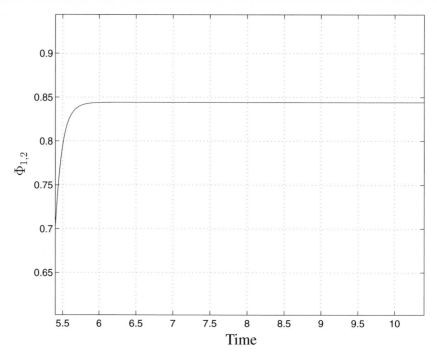

Fig. 7.5. *1:2 phase locking:* Time course of $\Phi_{1,2}(t)$ from (7.12) during the periodic stimulation (cf. Fig. 7.4).

7.2.3 Changes of the Synchronization Pattern

In the former two sections the cluster's state of synchronization did not change qualitatively in the course of the stimulation: In both examples a stimulus was applied to a one-cluster state which persisted during stimulation (Figs. 7.3 and 7.6). This section will point out that a stimulus may additionally induce transitions between qualitatively different patterns of synchronization, for instance, transitions from an m-cluster state to an n-cluster state, where $m \neq n$. This will be illustrated by considering a stimulation inducing a 1:2 entrainment which is accompanied by a transition from a one-cluster state to a symmetric two-cluster state. To this end we choose all model and stimulation parameters as in the former section except for a dimished stimulation intensity I_2. The parameters, thus, read

$$I_2 = 4, \; I_1 = I_3 = I_4 = \ldots = 0, \; \gamma_1 = \gamma_2 = \ldots = 0, \; Q = 0.4. \tag{7.15}$$

The time dependent stimulus is given by (7.12) and (7.13). The synchronization frequency of the cluster is $\Omega = 2\pi$, whereas the stimulus' frequency is $2\omega = 6\pi$.

Directly after the onset of the stimulation the cluster amplitudes $R_1, \ldots, R_4$ are excited. However, while R_2 and R_4 remain on a constant level, R_1

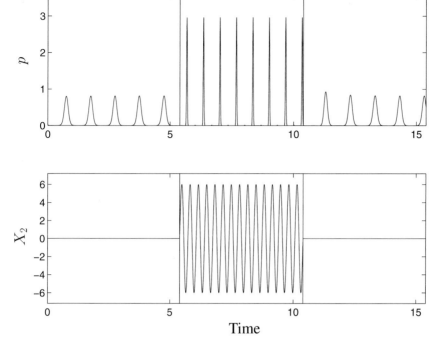

Fig. 7.6. *Entrained firing pattern:* Firing density $p(t) = n(0, t)$ and stimulus $X_2(t)$ from (7.12) are plotted versus time for the same simulation as in Fig. 7.4. Begin and end of the periodic stimulation are indicated by the vertical lines.

and R_3 decrease and finally vanish in the course of the stimulation (Fig. 7.7). For this reason the even modes dominate the synchronization pattern, and, correspondingly, a transition from a one-cluster state to a two-cluster state occurs (cf. Fig. 5.19). After the stimulus is turned off, the even modes are quickly damped by the nearly vanishing first mode, i.e. the order parameter. This effect is due to the slaving principle and was explained in the context of the excitation of higher order (Sect. 6.3.1, cf. Fig. 6.16). Finally, the first mode recovers, in this way resynchronizing the cluster, so that the initial one-cluster state emerges again (cf. Sects. 5.4.2 and 5.6.2).

As soon as the stimulation starts, the first cluster variable quickly becomes 1:2 phase locked to the stimulus, in this way adopting the frequency ω (Fig. 7.8). The transition from a one-cluster state to a two-cluster state is reflected by the firing pattern, too (Fig. 7.9). In the course of the stimulation the influence of the odd modes on the firing density $p(t)$ vanishes more and more (Fig. 7.7), and consequently a two-cluster state emerges, which finally tends to a symmetric configuration (Fig. 7.9). Accordingly, the dominant frequency of the firing activity turns from ω into 2ω. Hence, from the experimentalist's point of view the transition from the one-cluster state to the two-cluster state

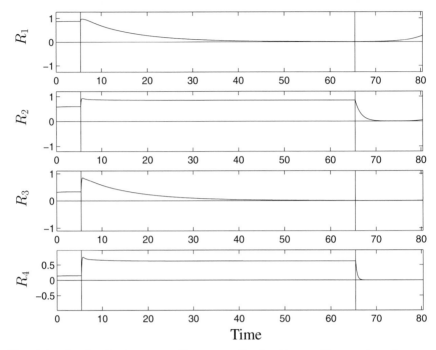

Fig. 7.7. *Amplitude dynamics:* Cluster amplitudes $R_1, \ldots, R_4$ are plotted versus time. Begin and end of the periodic stimulation are indicated by vertical lines. During the stimulation R_1 and R_3 slowly vanish, while R_2 and R_4 rapidly attain increased and constant values. Noise amplitude and coupling parameters as in the simulations of Figs. 7.1 and 7.4. The spontaneous synchronization frequency of the cluster is $\Omega = 2\pi$, and the frequency of the stimulus is $2\omega = 6\pi$. Stimulation parameters: $I_2 = 4$, $I_1 = I_3 = I_4 = \ldots = 0$, $\gamma_1 = \gamma_2 = \ldots = 0$.

shows up as a *frequency doubling in the course of the stimulation*. After the stimulation the increased even modes relax (Fig. 7.7), and, correspondingly, we observe a pronounced early response (Fig. 7.9, cf. Fig. 6.17). Finally the cluster resynchronizes in a way we are familiar with from Sects. 5.4.2 and 5.6.2.

7.3 Pulsatile Periodic Stimulation

The periodic stimulus need not be a smooth sinusoidal function. In physiological experiments trains of periodic pulses are frequently used, too (cf. Glass and Mackey 1988). Therefore it should be emphasized that our stochastic approach puts us in a position to study $n : m$ phase locking between pulsatile periodic stimuli and the cluster's firing activity. As an illustration of this kind of entrainment we consider a 1:1 locking caused by a train of twenty pulses following one another (Fig. 7.10). To model a pulse train we take into

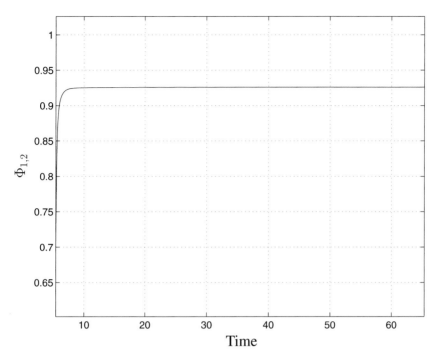

Fig. 7.8. *1:2 phase locking:* Time course of $\Phi_{1,2}(t)$ defined by (7.12) during the periodic stimulation (cf. Fig. 7.7).

account time dependent stimulation coefficients $I_1, I_2, \ldots$ in the stimulation mechanism (4.5). This yields

$$S(\psi, t) = \sum_{m=1}^{4} I_m(t) \cos(m\psi + \gamma_m) , \qquad (7.16)$$

where the series of M pulses is given by

$$I_m(t) = \begin{cases} \hat{I}_m & : \quad t \in [t_B + k\tau, t_B + k\tau + t_P] \\ 0 & : \quad \text{otherwise} \end{cases} \qquad (7.17)$$

with $k = 0, 1, \ldots, M-1$. The stimulation starts at time t_B, τ is the period of the periodic pulsatile stimulus, and t_P is the length of a single pulse, where $t_P < \tau$ has to fulfilled.

Let us apply such a stimulus to the one-cluster state considered in Sect. 7.2. To achieve a 1:1 entrainment it is sufficient to use a stimulus of first order, that means we put $\hat{I}_m = 0$ for $m > 1$. As visualized in Fig. 7.10 the stimulus quickly entrains the cluster variables, where the 1:1 locking can clearly be seen by comparing the time course of the first cluster variable and the stimulus. After the stimulation the cluster resynchronizes, so that the

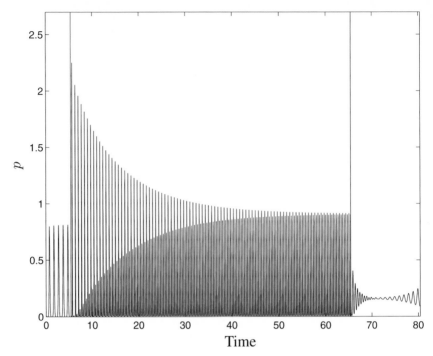

Fig. 7.9. *Entrained firing pattern:* The firing density $p(t) = n(0, t)$ is plotted versus time for the same simulation as in Fig. 7.7. Begin and end of the periodic stimulation are indicated by the vertical lines. During the stimulation a transition from a one-cluster state to a two-cluster state occurs, and consequently the dominant frequency of $p(t)$ turns from ω into 2ω.

initial one-cluster state is rapidly reestablished. The 1:1 entrainment as well as the cluster's recovery after stimulation also dominate the firing pattern (Fig. 7.11).

7.4 Annihilation of Rhythms

In the former sections we already saw that stimulation acts on both the phases and the amplitudes of the cluster variables. As a consequence of the stimulation the amplitudes may even be damped in such a way that the synchronized firing activity ceases. This is illustrated in Fig. 7.12, which shows how a pulsatile periodic stimulus annihilates the rhythmic firing activity of a one-cluster state. After the stimulation the cluster immediately resynchronizes and the rhythm restarts.

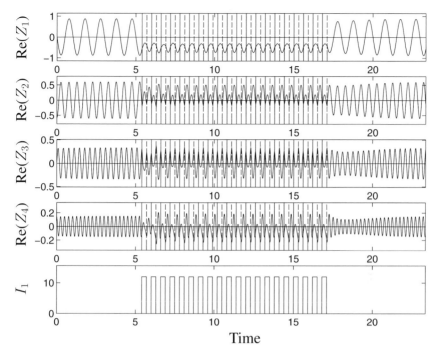

Fig. 7.10. *Amplitude dynamics:* Time course of the real parts of the cluster variables, $\text{Re}(Z_1), \ldots, \text{Re}(Z_4)$, as obtained by numerically integrating (4.21) with the pulse train (7.17), which is shown in the lowest plot, where $\hat{I}_1 = 12$, $t_P = 0.3$ and $\tau = 0.6$. The other stimulation parameters read $\hat{I}_m = 0$ for $m > 1$, $\gamma_1 = \pi/2$, and $\gamma_2 = \gamma_3 = \cdots = 0$. In the four upper plots begin and end of the single pulses are indicated by solid and dashed vertical lines, respectively. Coupling parameters: $K_1 = 1$, $K_2 = K_3 = K_4 = C_1 = \ldots = C_4 = 0$. Noise amplitude $Q = 0.4$.

7.5 Summary and Discussion

Stimulating a cluster of oscillators may serve quite different purposes. For instance, one may intend to annihilate rhythmic synchronized activity. Besides this desynchronizing effect a stimulus may be applied in order to modify an ongoing rhythm, for example, by changing the cluster's frequency or by reshaping the synchronization pattern.

A better understanding of effects of this kind is important for various reasons, for instance, in order to improve therapeutic stimulation techniques aiming at desynchronizing rhythmic activity (cf. Siegfried and Hood 1985, Volkmann and Sturm 1998, Volkmann, Sturm, Freund 1998). On the other hand in the brain rhythmic activity is processed by neuronal populations arranged in loops (cf. Creutzfeldt 1983). As yet, our knowledge concerning the dynamics of groups of rhythmically active clusters interacting in this

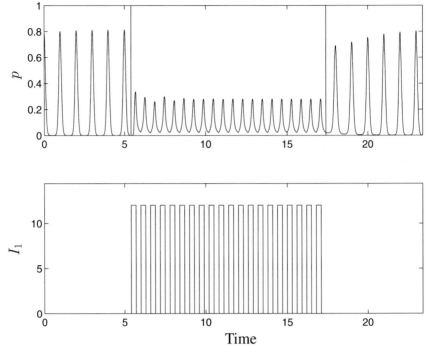

Fig. 7.11. *Entrained firing pattern:* The firing density $p(t) = n(0, t)$ is plotted versus time for the simulation of Fig. 7.10. In the upper plot begin and end of the periodic pulsatile stimulation are indicated by the vertical lines. The lower plot shows the pulse train (7.17).

way is sketchy. Hence, to cope with this complex issue in a first step it is appropriate to study how a cluster reacts on a rhythmic driving force.

Obviously it is a big advantage that the evolution equation of the average number density (4.21) also holds provided the stimulus is time dependent. In this way using our stochastic approach introduced in Chap. 4 we studied the impact of periodic stimuli on a cluster of interacting phase oscillators subject to random fluctuations. We saw that both smooth periodic and pulsatile periodic stimuli may entrain the cluster of oscillators in terms of an $n : m$ phase locking, where the strength of the cluster's synchronization may be changed (Fig. 7.3). Moreover the cluster may undergo stimulation induced qualitative modifications of the type of the synchronization pattern, for instance, we encountered transitions between different cluster states (Fig. 7.9). On the other hand as a consequence of a pronounced stimulus induced damping of the cluster amplitudes the cluster's rhythmic firing activity may vanish (Fig. 7.12).

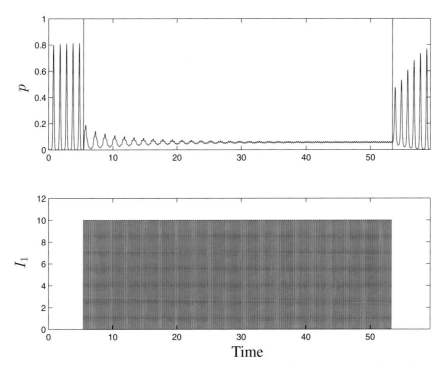

Fig. 7.12. *Annihilation of rhythmic firing activity:* The time course of the firing density $p(t) = n(0, t)$ (*upper plot*) is shown for a numerical integration of (4.21) with the pulse train (7.17) (*lower plot*) with parameters $\hat{I}_1 = 10$, $t_P = 0.15$ and $\tau = 0.3$, $\hat{I}_m = 0$ for $m > 1$, $\gamma_1 = \gamma_2 = \cdots = 0$. In the upper plot vertical lines indicate begin and end of the periodic pulsatile stimulation. Coupling parameters: $K_1 = 1$, $K_2 = K_3 = K_4 = C_1 = \ldots = C_4 = 0$. Noise amplitude $Q = 0.4$.

8. Data Analysis

8.1 Introductory Remarks

We should not get stuck in purely theoretical considerations concerning synchronization and desynchronization processes. Rather we should profit from the theory and use it as a basis for designing and evaluating phase resetting experiments. Consequently, this chapter is devoted to the data analysis tools throwing a bridge across the gap between modelling and experiment.

Section 8.2 is fundamental for all techniques presented in this chapter: We shall learn how to extract variables corresponding to cluster phases and cluster amplitudes out of experimental data. This method will be used for different purposes below where there are two main foci.

First, we shall dwell on desynchronization caused by a well-timed pulsatile stimulus. In Sect. 8.3 we shall elaborate the experimental procedure which makes it possible to find suitable parameters for a desynchronizing stimulation. This algorithm is important if we want to desynchronize synchronized pathological activity in patients suffering from movement disorders like the parkinsonian resting tremor. More details concerning the applications to neurological diseases will be presented in Chap. 10.

The second part of this chapter is devoted to MEG and EEG analysis. A variety of typical MEG and EEG experiments address neuronal responses evoked by sensory stimuli (cf. Hämäläinen et al. 1993, Hari and Salmelin 1997). Based on the theoretical results revealed in the former chapters several data analysis methods will be presented which make it possible to investigate stimulation induced neuronal dynamics: Triggered averaging, the standard technique to extract stereotyped neuronal responses to stimuli, is discussed in the context of phase resetting in Sect. 8.4.1. A modified, namely phase dependent version of this method will be presented in Sect. 8.4.2. This method is developed to investigate how the internal state of a particular rhythmically active population of neurons determines its reactions to a stimulus.

In Chap. 6 we saw that a pulsatile stimulus typically induces transient reponses. For this reason the idea suggests itself that coupled clusters of neurons may display stimulus induced responses which are somehow related. One possible relationship is that during certain epochs after the stimulation the phase difference of both clusters undergoes a stereotyped time course. In Sects. 8.4.3–8.4.7 new data analysis techniques will be presented which enable us to detect different types of stimulus locked epochs characterized by stereotyped patterns of the $n : m$ phase difference of two clusters. Favourably, the time resolution of these methods is only restricted by the sampling rate. Epochs of this kind can but do not need to be connected with $n : m$ phase synchronization.

A complementary approach is to detect $n : m$ phase synchronization by means of a single run analysis (Tass et al. 1998) as explained in Sect. 8.4.8. Since this tool uses a sliding window analysis the time resolution is restricted to windows corresponding to about eight periods of an oscillation. In Sect. 8.4.9 we shall compare this single run technique with the methods from Sects. 8.4.3–8.4.7. In this way we shall become aware of how to use both tools in a complementary way, for instance, to study whether there is a stimulus evoked flow of characteristic transient and synchronous epochs pouring out in different groups of brain areas (Sect. 8.4.10). We shall see that this approach is especially powerful if it is directly applied to the cerebral current density obtained with an appropriate inverse method. There are several methods of this kind, e.g., the minimum norm least squares (MNLS) method (Hämäläinen et al. 1993, Wang, Williamson, Kaufman 1995), the cortical current imaging (CCI) (Fuchs et al. 1995), the probabilistic reconstruction of multiple sources (PROMS) (Greenblatt 1993), the low resolution electromagnetic tomography (LORETA) (Pascual-Marqui, Michel, Lehmann 1994), or the magnetic field tomography (MFT) (Clarke and Janday 1989, Clarke, Ioannides, Bolton 1990, Ioannides, Bolton, Clarke 1990). By choosing an inverse algorithm suitable for the concrete experimental paradigm under consideration the spatial resolution will be remarkably enhanced.

The pitfalls in interpreting resetting processes by means of macrovariables, such as electroencephalography (EEG) or magnetoencephalography (MEG) data, are discussed in Sect. 8.4.11.

Finally, it should be mentioned that all data analysis tools explained in this chapter can also be applied to data which are not of neuronal origin. Since rhythmic oscillatory activity abounds in physiology and biology (cf. von Holst 1935, 1939, Winfree 1980, Hildebrandt 1982, 1987, Freund 1983, Glass and Mackey 1988, Steriade, Jones, Llinás 1990, Haken and Koepchen 1991, Schmid-Schönbein and Ziege 1991, Schmid-Schönbein et al. 1992, Perlitz et al. 1995, Hari and Salmelin 1997, Basar 1998a, 1998b) these analysis methods can certainly be used in the context of other physiological issues in a fruitful way.

8.2 Phases and Amplitudes

Let us put ourselves in the experimentalist's position. Applying a stimulus with certain parameters and measuring a signal corresponding to the firing density $p(t)$ defined by (5.95), the experimentalist wants to gain an insight into the cluster's dynamics. To this end Winfree (1980) introduced a discrete phase by means of marker events such as the signal's maxima. Elble and Koller (1990) used this type of phase in the context of tremor resetting experiments. Phase detection based on marker events is suitable for analyzing a single oscillator, in particular, a rigid rotator. But below we shall see that it is not an appropriate method for investigating the phase dynamics of a cluster of oscillators.

8.2.1 Marker Events

Considering a firing pattern as shown in Fig. 6.17 one might suggest that the maxima of $p(t)$ might be suitable marker events. However, since $p(t)$ is a firing density, experimental data are, of course, more noisy, and, thus, they typically do not show clear isolated maxima. For this reason in practice often a burst onset detection is performed. Denoting the timing points of the marker events, i.e. the bursts' onsets, by $t_1, t_2, \ldots$, we can introduce a phase ψ by putting

$$\psi(t_j) = 2\pi(j-1) , \tag{8.1}$$

so that the time intervals $[t_j, t_{j+1}]$ correspond to the oscillation's momentary period where $j = 1, \ldots, N$. In between the timing points of the marker events the phase ψ may be interpolated, e.g., with splines.

In Sect. 6.5.1 we saw that a stimulus can only desynchronize a cluster provided it is administered at or at least very close to a critical initial cluster phase. Accordingly, we need a method which estimates the phase as accurately as possible. For two reasons ψ from (8.1) is not an appropriate variable for the description of the relevant phase dynamics underlying the firing pattern:

1. The phase defined by (8.1) is shaky because it is difficult to find reliable marker events: In Sect. 2.8 we saw that stimuli with terms of higher order typically induce burst splitting, i.e. stimulation induced splitting of the cluster's bursts of collective activity (Fig. 2.26). On the other hand in stimulation experiments burst splitting is observed, for instance, in electromyography (EMG) data which originate from the synchronized activity of motor units controlling muscle activity (Hefter et al. 1992). Our investigation in Sect. 2.8 revealed that as a consequence of burst splitting both maxima and onset detection lead to artifacts and, therefore, do not reflect the dynamics of the ensemble phase appropriately (Fig. 2.26).

2. An even more devastating objection against defining a phase by means of marker events of the cluster's firing pattern is due to the relationship between cluster variables and firing density: All cluster variables contribute to the firing density $p(t)$ according to (6.18), whereas the phase of a particular cluster variable has to be extracted out of $p(t)$ in order to analyze the stimulus' impact on the phase dynamics of this particular mode. To illustrate this, let us consider the time course of the firing density in Fig. 6.17 and the underlying dynamics of the cluster amplitudes in Fig. 6.16. In Chap. 6

we saw that for given intensity parameters (I_1, I_2) one has to ensure that the stimulus is administered at a critical initial value $\varphi_{1,B}^{\text{crit}}$ of the first cluster phase for a critical duration T_{crit}. Otherwise, a partial desynchronization of the first mode cannot be achieved (Fig. 6.35).

To find the suitable parameters $\varphi_{1,B}^{\text{crit}}$ and T_{crit} the experimentalist has to perform a series of experiments in order to derive phase resetting curves which elicit the relationship between $\varphi_{1,B}$ and $\varphi_{1,E}$, that means between the first cluster phase before and after stimulation (cf. Fig. 6.8). This procedure will be explained in Sect. 8.3. In the present context, however, the very point to realize is that the experimentalist needs a tool in order to exclusively analyze the phase of the first cluster variable. Obviously, the marker event based phase detection with (8.1) fails, e.g., during a pronounced early response, when higher order modes dominate the firing and mask the dynamics of the first mode (Fig. 6.17).

8.2.2 Reconstruction of the Modes' Dynamics

It is possible to estimate the time course of the cluster amplitudes and the cluster phases from experimental data with a procedure, where first a *bandpass filtering* is performed to extract the activity of the leading modes. Second, by means of the *Hilbert transform* (see below) amplitude and phase of each mode are determined.

1. *Bandpass filtering:* A spectral analysis of the signal under consideration has to be carried out to reveal the fundamental peak and its harmonics. This is achieved by calculating the Fourier spectrum provided the data are stationary (before the stimulation). For nonstationary data the spectral analysis has to be performed, e.g., with autoregressive methods (cf. Haykin 1986). Bandpass filtering yields the fundamental mode (corresponding to the fundamental peak) and its harmonics (corresponding to the spectral harmonics), e.g., up to third order, where the band edges enclose the respective spectral peaks.

This procedure is illustrated by an example. Fig. 6.6 shows the firing density $p(t)$ before, during and after a stimulation causing a complete partial desynchronization of the first mode. As expected, a spectral analysis of $p(t)$ before the stimulation reveals a fundamental peak at 1 Hz and the first harmonic at 2 Hz, where the unit of the time t is set to be seconds. To visualize the quality of the bandpass filtering let us compare the second cluster variable, viz $\text{Re}(Z_2)$, with its reconstructed signal p_2. $\text{Re}(Z_2)$ is the real part of the second cluster variable directly obtained by the simulation, i.e. by integrating the Fourier transformed model equation (4.29) and taking into account (6.5), the relationship between cluster variables and Fourier modes. $p_2(t)$ is obtained by bandpass filtering $p(t)$ with band edges 1.5 Hz and 2.5 Hz, which enclose the first harmonic in the spectrum of $p(t)$. Figure 8.1 displays the time course of $\text{Re}(Z_2)$ and of p_2.

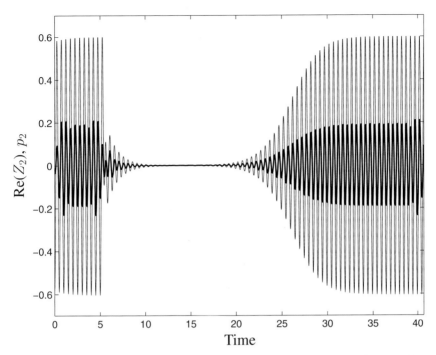

Fig. 8.1. *Reconstruction of the second mode:* The time course of Re(Z_2) and p_2 (*bold line*) is shown for the simulation from Figs. 6.5 and 6.6. Re(Z_2) is the numerically obtained real part of the second cluster variable, whereas p_2 is its reconstructed signal obtained via bandpass filtering $p(t)$ with band edges 1.5 Hz and 2.5 Hz.

Concerning the amplitude dynamics there is a good agreement between the reconstructed signals p_2 and the numerical signal Re(Z_2) except for a constant factor. This is not surprising since the cluster variables Z_j are subjected to a normalization (cf. (6.5)). With respect to the phase dynamics the reconstructed signal p_2 nicely agrees with the numerical signal Re(Z_2). This can be seen by comparing the timing points of the signals' corresponding maxima. The results of the reconstruction are invariant with respect to variations of the band edges (e.g., ±0.2 Hz on both band edges).

2. *Hilbert transform:* The *instantaneous* phase $\varphi(t)$ and amplitude $R(t)$ of an arbitrary signal $s(t)$ can be defined in a consistent and unique way based on the analytic signal approach (Panter 1965). To this end one has to determine the analytic signal $\zeta(t)$ which is defined by

$$\zeta(t) = s(t) + is_{\mathrm{H}}(t) = A(t)\exp[i\varphi(t)]\,, \tag{8.2}$$

where i is the complex unit, and $s_{\mathrm{H}}(t)$ is the *Hilbert transform* of $s(t)$ given by

$$s_{\mathrm{H}}(t) = \pi^{-1}\mathrm{P.V.}\int_{-\infty}^{\infty}\frac{s(\tau)}{t-\tau}d\tau \tag{8.3}$$

(where P.V. means that the integral is taken in the sense of the Cauchy principal value). In practice, the Hilbert transform is realized, e.g., by means of a filter which optimally approximates an ideal filter whose amplitude response is unity and whose phase response is a constant phase shift of $\pi/2$ for all frequencies (Panter 1965). Such filters are designed, e.g., by means of the Parks–McClellan algorithm (cf. Parks and Burrus 1987).

In several studies the phase determination by means of the Hilbert transform was used for the investigation of physiological data, for instance, for the analysis of tracking movement patterns (Tass et al. 1996), of human posture control (Rosenblum et al. 1998), of the interaction between heartbeat and ventilation (Schäfer et al. 1998), and for the detection of $n : m$ phase synchronization in MEG data (Tass et al. 1998).

In our phase resetting approach we apply the Hilbert transform to each of the leading modes $s_1, s_2, s_3, \ldots$ obtained by bandpass filtering of the measured signal $s(t)$ in the follwing way. For the fundamental mode, s_1, and the first and higher harmonics, $s_2, s_3, \ldots$, the analytic signal ζ_j is determined according to (8.2):

$$\zeta_j(t) = s_j(t) + is_{j,\mathrm{H}}(t) = A_j(t)\exp[i\varphi_j(t)]\,, \qquad (8.4)$$

where $s_{j,\mathrm{H}}$ is the Hilbert transform of s_j. In this way we obtain the phase φ_1 and the amplitude A_1 of the fundamental mode and the phases $\varphi_2, \varphi_3, \ldots$ and amplitudes $A_2, A_3, \ldots$ of the first and higher harmonics.

It should be emphasized that stimulation induced transient dynamics can be analyzed with the Hilbert transform because the latter does not require stationarity of the data (cf. Rosenblum and Kurths 1998). Correspondingly, the Hilbert transform was used, e.g., in mechanical engineering to identify elastic and damping properties of vibrating systems (Feldman 1985, 1994, Feldman and Rosenblum 1988).

8.2.3 Slaving Principle and Transients

We should be aware of the fact that the procedure presented in the former section extracts modes out of measured signals no matter whether they are order parameters or enslaved modes. Both sorts of modes are relevant in the context of phase resetting. On the one hand we have to analyze the stimulation induced phase dynamics of the order parameters to find suitable parameters for a desynchronizing stimulation. On the other hand we are interested in the higher order modes, for instance, to study whether or not a pronounced early response occurs. The latter has to be detected, so that its functional role and its physiological meaning can be analyzed.

There are other data analysis methods aiming exclusively at reconstructing the dynamics of order parameters. Along the lines of a top-down approach and based on Haken's (1977, 1983) order parameter concept spontaneously emerging spatio-temporal dynamics during a seizure of petit-mal epilepsy (Friedrich and Uhl 1992, Friedrich, Fuchs, Haken 1992) and during an α-EEG (Fuchs et al. 1987) were analyzed by determining phenomenological order parameter equations. In this case the order parameters are a few spatio-temporal modes contributing the major part of the measured signal.

A similar approach was suggested by Uhl et al. (1998) for the investigation of event related potentials which can be considered as stimulation induced transient MEG or EEG data. However, for *transient* dynamics the results in this book clearly show the dubiousness of decomposing a signal into certain modes in whatever way in order to define those few modes as order parameters which contribute to the signal the most, for instance, more than 90 %. Our results point out that a stimulus may excite enslaved modes, so that during stimulation induced transients they contribute to the signal to a relevant degree (cf. Sect. 6.3). For instance, during a pronounced early response the cluster's firing behavior is *quantitatively* dominated by enslaved modes although the order parameters govern the cluster's dynamics according to the slaving principle (Figs. 6.16 and 6.17). Hence, while a phenomenological reconstruction of an order parameter dynamics is an important data analysis tool for spontaneously emerging stationary dynamics, it is at best very problematic in the case of a stimulation induced transient dynamics.

8.3 Tracking Down the Black Holes

As explained in Chap. 6 a well-timed stimulus of a suitable intensity induces a partial desynchronization of the jth mode. That means that the stimulus has to be administered at a critical initial cluster phase $\varphi_{j,\mathrm{B}}^{\mathrm{crit}}$ for a critical duration T_{crit}. Moreover the stimulus' intensity parameters, $\{I_k\}$ and $\{\gamma_k\}$ from (6.20), have to be appropriate. The procedure the experimentalist has to perform to derive these critical parameters was explained in detail by Winfree (1980). For the analysis of the phase before and after the stimulation Winfree (1980) used a phase detection based on marker events. As explained in detail in Sect. 8.2.2 the signals we are interested in force us to detect the phase of a relevant mode by using *bandpass filtering* and *Hilbert transform*.

To find suitable parameters for a stimulation which induces a partial desynchronization of the jth mode the experimentalist has to proceed in the following way. A series of stimulation experiments is carried out, where the stimulus' intensity parameters, $\{I_k\}$ and $\{\gamma_k\}$, are kept constant. Obviously, the intensity has to be high enough to efficiently affect the cluster. For a fixed value of the stimulation duration T we perform a series of stimulations with the initial cluster phase $\varphi_{j,\mathrm{B}}$ equally spaced in $[0, 2\pi]$. To evaluate the stimulus' effect on the phase of the jth mode we determine the corresponding phase resetting curve, where $\varphi_{j,\mathrm{E}}$ is plotted over $\varphi_{j,\mathrm{B}}$ (cf. Sect. 6.2.3). $\varphi_{j,\mathrm{E}}$ is the phase of the jth mode at the end of the stimulation, i.e. $\varphi_{j,\mathrm{E}} = \varphi_j(t_{\mathrm{E}})$. Type 0 and type 1 resetting are connected with a winding number 0 and 1 of the phase resetting curve, respectively. For the next series of stimulations with equally spaced $\varphi_{j,\mathrm{B}}$ we choose a larger or smaller value of the duration

T if the winding number of the first series is 1 or 0, respectively. Also in the second series T is kept constant. In this way we iteratively run several series of stimulations adjusting the duration T to the type of resetting of the preceding series until, finally, we reveal the critical duration T_{crit}. The latter is connected with a transition from type 1 to type 0 resetting as shown, e.g., in Fig. 6.8.

The winding number of the phase resetting curves is invariant against a transformation $\varphi_{j,\text{E}} \rightarrow \varphi_{j,\text{E}} + \Delta\varphi$. Therefore, if it is not possible to evaluate φ_j directly at the end of the stimulation due to the experimental setup, we can evaluate φ_j after a constant time Δt after the end of the stimulation. Hence, in the phase resetting curve we simply replace $\varphi_{j,\text{E}}$ by $\varphi_j(t_{\text{E}}+\Delta t)$. Certainly, this is only allowed provided the oscillators' frequencies are sufficiently stationary after the stimulation.

In the same way we can fix the stimulation duration T and perform several series of stimulations where from series to series we adjust the stimulation intensity according to the type of resetting of the preceding series (cf. Winfree 1980). No matter how we iteratively approach the critical stimulation parameters we have to be aware of the fact that this procedure requires the stationarity of the oscillators' frequencies before and after the stimulation, at least to a certain extent. Otherwise, the phase resetting curves cannot be evaluated in a reliable way.

Comparing plots 3 and 4 in Fig. 6.8 it immediately turns out that the numbers of simulations within each series has to be large enough, so that $\varphi_{j,\text{B}}$ is sufficiently densely spaced in $[0, 2\pi]$. This is necessary because for the distinction between type 0 and type 1 resetting we have to resolve the critical region in the phase resetting curve, where $\varphi_{j,\text{B}} \approx \varphi_{j,\text{B}}^{\text{crit}}$, with enough accuracy. Thus, in practice we have to make a compromise between a high resolution of the phase resetting curves on the one hand and the above mentioned stationarity requirement: High resolution requires a large number of experiments, whereas stationarity is most likely fulfilled provided the experiment is as short as possible.

8.4 MEG and EEG Analysis

The electromagnetic field generated by synchronously active neurons in the human brain can be noninvasively assessed outside the head: the electric field by means of electroencephalography (EEG) (Berger 1929, cf. Freeman 1975, Cooper, Osselton, Shaw 1984, Niedermeyer and Lopes da Silva 1987, Jansen and Brandt 1993, Nunez 1995) and the magnetic field by means of magnetoencephalography (MEG) (Cohen 1972, cf. Ahonen et al. 1991, Hämäläinen et al. 1993). MEG and EEG signals are noisy. Hence, one needs strategies which make it possible to get rid of the noise, to extract the physiologically relevant features and to avoid artifacts.

8.4.1 Triggered Averaging

The decomposition of a signal into the amplitudes and the phases of its modes presented in Sect. 8.2.2 can also be used in the context of averaging. The latter is a standard technique for improving the signal-to-noise ratio of noisy data, such as EEG or MEG data (Hämäläinen et al. 1993). Averaging can only be used for repetitive phenomena such as stimulation induced *evoked responses*, where, e.g., an auditory stimulus like a sound is repeatedly administered while the MEG or EEG signals are registered. The assumption behind the triggered averaging is that the brain's reaction to the stimulus is stereotyped and to a certain extent masked by noise. Accordingly, an MEG or EEG signal $s(t)$ recorded, for instance, over the auditory cortex is assumed to be of the form

$$s(t) = e(t) + \xi(t) \, , \qquad (8.5)$$

where $e(t)$ is the stereotyped evoked response, and $\xi(t)$ is additive random Gaussian noise (cf. Hämäläinen et al. 1993). Let us introduce some notations: τ_k is the timing point of the onset of the kth stimulus. We evaluate $s(t)$ within windows attached to the stimuli. The windows are defined by $[\tau_k - t_-, \tau_k + t_+]$, where both t_- and t_+ are positive, so that each window encloses the corresponding stimulus. N_a denotes the number of averages. With that the averaged signal takes the form

$$\bar{S}(t) = \frac{1}{N_a} \sum_{k=1}^{N_a} s(t + \tau_k) \, , \qquad (8.6)$$

where $t \in [-t_-, +t_+]$. Averaging causes an improvement of the signal-to-noise ratio of $\sqrt{N_a}$, where N_a typically equals 20-300 (Hämäläinen et al. 1993). Hence, in the ideal case of perfectly replicable signals averaging extracts the stereotyped evoked response: $\bar{S}(t) \to e(t)$ for $N_a \to \infty$. Real signals originating from human brains are certainly not perfectly replicable, for example, as a consequence of fatigue. The latter takes place on a slow time scale compared to the signals under consideration, and, thus, in practice one has to make a compromise to choose an appropriate number of averages: On the one hand N_a has to be large enough so that the noise $\xi(t)$ is cancelled out and the stereotyped evoked response $e(t)$ becomes unmasked. On the other hand N_a must not be chosen too large since otherwise, e.g., long-term modulations may spoil the extraction of $e(t)$.

8.4.2 Phase Dependent Triggered Averaging

Aware of the results in this book one may question whether the crucial assumption (8.5) is valid in general. In terms of phase resetting (8.5) means that the stimulus causes the same reaction no matter at what initial phase it is administered. In other words, the stimulus gives rise to a type 0 resetting with a perfectly horizontal phase resetting curve as, e.g., shown in plots no. 6 of Figs. 6.8 and 6.11. Strictly speaking the averaging procedure (8.6) requires this strong version of type 0 resetting.

In contrast to the conventional averaging procedure (8.6) we can apply a modified approach, which addresses the issue as to whether the evoked response depends on the dynamical state of the neuronal cluster in the brain generating this particular electromagnetic field. Put otherwise, rather than being a reflex-like reaction the evoked response might depend on the intitial phase at which a stimulus is administered. Since power spectra of EEG and MEG signals contain more than one peak we have to use the procedure explained in Sect. 8.2.2 to extract the phase of that mode which is relevant in the concrete physiological context. To this end we have to filter each EEG or MEG signal $s(t)$ with one appropriate bandpass filter, in this way obtaining $s_j(t)$, where the subscript j refers to the relevant mode, i.e. the relevant peak in the power spectrum of $s(t)$. By means of the Hilbert transform we extract the phase $\varphi_j(t)$ of $s_j(t)$. Let us subdivide $[0, 2\pi]$ in N_B bins B_l of equal width $\Delta\phi$, where $l = 1, \ldots, N_B$. The lth bin is a phase interval defined by $B_l = [\phi_l - \Delta\phi/2, \phi_l + \Delta\phi/2]$. Considering $\varphi_j(t)$ modulo 2π we can, thus, assign the initial phase $\varphi_{j,B}$ of each response to a certain bin. This enables us to selectively average all responses for which $\varphi_{j,B}$ is contained in a particular bin. Our theoretical results clearly indicate that the bin width $\Delta\phi$ has to be small enough because, for example, desynchronizing effects only occur provided $\varphi_{j,B}$ is close to a critical value $\varphi_{j,B}^{\mathrm{crit}}$ (cf. Sect. 6.5).

Phase dependent triggered averaging means that for each bin we average the responses seperately. Hence, denoting the number of responses belonging to the bin B_l as N_l, for this particular bin we obtain the phase dependent average

$$\bar{s}(t; l) = \frac{1}{N_l} \sum_{k(l)} s(t + \tau_k) \,, \tag{8.7}$$

where $t \in [-t_-, +t_+]$, and $\sum_{k(l)}$ denotes the summation over all responses with $\varphi_{j,B}$ contained in B_l. Assuming that for each bin the same number of responses is averaged, i.e.

$$N_l = N_k \quad (k, l = 1, \ldots, N_B) \,, \tag{8.8}$$

we obtain

$$N_a = N_l N_B \tag{8.9}$$

for all $l = 1, \ldots, N_B$. Comparing (8.6) with (8.7) and using (8.9) immediately yields the relationship between the phase dependent triggered averaging and the conventional triggered averaging:

$$\bar{S}(t) = \frac{1}{N_B} \sum_{l=1}^{N_B} \bar{s}(t; l) \,, \tag{8.10}$$

where $t \in [-t_-, +t_+]$. The conventional averaging procedure, thus, averages over all initial phases at which the stimulus is administered. That is why phase dependent features of stimulation induced responses can be missed. If (8.8) is not fulfilled, for instance, because the majority of responses belongs to a particular bin B_k, the conventional triggered average is dominated by the responses of B_k.

8.4.3 Stimulus Locked $n : m$ Transients

This and the following four sections are dedicated to the detection of stimulus locked epochs displaying particular patterns of the $n : m$ phase difference. Let us first clarify some expressions in order not to become confused by the terminology. The word locking has several meanings. In physics locking is a shortform for *frequency locking* or $n : m$ *phase locking* where $n : m$ phase locking is a synonym for $n : m$ *phase synchronization*. Both frequency and phase locking will be defined in detail in this and, especially, in Sect. 8.4.8. For the time being it is important that apart from these two meanings in the field of neuroscience the term locking is additionally used in another sense: An event is *stimulus locked* provided this event occurs with a certain and fixed delay after the stimulus onset. So, in this case locking means that between two events there is a particular relationship in time. In this chapter we shall speak of stimulus locked events, activity etc in the sense of this definition.

Whether or not a signal is stimulus locked crucially depends on the signal's phase dynamics. Considering, e.g., (8.6) we easily realize that a nonvanishing average $\bar{S}(t)$ does only occur provided the time course of the amplitude and, in particular, the phase dynamics of the responses is tightly stimulus locked. Likewise one obtains a nonvanishing phase dependent average $\bar{s}(t; \phi_l)$ only provided the responses' phase dynamics is tightly stimulus locked (cf. (8.7)).

In animal experiments it was shown that a visual stimulus induces two types of dynamical patterns observed in local field potentials (LFP): (a) a visual evoked cortical potential (VECP) occurring after about 50 msec, and (b) oscillatory activity in the γ-band range (> 30 Hz) observed after approximately 500 msec (Singer 1989, Eckhorn et al. 1990). Triggered averaging extracts the VECP indicating that the VECP is tightly stimulus locked (cf. Sect. 8.4.1). In contrast, the oscillatory γ-band responses are typically averaged out, since they are not tightly stimulus locked (Singer 1989, Eckhorn et al. 1990).

One typically observes phase-locked γ-band oscillations in different parts of the visual cortices, where the in-phase synchronization of these oscillations appears to be the binding mechanism which combines seperate visual features belonging to the same visual pattern (Gray and Singer 1987, 1989, Eckhorn et al. 1988). For more details I refer to the discussion of the binding problem in Sect. 3.2.

This issue illustrates that we need a tool for the analysis of bivariate data which enables us to detect tightly stimulus locked epochs of $n : m$ phase synchronized activity in MEG or EEG data. The idea behind this approach is that an appropriate stimulus might induce neuronal activity displaying stereotyped synchronization patterns while the time course of the single responses does not need to be stereotyped at all. In other words, the time course of the $n : m$ phase difference between certain MEG/EEG signals or between the latter and peripheral signals (e.g., EMG signals) may be tightly stimulus locked, while, in contrast, there does not need to be a tight relationship between the time course of the single signals and the stimulus onset.

How might the phase dynamics of both signals during these short-term epochs look like? From the outset we cannot expect that a stimulus induces perfectly synchronized activity, connected with a plateau of the $n : m$ phase difference. On the contrary, the simulations shown especially in Chap. 6 underline that a stimulus typically induces transient dynamics before a stationary state reappears. Hence, if a single cluster exhibits transient responses to a stimulus one might also expect that interacting clusters undergo stimulation induced transient dynamics. Such transients may be rather short, for example, of a length corresponding to, say, one or two periods of an oscillation. Accordingly, it may be rather difficult to decide whether or not $n : m$ phase synchronization can be observed.

Therefore in this section we choose a different approach. With this aim in view it is appropriate to additionally introduce the expression *stimulus locked* $n : m$ *transient*. The latter is defined as a stimulus locked epoch exhibiting a stereotyped time course of the $n : m$ phase difference of both signals under consideration. In other words, during a time interval following a stimulus with a fixed delay the $n : m$ phase difference (modulo 2π) between the two signals displays always the same time course except for small or moderate variations caused by noise. Note that this stereotyped time course need not be a horizontal plateau corresponding to perfect $n : m$ phase synchronization. Consequently, the term stimulus locked $n : m$ transient is not restricted to $n : m$ phase synchronized activity. A stimulus locked epoch of $n : m$ phase synchronized activity is just a special case of a stimulus locked $n : m$ transient.

To detect such transients we use the phase extraction method explained in Sect. 8.2.2: The MEG/EEG signals of all channels are filtered with the same bandpass, where the latter is relevant in the concrete physiological context. I sketch the analysis approach for one channel, where $s_j(t)$ denotes the bandpass filtered signal of this particular channel, and the subscript j refers to the mode, i.e. the bandpass, e.g., the γ-band range (cf. Sect. 8.2.2). To investigate synchronization processes we choose a reference signal denoted by $s_r(t)$, which is a suitably bandpass filtered MEG/EEG signal or a bandpass filtered peripheral signal, such as an EMG signal. The bandpass of $s_r(t)$ need not be the same as that of $s_j(t)$. In this way we can study $n : m$ synchronization processes between an MEG reference channel and all other MEG channels in order to reveal stimulation induced $n : m$ phase synchronization between the activity in different brain areas.

By means of the Hilbert transform we prepare the analytic signal of both signals as defined by (8.4). This means that we decompose the signals into their instantaneous real amplitudes, $A_j(t)$, $A_r(t)$, and real phases, $\varphi_j(t)$, $\varphi_r(t)$, according to

$$\zeta_j(t) = s_j(t) + is_{j,\mathrm{H}}(t) = A_j(t)\exp[i\varphi_j(t)] \ , \tag{8.11}$$

$$\zeta_r(t) = s_r(t) + is_{r,\mathrm{H}}(t) = A_r(t)\exp[i\varphi_r(t)] \tag{8.12}$$

(cf. Sect. 8.2.2). To extract that part of $s_j(t)$ which is $n : m$ phase locked to the reference channel we introduce the $n : m$ phase difference between $s_j(t)$ and $s_r(t)$ by putting

$$\phi_{n,m}^{(j,r)}(t) = n\varphi_j(t) - m\varphi_r(t) \tag{8.13}$$

and we prepare the signal

$$\zeta_{j,r}^{(n,m)}(t) = A_j(t)\exp\left\{i\left[\phi_{n,m}^{(j,r)}(t)\right]\right\} \ , \tag{8.14}$$

which reflects both the time course of the amplitude $A_j(t)$ and of the $n : m$ phase difference $n\varphi_j(t) - m\varphi_r(t)$. In this way we want to detect coordinated dynamical epochs during which the ratio of the frequencies of $s_r(t)$ versus $s_j(t)$ is approximately equal to n/m.

In the spirit of Sect. 8.4.2 we perform a phase dependent triggered averaging. To this end we divide $[0, 2\pi]$ in N_B equally wide bins $B_l = [\phi_l - \Delta\phi/2, \phi_l + \Delta\phi/2]$ with width $\Delta\phi$, where $l = 1, \ldots, N_\mathrm{B}$. Similar to Sect. 8.4.2 we assign the initial phase $\varphi_{j,\mathrm{B}}$ modulo 2π of each response to the corresponding bin. In this way we can selectively average all responses for which $\varphi_{j,\mathrm{B}}$ belongs to a particular bin. Accordingly, the stimulation induced activity of the MEG/EEG channel under consideration which displays a stereotyped response concerning the $n : m$ phase relation to the reference channel is obtained by calculating

$$\bar\zeta_{j,r}^{(n,m)}(t; l) = \frac{1}{N_l}\sum_{k(l)}\zeta_{j,r}^{(n,m)}(t + \tau_k) \ , \tag{8.15}$$

where $t \in [-t_-, +t_+]$, the latter denoting the time window enclosing the stimulus onset. N_l is the number of responses corresponding to the bin B_l. Similar to (8.7) $\bar\zeta_{j,r}^{(n,m)}(t; l)$ averages all responses with $\varphi_{j,\mathrm{B}}$ belonging to the bin B_l. It is necessary to determine the phase dependent average (8.15) because the initial phase of a cluster might not only influence the cluster's response. It might, moreover, influence the interaction of this particular cluster with other neuronal populations.

In contrast to (8.15) a triggered averaging which is not phase dependent is achieved by calculating

$$\bar{X}_{j,\mathrm{r}}^{(n,m)}(t) = \frac{1}{N_\mathrm{a}} \sum_{k=1}^{N_\mathrm{a}} \zeta_{j,\mathrm{r}}^{(n,m)}(t + \tau_k) \,, \tag{8.16}$$

where $t \in [-t_-, +t_+]$, and N_a denotes the number of all averaged responses.

If we assume that for each bin the same number of responses is averaged, (8.8) and (8.9) hold, and we obtain

$$\bar{X}_{j,\mathrm{r}}^{(n,m)}(t) = \frac{1}{N_\mathrm{B}} \sum_{l=1}^{N_\mathrm{B}} \bar{\zeta}_{j,\mathrm{r}}^{(n,m)}(t; l) \,, \tag{8.17}$$

where $t \in [-t_-, +t_+]$.

We apply the Hilbert transform to investigate the amplitude and phase dynamics of the $n : m$ phase locked responses by determining the amplitudes and phases of (8.15) and (8.16) which yields

$$\bar{\zeta}_{j,\mathrm{r}}^{(n,m)}(t; l) = \bar{a}_{j,\mathrm{r}}^{(n,m)}(t; l) \exp\left[\mathrm{i}\bar{\phi}_{j,\mathrm{r}}^{(n,m)}(t; l)\right] \tag{8.18}$$

and

$$\bar{X}_{j,\mathrm{r}}^{(n,m)}(t) = \bar{A}_{j,\mathrm{r}}^{(n,m)}(t) \exp\left[\mathrm{i}\bar{\Phi}_{j,\mathrm{r}}^{(n,m)}(t)\right] \,, \tag{8.19}$$

where $t \in [-t_-, +t_+]$. $\bar{a}_{j,\mathrm{r}}^{(n,m)}(t; l)$ and $\bar{A}_{j,\mathrm{r}}^{(n,m)}(t)$ provide us with the time course of the real amplitudes of the averaged $n : m$ locked responses (8.15) and (8.16), whereas $\bar{\phi}_{j,\mathrm{r}}^{(n,m)}(t; l)$ and $\bar{\Phi}_{j,\mathrm{r}}^{(n,m)}(t)$ are the corresponding time dependent averaged $n : m$ phase differences. It is important to distinguish between the notations of the different types of $n : m$ phase differences. As indicated by the bars $\bar{\phi}_{j,\mathrm{r}}^{(n,m)}(t; l)$ and $\bar{\Phi}_{j,\mathrm{r}}^{(n,m)}(t)$ refer to the averaged responses according to (8.18) and (8.19), whereas $\phi_{n,m}^{(j,r)}(t)$ from (8.13) denotes the $n : m$ phase difference between the nonaveraged signals $s_j(t)$ and $s_\mathrm{r}(t)$.

Nonvanishing amplitudes $\bar{a}_{j,\mathrm{r}}^{(n,m)}(t; l)$ and $\bar{A}_{j,\mathrm{r}}^{(n,m)}(t)$ indicate that during that particular period of time the stimulation induced activity of the MEG/EEG channel under consideration undergoes a stereotyped time course of the $n : m$ phase realtionship with the reference signal. Moreover, $\bar{\phi}_{j,\mathrm{r}}^{(n,m)}(t; l)$ and $\bar{\Phi}_{j,\mathrm{r}}^{(n,m)}(t)$ allow us to characterize this phase relationship more precisely. In this way we can show that the activity of this particular channel is $n : m$ phase locked to the reference signal during that epoch. In this case $\bar{\phi}_{j,\mathrm{r}}^{(n,m)}(t; l)$ and $\bar{\Phi}_{j,\mathrm{r}}^{(n,m)}(t)$ will display a horizontal plateau. In more complicated cases one might expect that $\bar{\phi}_{j,\mathrm{r}}^{(n,m)}(t; l)$ and $\bar{\Phi}_{j,\mathrm{r}}^{(n,m)}(t)$ switch between different preferred values in a stereotyped way so that a characteristic sequence of different plateaus is observed.

In addition to (8.14) we can also introduce a normalized variable according to

$$\eta_{j,\mathrm{r}}^{(n,m)}(t) = \exp\left\{\mathrm{i}\left[\phi_{n,m}^{(j,\mathrm{r})}(t)\right]\right\} \,, \tag{8.20}$$

where $\phi_{n,m}^{(j,r)}(t)$ is the $n : m$ phase difference between $s_j(t)$ and $s_r(t)$ defined by (8.13). This means that compared to $\zeta_{j,r}^{(n,m)}(t)$ defined by (8.14) one simply drops the amplitude $A_j(t)$. Such a signal is more appropriate if one is exclusively interested in the locking behavior while the amplitude $A_j(t)$ is less important. Let us briefly dwell on this point since it is important to be aware of the difference between (8.14) and (8.20).

The magnetic and the electric field of the brain measured with MEG and EEG are produced by the cerebral neurons, in particular, in large part by synaptic current flow (Hämäläinen et al. 1993). At least a million synapses have to be synchronously active to produce a typical evoked response as detected by means of the conventional averaging (8.6) (Hämäläinen et al. 1993). In a complex way the amplitude of such a response depends on several factors like the number of synchronously active neurons, the strength of the neuronal synchronization, the arrangement of the neuronal population etc. Using the normalized signal (8.20) makes it possible to focus on the synchronization between different areas and to counteract possibly masking effects of the amplitudes. Obviously, both the non-normalized signal (8.14) as well as the normalized signal (8.20) provide us with complementary information.

A phase dependent triggered averaging for (8.20) can be performed in a similar way as for the non-normalized signals by computing

$$\bar{\eta}_{j,r}^{(n,m)}(t;l) = \frac{1}{N_l} \sum_{k(l)} \eta_{j,r}^{(n,m)}(t + \tau_k) , \tag{8.21}$$

where $t \in [-t_-, +t_+]$ (cf. (8.15)). Correspondingly, a triggered averaging which is not phase dependent reads

$$\bar{Y}_{j,r}^{(n,m)}(t) = \frac{1}{N_a} \sum_{k=1}^{N_a} \eta_{j,r}^{(n,m)}(t + \tau_k) \tag{8.22}$$

with $t \in [-t_-, +t_+]$ and N_a denoting the number of all averaged responses (cf. (8.16)). Assuming that for each bin the same number of responses is averaged, (8.8) and (8.9) hold. This yields

$$\bar{Y}_{j,r}^{(n,m)}(t) = \frac{1}{N_B} \sum_{l=1}^{N_B} \bar{\eta}_{j,r}^{(n,m)}(t;l) , \tag{8.23}$$

where $t \in [-t_-, +t_+]$ (cf. (8.17)).

We apply the Hilbert transform to investigate the amplitude and phase dynamics of the $n : m$ phase locked responses by determining the amplitudes and phases of (8.15) and (8.16) which yields

$$\bar{\eta}_{j,r}^{(n,m)}(t;l) = \bar{b}_{j,r}^{(n,m)}(t;l) \exp\left[i\bar{\psi}_{j,r}^{(n,m)}(t;l) \right] \tag{8.24}$$

and

$$\bar{Y}_{j,r}^{(n,m)}(t) = \bar{B}_{j,r}^{(n,m)}(t) \exp\left[i\bar{\Psi}_{j,r}^{(n,m)}(t)\right] , \qquad (8.25)$$

where $t \in [-t_-, +t_+]$ (cf. (8.18) and (8.19)). Due to the normalization achieved by (8.20) the properties

$$0 \le \bar{b}_{j,r}^{(n,m)}(t;l) \le 1 , \quad 0 \le \bar{B}_{j,r}^{(n,m)}(t) \le 1 \qquad (8.26)$$

hold for all times $t \in [-t_-, +t_+]$. In this way we can easily estimate the strength of the stimulus locking of the $n : m$ phase synchronized activity. Vanishing $\bar{b}_{j,r}^{(n,m)}(t;l)$ and $\bar{B}_{j,r}^{(n,m)}(t)$ mean that at time t there is no stimulus locked $n : m$ phase synchronized activity. On the contrary $\bar{b}_{j,r}^{(n,m)}(t;l) = 1$ and $\bar{B}_{j,r}^{(n,m)}(t) = 1$ only occur if at time t the $n : m$ phase difference between both signals is perfectly stimulus locked. By comparing $\bar{b}_{j,r}^{(n,m)}(t;l)$ with $\bar{B}_{j,r}^{(n,m)}(t)$ we can study whether certain stimulus locked $n : m$ transients preferentially or exclusively occur provided the stimulus is administered at certain initial phases $\varphi_{j,B}$ of the signal s_j.

$\bar{\psi}_{j,r}^{(n,m)}(t;l)$ and $\bar{\Psi}_{j,r}^{(n,m)}(t)$ provide us with the time course of the $n : m$ phase difference during the stimulus locked epochs. A plateau of $\bar{\psi}_{j,r}^{(n,m)}(t;l)$ and $\bar{\Psi}_{j,r}^{(n,m)}(t)$ would correspond to a perfectly $n : m$ phase locked epoch. Of course, other patterns of phase differences might also be observed, for example, a stereotyped transition between two $n : m$ phase locked states which would correspond to two plateaus following one another.

We can avoid a spurious detection of stimulus locked $n : m$ transients which may be due to noise or bandpass filtering by deriving time independent significance levels $\bar{a}_{j,r,S}^{(n,m)}(l)$, $\bar{A}_{j,r,S}^{(n,m)}$ $\bar{b}_{j,r,S}^{(n,m)}(l)$, and $\bar{B}_{j,r,S}^{(n,m)}$ of the time dependent amplitudes $\bar{a}_{j,r}^{(n,m)}(t;l)$, $\bar{A}_{j,r}^{(n,m)}(t)$, $\bar{b}_{j,r}^{(n,m)}(t;l)$, and $\bar{B}_{j,r}^{(n,m)}(t)$, respectively. For this purpose, we simply apply the same analysis as described in this section to surrogates which are, e.g., white noise signals filtered exactly in the same way as the original signals. Such surrogates turned out to be appropriate in a similar context as explained in the following section. The 95th percentile of the distribution of the amplitudes of the surrogates serves as significance levels denoted by $\bar{a}_{j,r,S}^{(n,m)}(l)$, $\bar{A}_{j,r,S}^{(n,m)}$, $\bar{b}_{j,r,S}^{(n,m)}(l)$, and $\bar{B}_{j,r,S}^{(n,m)}$, respectively. To take into account only relevant values of the amplitudes we introduce the significant amplitudes by setting

$$\bar{r}_{j,r}^{(n,m)}(t;l) = \max\left\{\bar{a}_{j,r}^{(n,m)}(t;l) - \bar{a}_{j,r,S}^{(n,m)}(l), 0\right\} , \qquad (8.27)$$

$$\bar{R}_{j,r}^{(n,m)}(t) = \max\left\{\bar{A}_{j,r}^{(n,m)}(t) - \bar{A}_{j,r,S}^{(n,m)}, 0\right\} , \qquad (8.28)$$

$$\bar{e}_{j,r}^{(n,m)}(t;l) = \max\left\{\bar{b}_{j,r}^{(n,m)}(t;l) - \bar{b}_{j,r,S}^{(n,m)}(l), 0\right\} , \qquad (8.29)$$

$$\bar{E}_{j,r}^{(n,m)}(t) = \max\left\{\bar{B}_{j,r}^{(n,m)}(t) - \bar{B}_{j,r,S}^{(n,m)}, 0\right\} , \qquad (8.30)$$

respectively. $\bar{a}_{j,\mathrm{r}}^{(n,m)}(t;l)$ and $\bar{b}_{j,\mathrm{r}}^{(n,m)}(t;l)$ may be called *phase dependent* $n:m$ *stimulus locking indices*, whereas $\bar{A}_{j,\mathrm{r}}^{(n,m)}(t;l)$ and $\bar{B}_{j,\mathrm{r}}^{(n,m)}(t;l)$ may be called $n:m$ *stimulus locking indices*. Correspondingly, we may denote $\bar{r}_{j,\mathrm{r}}^{(n,m)}(t;l)$ and $\bar{e}_{j,\mathrm{r}}^{(n,m)}(t;l)$ as *significant phase dependent* $n:m$ *stimulus locking indices*, and $\bar{R}_{j,\mathrm{r}}^{(n,m)}(t;l)$ as well as $\bar{E}_{j,\mathrm{r}}^{(n,m)}(t;l)$ as *significant* $n:m$ *stimulus locking indices*.

In summary, by means of this averaging procedure stimulus locked $n:m$ transients are analyzed in the following way:

1. *Occurrence of* $n:m$ *stimulus locked transients:* We have to decide whether and at what times these epochs occur. To this end we have to detect the periods with nonvanishing significant amplitudes defined by (8.27) to (8.30). $\bar{r}_{j,\mathrm{r}}^{(n,m)}(t;l)$ and $\bar{R}_{j,\mathrm{r}}^{(n,m)}(t)$ take into account the amplitude of s_j, the particular signal under consideration, whereas the normalized amplitudes $\bar{e}_{j,\mathrm{r}}^{(n,m)}(t;l)$ and $\bar{E}_{j,\mathrm{r}}^{(n,m)}(t)$ only account for stimulus locked patterns of the $n:m$ phase difference between the signal s_j and the reference signal s_r. Depending on the concrete experimental issue the normalized or the non-normalized amplitudes may be more appropriate, and often they may provide us with complementary information.

2. *Phase dependence of the stimulation's outcome:* A stimulus locked $n:m$ transient may only occur provided the stimulus is applied for certain initial phases. This can straightforwardly be estimated by comparing the amplitudes of the phase dependent averages $\bar{r}_{j,\mathrm{r}}^{(n,m)}(t;l)$ and $\bar{e}_{j,\mathrm{r}}^{(n,m)}(t;l)$ with the amplitudes of the averages which are not phase dependent, namely $\bar{R}_{j,\mathrm{r}}^{(n,m)}(t)$ and $\bar{E}_{j,\mathrm{r}}^{(n,m)}(t)$. The stimulus locked $n:m$ transient might be particularly strong provided the stimulus is administered at an initial phase $\varphi_{j,\mathrm{B}}$ (modulo 2π) lying within a particular bin $B_{\tilde{l}}$. In this case the corresponding phase dependent averages $\bar{r}_{j,\mathrm{r}}^{(n,m)}(t;\tilde{l})$ and $\bar{e}_{j,\mathrm{r}}^{(n,m)}(t;\tilde{l})$ would be largest compared to those of the other bins. Apart from such a gradual amplitude effect the whole time course of the amplitudes may qualitatively depend on the stimulation's initial phase $\varphi_{j,\mathrm{B}}$. For example, for $\varphi_{j,\mathrm{B}}$ contained in particular bins a stimulus locked $n:m$ transient might occur shortly after the stimulus' end, whereas for $\varphi_{j,\mathrm{B}}$ within other bins a stimulus locked $n:m$ transient might occur later.

3. *Stereotyped pattern of the* $n:m$ *phase difference:* If the amplitudes reveal significant stimulus locked epochs we can assess and quantify the strength of the stimulus locking according to (8.26). Finally, the time course of the $n:m$ phase difference is studied by considering $\bar{\phi}_{j,\mathrm{r}}^{(n,m)}(t;l)$ or $\bar{\Phi}_{j,\mathrm{r}}^{(n,m)}(t)$ and, in particular, $\bar{\psi}_{j,\mathrm{r}}^{(n,m)}(t;l)$ or $\bar{\Psi}_{j,\mathrm{r}}^{(n,m)}(t)$, depending on whether or not the epochs occur only for certain initial phases (cf. (8.18), (8.19), (8.24), (8.25)). By this one might reveal a stimulus locked epoch exhibiting $n:m$ phase synchronized activity which corresponds to a horizontal plateau.

The time course of the $n : m$ phase difference might crucially depend on the stimulation's initial phase $\varphi_{j,B}$ in a similar way as discussed for the amplitudes above.

Of course, it is necessary to perform the analysis with suitable values of n and m. To this end we might systematically vary n and m and pick up those values which are related to the largest stimulus locking indices. However, the range of possible values of n and m is restricted due to both the spectral properties of the systems under consideration and due to the bandpass filtering. For this reason in practice n and m need not be varied systematically since suitable values of n and m can be estimated by means of visual inspection of the time series and their spectra.

8.4.4 Stimulus Locked $n : m$ Transients with Delay

Imagine the following scenario: A sensory stimulation gives rise to a response in a certain brain area, which processes this incoming information and sends it to another brain area. Due to a conduction delay the response of the second area shows up after the response of the first area has vanished. Although both responses do not coincide the time course of their phase difference may be stereotyped except for the conduction delay.

To detect such effects we have to take into account the impact of conduction delays on short-term transients. This can be done by means of

$$\phi_{n,m}^{(j,r)}(t; T_\delta) = n\varphi_j(t - T_\delta) - m\varphi_r(t) , \qquad (8.31)$$

the time-delayed $n : m$ phase difference between the particular channel under consideration $s_j(t)$ and the reference signal $s_r(t)$. The time delay is denoted by T_δ, where $\varphi_j(t)$ and $\varphi_r(t)$ are the phases of $s_j(t)$ and $s_r(t)$, respectively. With this time-delayed phase difference we can define the time-delayed versions of (8.14) and (8.20) which read

$$\zeta_{j,r}^{(n,m)}(t; T_\delta) = A_j(t - T_\delta) \exp\left\{i\left[\phi_{n,m}^{(j,r)}(t; T_\delta)\right]\right\} \qquad (8.32)$$

and

$$\eta_{j,r}^{(n,m)}(t; T_\delta) = \exp\left\{i\left[\phi_{n,m}^{(j,r)}(t; T_\delta)\right]\right\} , \qquad (8.33)$$

respectively. Averaging (8.32) and (8.33) in the same way as in (8.15), (8.16), (8.21), and (8.22) yields the quantities

$$\bar{\zeta}_{j,r}^{(n,m)}(t; l, T_\delta) = \frac{1}{N_l} \sum_{k(l)} \zeta_{j,r}^{(n,m)}(t + \tau_k; T_\delta) , \qquad (8.34)$$

$$\bar{X}_{j,r}^{(n,m)}(t; T_\delta) = \frac{1}{N_a} \sum_{k=1}^{N_a} \zeta_{j,r}^{(n,m)}(t + \tau_k; T_\delta) , \qquad (8.35)$$

$$\bar{\eta}_{j,\mathrm{r}}^{(n,m)}(t;l,T_\delta) = \frac{1}{N_l}\sum_{k(l)}\eta_{j,\mathrm{r}}^{(n,m)}(t+\tau_k;T_\delta)\,, \tag{8.36}$$

$$\bar{Y}_{j,\mathrm{r}}^{(n,m)}(t;T_\delta) = \frac{1}{N_\mathrm{a}}\sum_{k=1}^{N_\mathrm{a}}\eta_{j,\mathrm{r}}^{(n,m)}(t+\tau_k;T_\delta)\,, \tag{8.37}$$

where $t \in [-t_-,+t_+]$. With the Hilbert transform we determine the amplitudes and phases of these averaged signals in this way obtaining

$$\bar{\zeta}_{j,\mathrm{r}}^{(n,m)}(t;l,T_\delta) = \bar{a}_{j,\mathrm{r}}^{(n,m)}(t;l,T_\delta)\exp\left[\mathrm{i}\bar{\phi}_{j,\mathrm{r}}^{(n,m)}(t;l,T_\delta)\right]\,, \tag{8.38}$$

$$\bar{X}_{j,\mathrm{r}}^{(n,m)}(t;T_\delta) = \bar{A}_{j,\mathrm{r}}^{(n,m)}(t;T_\delta)\exp\left[\mathrm{i}\bar{\Phi}_{j,\mathrm{r}}^{(n,m)}(t;T_\delta)\right]\,, \tag{8.39}$$

$$\bar{\eta}_{j,\mathrm{r}}^{(n,m)}(t;l,T_\delta) = \bar{b}_{j,\mathrm{r}}^{(n,m)}(t;l,T_\delta)\exp\left[\mathrm{i}\bar{\psi}_{j,\mathrm{r}}^{(n,m)}(t;l,T_\delta)\right]\,, \tag{8.40}$$

$$\bar{Y}_{j,\mathrm{r}}^{(n,m)}(t;T_\delta) = \bar{B}_{j,\mathrm{r}}^{(n,m)}(t;T_\delta)\exp\left[\mathrm{i}\bar{\Psi}_{j,\mathrm{r}}^{(n,m)}(t;T_\delta)\right]\,, \tag{8.41}$$

where

$$0 \le \bar{b}_{j,\mathrm{r}}^{(n,m)}(t;l,T_\delta) \le 1\,,\quad 0 \le \bar{B}_{j,\mathrm{r}}^{(n,m)}(t;T_\delta) \le 1 \tag{8.42}$$

hold for all times $t \in [-t_-,+t_+]$ as a consequence of the normalization by (8.33).

Artifacts due to noise or bandpass filtering can be avoided with time independent significance levels obtained by applying the just mentioned analysis to surrogates, i.e. white noise signals filtered exactly as the original signals. The 95th percentile of the distribution of the amplitudes of the surrogates is taken as significance level, respectively. There is a more convenient and, especially, computer time saving way of deriving these significance levels: As the time delay T_δ does obviously not change the values of the time independent significance levels we are allowed to simply take the significance levels $\bar{a}_{j,\mathrm{r},\mathrm{S}}^{(n,m)}(l)$, $\bar{A}_{j,\mathrm{r},\mathrm{S}}^{(n,m)}$, $\bar{b}_{j,\mathrm{r},\mathrm{S}}^{(n,m)}(l)$, and $\bar{B}_{j,\mathrm{r},\mathrm{S}}^{(n,m)}$, used for the calculation of the significant amplitudes (8.27) to (8.30) which were determined for $T_\delta = 0$.

To reveal only significant values of the amplitudes we introduce the *significant $n:m$ stimulus locking indices* by putting

$$\bar{r}_{j,\mathrm{r}}^{(n,m)}(t;l,T_\delta) = \max\left\{\bar{a}_{j,\mathrm{r}}^{(n,m)}(t;l,T_\delta) - \bar{a}_{j,\mathrm{r},\mathrm{S}}^{(n,m)}(l),0\right\}\,, \tag{8.43}$$

$$\bar{R}_{j,\mathrm{r}}^{(n,m)}(t;T_\delta) = \max\left\{\bar{A}_{j,\mathrm{r}}^{(n,m)}(t;T_\delta) - \bar{A}_{j,\mathrm{r},\mathrm{S}}^{(n,m)},0\right\}\,, \tag{8.44}$$

$$\bar{e}_{j,\mathrm{r}}^{(n,m)}(t;l,T_\delta) = \max\left\{\bar{b}_{j,\mathrm{r}}^{(n,m)}(t;l,T_\delta) - \bar{b}_{j,\mathrm{r},\mathrm{S}}^{(n,m)}(l),0\right\}\,, \tag{8.45}$$

$$\bar{E}_{j,\mathrm{r}}^{(n,m)}(t;T_\delta) = \max\left\{\bar{B}_{j,\mathrm{r}}^{(n,m)}(t;T_\delta) - \bar{B}_{j,\mathrm{r},\mathrm{S}}^{(n,m)},0\right\}\,. \tag{8.46}$$

$\bar{\phi}_{j,\mathrm{r}}^{(n,m)}(t;l,T_\delta)$, $\bar{\Phi}_{j,\mathrm{r}}^{(n,m)}(t;l,T_\delta)$, $\bar{\psi}_{j,\mathrm{r}}^{(n,m)}(t;l,T_\delta)$, and $\bar{\Psi}_{j,\mathrm{r}}^{(n,m)}(t;l,T_\delta)$ finally provide us with the time course of the averaged $n:m$ phase difference during the significant epochs.

The analysis by means of this approach is more extensive compared to the calculation of the significant amplitudes (8.27) to (8.30). Here, we have to scan a range of physiologically possible delays T_δ, e.g., for T_δ between 0 and 100 ms. In this way we can detect a certain delay which yields optimal values of the significant amplitudes of the $n : m$ locked epochs. On the other hand, in particular, in the case of short-term epochs it is also possible that different ranges of T_δ can be related to qualitatively different stimulus locked transients.

8.4.5 Multiple Stimulus Locked $n : m$ Transients

The reactions of two interacting clusters of neurons may, however, be even more complex compared to the transient dynamics analyzed in the last two sections. For instance, there may be a variability of the clusters' reactions to the same and repetitively administered stimulus. Especially, we may expect a switching between different neuronal reactions in the course of the repetitive application of the stimulus. To illustrate transient phenomena of this kind let us introduce $\Psi_{n,m}^{(j,r)}$, the $n : m$ phase difference between the jth signal $s_j(t)$ and the reference signal $s_r(t)$ modulo 2π, by setting

$$\Psi_{n,m}^{(j,r)}(t) = \left[n\varphi_j(t) - m\varphi_r(t) \right] \bmod 2\pi \, , \tag{8.47}$$

where $\varphi_j(t)$ and $\varphi_r(t)$ are the phases of $s_j(t)$ and $s_r(t)$, respectively.

Figure 8.2a shows a schematic plot of a stimulus locked $n : m$ transient. The time axis is chosen in a way that the stimulation ends at $t = 0$. The figure displays the time courses of $\Psi_{n,m}^{(j,r)}(t)$ of 20 different (simulated) trials. Before and after the transient epoch the alignment of the 20 trajectories appears to be at random. During the epoch the trajectories are joined into a bundle. For this reason we observe a stimulus locked stereotyped pattern of the time course of $\Psi_{n,m}^{(j,r)}(t)$. Such a transient can be analyzed by means of the averaging technique explained in Sect. 8.4.3. The tighter the trajectories are bundled the larger are the corresponding values of the indices $\bar{e}_{j,r}^{(n,m)}(t; l)$ and $\bar{E}_{j,r}^{(n,m)}(t)$, respectively (cf. (8.29) and (8.30)).

A different situation occurs provided the two interacting neuronal clusters may react on the same stimulus in a few qualitatively different ways concerning the time course of their stimulus locked $n : m$ phase difference. For example, during an experiment in, say, 60 % of the trials we observe a certain stimulus locked $n : m$ transient, whereas in 40 % we observe a stimulus locked $n : m$ transient of a qualitatively different type. Figures 8.2b–d schematically illustrate possible realizations of such a situation. In Fig. 8.2b 60 % of the trajectories are joined into the lower bundle, whereas the other 40 % belong to the upper bundle. Such a bundle may also be called a *branch*, and stimulus locked $n : m$ transients consisting of more than one branch will be denoted as *multiple stimulus locked $n : m$ transients*.

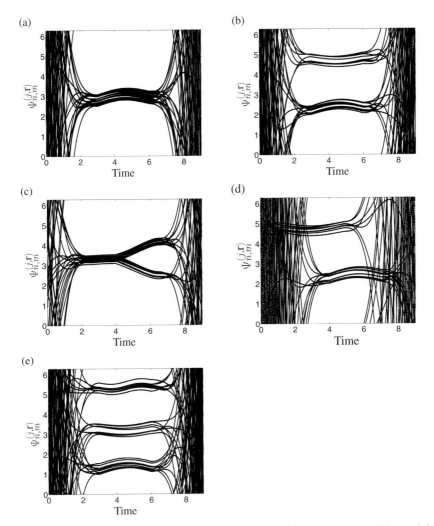

Fig. 8.2a–e. *Stimulus locked $n : m$ transients:* (**a**) to (**e**) show superpositions of the stimulation induced time courses of $\Psi_{n,m}^{(j,\mathrm{r})}(t)$ of different simulated experiments. In each experiment a stimulus is repetitively administered. $\tau_1', \tau_2', \ldots, \tau_{N_\mathrm{a}}'$ denote the timing points at which the respective stimulus ends. Each plot displays $\Psi_{n,m}^{(j,\mathrm{r})}(t + \tau_k')$ for $k = 1, \ldots, N_\mathrm{a}$. The time t runs in the window $[0, 9]$ following each stimulus, respectively (with arbitrary time units), where at $t = 0$ the stimulation ends. (**a**) shows $N_\mathrm{a} = 20$ superimposed responses of $\Psi_{n,m}^{(j,\mathrm{r})}$ which exhibit a stimulus locked $n : m$ transient consisting of one branch (for $3 \leq t \leq 6$). Multiple stimulus locked $n : m$ transients with two branches are displayed in (**b**) to (**d**). In (**b**) and (**c**) both branches occur at the same time (for $3 \leq t \leq 6$ in (**b**) and for $2.5 \leq t \leq 6.5$ in (**c**)), whereas in (**d**) the two branches do not coincide. The two branches partially overlap in (**c**). (**e**) illustrates a multiple stimulus locked $n : m$ transient consisting of three branches.

We may expect that there is a huge variety of different realizations of transients of this kind. Both branches may partially overlap in time as shown in Fig. 8.2c. Begin and end of the different branches need not coincide (Fig. 8.2d). Moreover, multiple stimulus locked $n : m$ transients may consist of more than two branches (Fig. 8.2e).

Obviously, it is very difficult to detect multiple stimulus locked $n : m$ transients with the averaging procedure explained in Sect. 8.4.3. For instance, two antiphase branches of a stimulus locked epoch may be averaged out by calculating the indices $\bar{R}_{j,r}^{(n,m)}(t)$ and $\bar{E}_{j,r}^{(n,m)}(t)$ (cf. 8.28 and 8.30). By means of the averaging technique of Sect. 8.4.3 we can only detect multiple stimulus locked $n : m$ transients in a reliable way provided each branch corresponds to a different range of the initial phase $\varphi_{j,B}$ of a particular rhythm. We may imagine an example where for $\varphi_{j,B} \in [0, \pi/2]$ all trajectories of $\Psi_{n,m}^{(j,r)}(t)$ run through the upper branch in Fig. 8.2b, whereas the trajectories for $\varphi_{j,B} \in]\pi/2, 2\pi[$ are joined into the lower bundle. In such a simple case the different branches can be resolved and detected by determining the phase dependent $n : m$ stimulus locking indices $\bar{r}_{j,r}^{(n,m)}(t; l)$ and $\bar{e}_{j,r}^{(n,m)}(t; l)$ (cf. (8.27) and (8.29)).

However, the dynamics may be more complex. On the one hand the initial state of *several* rhythms may determine through which branch a particular trajectory of $\Psi_{n,m}^{(j,r)}(t)$ runs. On the other hand apart from initial conditions at the beginning of the stimulation other dynamical features may be important, too. Memory effects or experimentally induced shifts of the subject's attention may influence the choice of a particular branch. Also the interactions between different clusters during a multiple stimulus locked $n : m$ transient may determine which branch a particular trajectory of $\Psi_{n,m}^{(j,r)}(t)$ chooses. Partially overlapping branches which diverge in the course of the stimulus locked epoch (Fig. 8.2c) might correspond to a neuronal pattern recognition process. The two diverging branches might, thus, be related to, e.g., two different perceptions or two different sensations. Anyway, studying multiple stimulus locked $n : m$ transients enables us to probe the brain's internal variablility of reactions to external stimuli. Due to the pitfalls of the averaging technique we, hence, need a different approach.

8.4.6 Detection of Multiple Stimulus Locked $n : m$ Transients

As in the former section let us imagine an experiment where a series of identical stimuli is administered. Whether or not multiple stimulus locked $n : m$ transients occur can be checked in the following way. To compare the phase difference $\Psi_{n,m}^{(j,r)}(t)$ typically rushing ahead of and following the stimulus we consider the distribution given by

$$\left\{ \Psi_{n,m}^{(j,r)}(t + \tau_k) \right\}_{k=1,\ldots,N_a} \tag{8.48}$$

for each time $t \in [-t_-, t_+]$. τ_k is the timing point of the onset of the kth stimulus, and N_a denotes the number of all trials or, in other words, the number of the repetitively administered stimuli. The time t runs in the window $[-t_-, t_+]$ which encloses the stimulus applied at time $t = 0$ $(t_-, t_+ > 0)$. So, (8.48) consists of all values of $\Psi_{n,m}^{(j,r)}$ which are ahead of the stimulus with a lead given by t (< 0) or which follow the stimulus after a particular time t (> 0) in all stimulation trials, respectively.

If the $n : m$ phase difference of both signals is perfectly stimulus locked at time t we measure the same value of the $n : m$ phase difference in each trial, so that we obtain $\Psi_{n,m}^{(j,r)}(t+\tau_k) = \Psi_{n,m}^{(j,r)}(t+\tau_l)$ for all $k, l = 1, \ldots, N_{\mathrm{a}}$. In other words, in this case (8.48) is a Dirac-like distribution. If the phase difference of both signals is not at all stimulus locked at time t (8.48) is a uniform distribution. On the other hand we may also expect that the distribution (8.48) may have two or more distinct and pronounced peaks. For example, there may be two equally large peaks with a phase difference of approximately π. As a consequence of calculating the normalized varibale $\eta_{j,r}^{(n,m)}(t)$ according to (8.20) the two peaks may be averaged out. The existence of multiple bumps has to be checked experimentally, e.g., by simply plotting the time course of the distribution (8.48). If we encounter such a situation we have to perform a different analysis:

For each time t we characterize the distribution (8.48) in a statistical sense by comparing it with the uniform distribution $1/(2\pi)$. This can be achieved by introducing the $n : m$ stimulus locking index $\tilde{\mu}_{n,m}(t)$ which provides us with the strength of the stimulus locking no matter whether the distribution has only one or more bumps. We can define $\tilde{\mu}_{n,m}(t)$ by putting

$$\tilde{\mu}_{n,m}(t) = \frac{S_{\mathrm{max}} - S(t)}{S_{\mathrm{max}}} , \quad \text{where } S(t) = -\sum_{k=1}^{N_{\mathrm{B}}} p_k \ln p_k \qquad (8.49)$$

is the entropy of the distribution (8.48) for time t, and $S_{\mathrm{max}} = \ln N_{\mathrm{B}}$, where N_{B} is the number of bins which is optimally given by $N_{\mathrm{B}} = \exp[0.626 + 0.4\ln(N_{\mathrm{a}}-1)]$, where N_{a} denotes the number of all trials (Otnes and Enochson 1972). The relative frequency of finding $\Psi_{n,m}^{(j,r)}$ within the kth bin is denoted by p_k. Due to this normalization $0 \le \tilde{\mu}_{n,m} \le 1$ is fulfilled. $\tilde{\mu}_{n,m} = 0$ corresponds to a uniform distribution without any stimulus locking at time t, whereas $\tilde{\mu}_{nm} = 1$ corresponds to a Dirac-like distribution, that means to a perfect stimulus locking at time t. Multiple stimulus locked $n : m$ transients correspond to two or more pronounced bumps which are associated with large values of $\tilde{\mu}_{n,m}(t)$ (< 1). Instead of (8.49) we may also use a different statistical parameter in order to characterize the difference between (8.48) and a uniform distribution. We might, for example, use an index based on the conditional probability as defined in detail in Sect. 8.4.8 (cf. (8.60)).

To eliminate effects which are caused by the bandpass filtering and noise we determine a significance level $\tilde{\mu}_{n,m}^{\mathrm{s}}$ by calculating (8.49) for surrogate data which are, e.g., white noise signals filtered exactly in the same way as the original signals $s_j(t)$ and $s_{\mathrm{r}}(t)$. We choose the 95th percentile of the distribution of $\tilde{\mu}_{n,m}$ of the surrogates as significance level $\tilde{\mu}_{n,m}^{\mathrm{s}}$. The significant values are extracted by introducing the *significant $n : m$ stimulus locking index*

$$\mu_{n,m}(t) = \max\{\tilde{\mu}_{n,m}(t) - \tilde{\mu}_{n,m}^{\mathrm{s}}, 0\} . \qquad (8.50)$$

Certainly, this approach based on a statistical characterization of the distribution (8.48) is more envolved compared to the averaging analysis discussed above (cf. (8.27) to (8.30)). We can expect that both approaches will reveal results which qualitatively agree provided there are no multiple stimulus locked $n : m$ transients. In this case the distribution (8.48) has at most one bump for all times t during the stimulus locked epoch. It should be emphasized again that we are only allowed to apply the averaging technique from Sect. 8.4.3 provided there are no multiple branches. This can be tested in a convenient way, e.g., by plotting the time course of the distribution (8.48) for $t \in [-t_-, t_+]$. However, if there are, indeed, multiple stimulus locked $n : m$ transients we have to perform the statistical analysis by determining (8.50).

In this way applying both the averaging and the statistical technique to experimental data will show us which approach is more appropriate and more sensitive in the context of the particular neurophysiological issue under consideration. On the other hand if we want to perform a screening analysis to detect all sorts of single and multiple stimulus locked $n : m$ transients we can, first, apply the statistical analysis explained in this section. To judge how many branches the stimulus locked epochs contain we, second, plot the time course of the distribution (8.48).

8.4.7 Multiple Stimulus Locked $n : m$ Transients With Delay

During a repetitive application of stimuli we may also encounter a switching between qualitatively different types of stimulus locked $n : m$ transients with delay. As an example let us imagine a stimulation experiment where, say, 60 % of the stimuli induce particular responses in two brain areas which do not coincide. Nevertheless, the time course of the responses' phase difference is stereotyped except for the conduction delay. In the other 40 % we measure a qualitatively different type of time-delayed coordinated responses of both areas. To study this problem we can perform a similar analysis as in the former section. We merely have to introduce the time-delayed $n : m$ phase difference modulo 2π

$$\Psi_{n,m}^{(j,\mathrm{r})}(t; T_\delta) = \left[n\varphi_j(t - T_\delta) - m\varphi_{\mathrm{r}}(t)\right] \bmod 2\pi \qquad (8.51)$$

between the jth signal $s_j(t)$ and the reference signal $s_{\mathrm{r}}(t)$, where $\varphi_j(t)$ and $\varphi_{\mathrm{r}}(t)$ are the phases of $s_j(t)$ and $s_{\mathrm{r}}(t)$, respectively. The time delay is denoted as T_δ. We then determine the distribution

$$\left\{\Psi_{n,m}^{(j,\mathrm{r})}(t + \tau_k; T_\delta)\right\}_{k=1,\dots,N_{\mathrm{a}}} \qquad (8.52)$$

for each time t within the window $[-t_-, t_+]$ enclosing the stimulus applied at time $t = 0$. τ_k is the time of the onset of the kth stimulus. N_a denotes the number of stimuli.

A stimulus locked $n : m$ transient with L branches is associated with L bumps of the distribution (8.52) for all times t during that stimulus locked epoch. We detect such epochs by comparing the distribution (8.52) with the uniform distribution $1/(2\pi)$ for each time t. With this aim in view we introduce the $n : m$ stimulus locking index $\tilde{\mu}_{n,m}(t; T_\delta)$ which gives us the strength of the stimulus locking irrespective of the particular number of bumps. According to the former section we define $\tilde{\mu}_{n,m}(t; T_\delta)$ by setting

$$\tilde{\mu}_{n,m}(t; T_\delta) = \frac{S_{\max} - S(t; T_\delta)}{S_{\max}} \ , \quad \text{where} \ \ S(t; T_\delta) = -\sum_{k=1}^{N_B} p_k \ln p_k \qquad (8.53)$$

provides us with the entropy of the distribution (8.52) for time t. The maximal entropy is given by $S_{\max} = \ln N_B$, where the optimal number of bins is given by $N_B = \exp[0.626 + 0.4 \ln(N_a - 1)]$, and N_a denotes the number of all trials (Otnes and Enochson 1972). p_k is the relative frequency of finding $\Psi_{n,m}^{(j,r)}(t; T_\delta)$ within the kth bin. From the definition (8.53) it follows that $0 \leq \tilde{\mu}_{n,m}(t; T_\delta) \leq 1$ holds. Absence of any stimulus locking at time t is related to $\tilde{\mu}_{n,m}(t; T_\delta) = 0$. On the contrary, a stimulus locked $n : m$ transient consisting of one perfectly thin branch is associated with $\tilde{\mu}_{nm} = 1$ at time t. Accordingly, multiple stimulus locked $n : m$ transients correspond to large values of $\tilde{\mu}_{n,m}(t)$ (< 1). As already mentioned in the former section instead of (8.53) for the quantitative description of the distribution we may also use different statistical indices like that based on the conditional probability (cf. (8.60)).

Next, we have to avoid artifacts which are caused by the bandpass filtering and noise. This can be achieved by determining a significance level $\tilde{\mu}_{n,m}^s$. To this end (8.53) is calculated for surrogate data, i.e., for example, white noise signals which are filtered with exactly the same filters as the original signals $s_j(t)$ and $s_r(t)$. Obviously, $\tilde{\mu}_{n,m}^s$ does not depend on the delay T_δ. For this reason it is sufficient to determine the significance level for one value of the delay, where we conveniently choose $T_\delta = 0$. The 95th percentile of the distribution of $\tilde{\mu}_{n,m}$ of the surrogates serves as significance level $\tilde{\mu}_{n,m}^s$. To determine the significant values we define the *significant $n : m$ stimulus locking index*

$$\mu_{n,m}(t; T_\delta) = \max\{\tilde{\mu}_{n,m}(t; T_\delta) - \tilde{\mu}_{n,m}^s, 0\} \ . \qquad (8.54)$$

We cannot reliably analyze multiple stimulus locked $n : m$ transients with delay by means of the averaging technique explained in Sect. 8.4.4. As already discussed in Sect. 8.4.5 multiple branches can easily be averaged out. Thus, before we are allowed to apply the averaging method we have to judge how many branches the stimulus locked epochs contain, e.g., by means of simply plotting the time course of the distribution (8.52). If the stimulus locked epochs do not consist of more than one branch, respectively, we can apply the averaging technique as well as the statistical method explained in this section. Probably the results will be qualitatively the same. Nevertheless, there may be some minor differences concerning, e.g., the methods' sensitivity. Therefore it has to checked which method is most appropriate for the particular problem under consideration.

However, it is important to note that we have to perform our analysis for a range of physiologically relevant delays so that T_δ is varied, e.g., between 0 and 100 ms. For this reason it is much more convenient to generally use the statistical method which can be applied to stimulus locked $n : m$ transients consisting of an arbitrary number of branches. In this way we can perform an automatic screening since we can avoid the time consuming visual inspection which is necessary to rule out mutliple branches.

After reflecting on stimulus locked transient dynamics let us now turn to the question as to how $n : m$ phase synchronization of continuously interacting oscillators can be defined.

8.4.8 $n : m$ Phase Synchronization

Classically synchronization of two periodic non-identical interacting oscillators in the absence of noise means that both oscillators are able to adjust their rhythms. Correspondingly, the *phase locking* condition reads

$$c_1 < \phi_{n,m} < c_2 \ , \ \text{where} \ \phi_{n,m} = n\varphi_1 - m\varphi_2 \ , \qquad (8.55)$$

c_1 and c_2 are constants, n and m are integers, φ_1, φ_2 are the phases of the two oscillators and $\phi_{n,m}$ is the $n : m$ phase difference, or relative phase. Note that in (8.55) the phases φ_1 and φ_2 are defined on the whole real line, i.e. on $\mathbb{R}$. In this simplest case both oscillators are also frequency locked whenever they are phase locked, that means whenever condition (8.55) is fulfilled. *Frequency locking* is defined by

$$n\Omega_1 = m\Omega_2 \ , \ \text{where} \ \Omega_j = \langle \dot{\varphi}_j \rangle \qquad (8.56)$$

is the average frequency of the jth oscillator, and brackets denote time averaging. Note that for the determination of phase synchronized states it is irrelevant whether or not the amplitudes of both oscillators are different. Rather one only considers the phases of both oscillators.

The definition of synchronization in *noisy and/or chaotic systems* is more involved. In this context it is important that the notion of phase can also be introduced for chaotic systems, and that phase locking as defined by (8.55) can be observed in this case, too (Rosenblum, Pikovsky, Kurths 1996, Pikovsky, Rosenblum, Kurths 1996): In the synchronized states of chaotic oscillators their amplitudes remain chaotic and act on the phase dynamics qualitatively in the same way as external noise. For this reason we are allowed to consider synchronization processes of noisy periodic oscillators as well as chaotic oscillators within a common framework.

If the random or the purely deterministic (chaotic) perturbations are weak, the frequency locking condition (8.56) is fulfilled only approximately and the relative phase fluctuates in a random way. Moreover, if these perturbations are strong one observes phase slips which are rapid 2π jumps of the relative phase $\varphi_{n,m}$. Put otherwise: Both oscillators tend to maintain a preferred phase relationship except for small fluctuations. However, due to the perturbations the oscillators are kicked out of this preferred state, and one oscillator passes the other one so that a phase slip occurs. Nevertheless, by means of their synchronizing interactions both oscillators manage to reestablish their preferred phase relationship. As a consequence of several phase slips $\phi_{n,m}$ will undergo large variations. Consequently, the phase locking condition (8.55) fails in the presence of noise. Rather we have to describe the oscillators' tendency to continually synchronize their rhythms by resisting both the random and the purely deterministic perturbations.

Therefore we have to understand phase synchronization in terms of statistical physics. More precisely, we have to understand synchronization of noisy and/or chaotic systems as appearance of peaks in the distribution of the cyclic relative phase

$$\Psi_{n,m}(t) = [\phi_{n,m}(t)] \bmod 2\pi .\qquad(8.57)$$

To understand why we have to introduce the cyclic relative phase let us focus on the phase slips. Depending on the oscillators frequency detuning the probability of these upward and downward phase jumps may either be equal or different so that the relative phase performs either an unbiased or a biased random walk. In the first case we observe a frequency locking as defined by (8.56), whereas in the second case there is no frequency locking. For this reason in a statistical sense phase synchronization is characterized by the existence of one or a few preferred values of $\Psi_{n,m}$, no matter whether the oscillators' averaged frequencies are equal or different.

Let me now explain the method of Tass et al. (1998) for the detection of epochs of $n : m$ phase synchronization in noisy non-stationary data. This data analysis tool is a single run analysis, in contrast to the methods presented in the former sections. This means that $n : m$ phase locking between two signals $s_1(t)$ and $s_2(t)$ is studied without any averaging procedure. The analysis starts by extracting the signals' phases with the method described in Sect. 8.2.2. For this purpose first a bandpass filtering of the MEG/EEG signals and the related peripheral, e.g. EMG, signals has to be carried out. The band edges of the filters depend on the neurophysiological process under consideration and can be determined by visual inspection of the signals' power spectra. Applying the Hilbert transform to the bandpass filtered signals yields the instantaneous phases $\varphi_1(t)$ and $\varphi_2(t)$ of the signals $s_1(t)$ and $s_2(t)$ (cf.

(8.11)). Next, $\Psi_{n,m}$, the $n : m$ phase difference modulo 2π, is calculated according to (8.55) and (8.57).

Data in medicine and biology like MEG or EEG data are inevitably noisy. Correspondingly, $n : m$ phase synchronization is understood in terms of statistical physics (cf. Stratonovich 1963, Haken 1983). Accordingly, $n : m$ phase synchronization is defined as appearance of one or more peaks in the distribution of $\Psi_{n,m}$. The strength of the $n : m$ locking can, thus, be measured by comparing the actual distribution of $\Psi_{n,m}$ with a uniform distribution corresponding to an absence of locking. This is realized by means of two $n : m$ *synchronization indices*:

1. The *index based on the Shannon entropy* is defined as

$$\tilde{\rho}_{nm} = \frac{S_{\max} - S}{S_{\max}} , \quad \text{where} \quad S = -\sum_{k=1}^{N_{\mathrm{B}}} p_k \ln p_k \qquad (8.58)$$

is the entropy of the distribution of $\Psi_{n,m}$ and $S_{\max} = \ln N_{\mathrm{B}}$, where N_{B} is the number of bins. p_k denotes the relative frequency of finding Ψ_{nm} within the kth bin. The optimal number of bins is given by $N_{\mathrm{B}} = \exp[0.626 + 0.4 \ln(M_{\mathrm{S}} - 1)]$, where M_{S} denotes the number of samples (Otnes and Enochson 1972). As a consequence of this normalization $0 \leq \tilde{\rho}_{nm} \leq 1$ holds, where $\tilde{\rho}_{nm} = 0$ corresponds to a uniform distribution without any synchronization, whereas $\tilde{\rho}_{nm} = 1$ corresponds to a Dirac-like distribution, that means to a perfect synchronization.

2. *Index based on conditional probability:* Suppose the two phases $\varphi_1(t_k)$ and $\varphi_2(t_k)$ are defined on the intervals $[0, n2\pi]$ and $[0, m2\pi]$, respectively. This can easily be achieved with a modulo operation, e.g., $\varphi_1(t_k) \rightarrow \varphi_1(t_k)$ mod $n2\pi$. The index k in t_k refers to the timing points where $k = 1, \dots, M_{\mathrm{P}}$. We divide each interval into N_{B} bins $B_l = [\phi_l - \Delta\phi/2, \phi_l + \Delta\phi/2]$ of equal width $\Delta\phi$, where $l = 1, \dots, N_{\mathrm{B}}$, and N_{B} is defined as above. For each bin B_l we calculate

$$y_l = \frac{1}{M_l} \sum_{k=1}^{M_{\mathrm{P}}} \exp[i\varphi_2(t_k)] \qquad (8.59)$$

such that $\varphi_1(t_k)$ belongs to this particular bin B_l, and M_l is the number of points within this bin. $|y_l| = 1$ provided there is a complete dependence between both phases, whereas $|y_l|$ vanishes if there is no dependence at all. Finally, we calculate the average over all bins according to

$$\tilde{\lambda}_{n,m} = \frac{1}{N_{\mathrm{B}}} \sum_{l=1}^{N_{\mathrm{B}}} |y_l| . \qquad (8.60)$$

Correspondingly, $\tilde{\lambda}_{n,m}$ measures the conditional probability for φ_2 to have a certain value provided φ_1 is in a certain bin.

Since data in medicine and biology are typically nonstationary this quantification of the strength of $n : m$ synchronization is performed in a sliding window. In other words, for each timing point t one calculates $\tilde{\rho}_{n,m}(t)$ and $\tilde{\lambda}_{n,m}(t)$ within the corresponding window $[t - T/2, t + T/2]$ of width T.

To extract only significant epochs of $n : m$ phase synchronization and to avoid artefacts caused by the bandpass filtering, significance levels $\tilde{\rho}_{n,m}^{s}$ and $\tilde{\lambda}_{n,m}^{s}$ are derived for each $n : m$ synchronization index $\tilde{\rho}_{n,m}$ and $\tilde{\lambda}_{n,m}$ by applying the analysis to surrogate data. The latter are white noise signals filtered with the same bandpass filters as the original signals under consideration. The 95th percentile of the distribution of the $n : m$ synchronization indices, $\tilde{\rho}_{n,m}$ or $\tilde{\lambda}_{n,m}$, of the surrogates serves as significance level $\tilde{\rho}_{n,m}^{s}$ or $\tilde{\lambda}_{n,m}^{s}$, respectively. Only relevant values of the $n : m$ synchronization indices are taken into account by introducing the *significant $n : m$ synchronization indices*

$$\rho_{n,m}(t) = \max\{\tilde{\rho}_{n,m}(t) - \tilde{\rho}_{n,m}^{s}, 0\} \ , \quad \lambda_{n,m}(t) = \max\{\tilde{\lambda}_{n,m}(t) - \tilde{\lambda}_{n,m}^{s}, 0\} \ . \tag{8.61}$$

The results obtained by means of calculating $\rho_{n,m}(t)$ and $\lambda_{n,m}(t)$ are robust with respect to variations of the algorithm's parameters, namely the window length and the band edges of all filters used. Computation of both indices yields consistent results (Tass et al. 1998). However, $\lambda_{n,m}$ appears to be more sensitive to intermittent $n : m$ phase synchronization compared to $\rho_{n,m}$. Future studies will address the differences between both indices in this way clarifying the neurophysiological role of different types of synchronization processes like strong phase synchronization versus intermittent phase synchronization.

In order to find suitable values of n and m one tries different values and picks up those that yield the largest indices. Of course, the range of possible values of n and m is restricted due to the properties of the particular systems under consideration and, in particular, due to the bandpass filtering. For this reason suitable values of n and m can initially be guessed by means of visual inspection of the time series and their spectra.

Applying this method to MEG data and EMG data of a Parkinsonian patient revealed that the temporal evolution of the peripheral tremor rhythms directly reflects the time course of the synchronization of abnormal activity between cortical motor areas (Tass et al. 1998). These results will be discussed in detail in Sect. 10.2.1.

It is very important to stress that phase synchronization is not equivalent to correlation and coherence. In particular, oscillatory signals may be coherent although there are not phase synchronized. Let us briefly recall that the magnitude-squared coherence between two signals $x(t)$ and $y(t)$ is defined by

$$C_{xy}(\omega) = \frac{|P_{xy}(\omega)|^2}{P_{xx}(\omega)P_{yy}(\omega)} \ , \tag{8.62}$$

where P_{xx} and P_{yy} denote the power spectral density function of x and y, whereas C_{xy} is the cross spectral density function of x and y. Correspondingly,

(8.62) is a real number between 0 and 1 which estimates the correlation between $x(t)$ and $y(t)$ at the frequency ω.

To illustrate the difference between phase synchronization and coherence let us consider two chaotic oscillatory signals x_1 and x_2 generated by, e.g., Rössler systems. Assume that both oscillators and, thus, x_1 and x_2 are not phase synchronized. To mimic the real situation during an MEG or EEG measurement we construct the signals $u = (1-\varepsilon)x_1 + \varepsilon x_2$ and $w = \varepsilon x_1 + (1-\varepsilon)x_2$, where ε is positive and small. u and w correspond to the signals measured by the first and second sensor, respectively. The first sensor is close to the first cerebral current which generates a field x_1, whereas the second sensor is located near the second current and, accordingly, mainly picks up the corresponding second field x_2. As MEG and EEG sensors measure signals originating from more than one cerebral area, the measured signals are typically mixtures like u and w. Even if the distance between both cerebral currents is large the first sensor (u) picks up a bit of the field of the second current (modelled by εx_2) and vice versa. u and w are not phase synchronized since x_1 and x_2 are not phase synchronized. On the contrary, u and w are typically significantly coherent as revealed by a cross-spectrum analysis by means of the Welch technique with the Bartlett window (Tass et al. 1998). This means that the usage of coherence can lead to an enormous spatial overestimation of cortico-cortical interactions.

8.4.9 Self-Synchronization Versus Transients

Let us compare the data analysis tool for the detection of $n : m$ phase synchronization (Sect. 8.4.8) with those methods for the study of single or multiple stimulus locked $n : m$ transients (Sects. 8.4.3 and 8.4.6).

The method by Tass et al. (1998) is a single run analysis which works without any averaging. Thus, applied to stimulation experiments this method enables us to detect epochs of $n : m$ phase synchronization also if they are not tightly stimulus locked and if the phase difference between the two signals varies from trial to trial. This is a crucial advantage.

However, the drawback of the method by Tass et al. (1998) is its limited time resolution because it is a sliding window analysis and one needs a window length T corresponding to at least, say, eight to ten periods of the oscillation. For example, for a tremor activity of about 5 Hz one needs a window length of about 1.5 to 2 seconds minimum. In contrast, the time resolution of the methods presented in Sects. 8.4.3–8.4.7 is merely restricted by the sampling rate. Correspondingly, these techniques are associated with a remarkably enhanced time resolution.

The methods for detecting stimulus locked $n : m$ transients from Sects. 8.4.3 and 8.4.6 are developed for detecting transient patterns of the $n : m$ phase difference between two signals which occur with a certain delay after the stimulation. These patterns can only be detected if they are stereotyped, and, thus, an averaging (Sect. 8.4.3) or a statistical analysis (Sect. 8.4.6) has to be performed. Stimulus locked $n : m$ transients can but do not need to display horizontal plateaus of the $n : m$ phase difference. If they do, these transients are epochs of $n : m$ phase synchronized activity.

Future studies with experimental data will reveal under which conditions one may profit from the single run, the averaging or the statistical method, respectively. As far as the time resolution is concerned the tools for the analysis of stimulus locked $n : m$ transients with one (Sect. 8.4.3) or arbitrarily many branches (Sect. 8.4.6) may be advantageous for the detection of short-term eruptions of synchronous stimulus locked γ-band activity (> 30 Hz). On the other hand the method by Tass et al. (1998) may be particularly appropriate for the study of ongoing cerebral rhythms, e.g., in the context of the coordination of alternating bimanual movements.

8.4.10 The Flow of Synchronized Cerebral Activity

According to animal experiments short-term epochs of synchronized cerebral activity appear to be of great importance for neuronal information processing, in particular, concerning the binding problem (for a review see Singer and Gray 1995). The results of these experiments were already discussed in Sect. 3.2. Let us now think about how to analyze synchronization processes of this kind. Exemplarily we imagine a scenario of successively emerging epochs of synchronized cerebral activity. Assume that a stimulus, e.g., a particular visual pattern is presented to a human subject. Induced by this visual stimulation a group of cortical areas displays a short epoch of 1:1 synchronized γ-band activity. After this eruption of synchronized activity vanishes a different group of cortical areas exhibits another epoch of 1:1 synchronized γ-band activity. The ceasing and reemerging of synchronous epochs may occur several times where during each epoch different cortical areas may be involved. Such a scenario may correspond to a flow of cortical information processing. Groups of cortical areas become successively active, while within each group there is a parallel processing of the information realized by the synchronization.

To study such a process we can apply the single run method from Sect. 8.4.8 provided the length of these epochs corresponds to at least eight to ten periods of an oscillation which is possible, e.g., in the case of γ-band activity. The averaging (Sect. 8.4.3) and the statistical technique (Sect. 8.4.6), however, are certainly predestined for the analysis of stimulation induced short-term eruptions of synchronized activity. Nevertheless, we have to keep in mind that they are designed to detect stimulus locked epochs. In other words, if the flow of synchronized cerebral activity runs in a stereotyped way the techniques for the detection of stimulus locked $n : m$ are perfectly appropriate.

It is also very interesting to perform a phase dependent averaging by means of (8.27) and (8.29) to study whether the flow of neuronal processing depends on the initial phase of a relevant neuronal rhythm when the stimulation starts. Moreover, the analysis of multiple stimulus locked $n : m$ transients (Sect. 8.4.6) makes it possible to study the variability of neuronal reactions to external stimuli.

In summary, we may profit from studying both $n : m$ phase synchronization and stimulus locked $n : m$ transients. In this way we can distinguish those aspects and features of the flow of synchronized cerebral activity which are stimulus locked from those which are not stimulus locked. This may be relevant since both parts might be related to different physiological functions, respectively.

8.4.11 Inverse Problems

According to definition (6.18) the firing density $p(t)$ provides us with the density of neurons most likely firing or bursting at time t depending on whether the phase oscillators model firing or bursting neurons (cf. Sect. 3.4.1). A synchronous cluster of neurons generates an electromagnetic field (cf. Hämäläinen et al. 1993), which is why the time course of the firing density is related to the time course of the corresponding electromagnetic field. To interpret the neuronal processes giving rise to the electromagnetic activity one would like to identify the generating currents within the brain. Unfortunately, this issue is inextricably linked with a fundamental inverse problem: Helmholtz (1853) showed that a current distribution inside a conductor cannot be retrieved uniquely from the measured electromagnetic field outside (cf. Hämäläinen et al. 1993). Therefore, for instance, dynamical properties of the measured EEG or MEG signals were used in order to derive information about the neuronal clusters which generate the measured field.

Before we turn to one of these approaches let us first imagine an experiment in which a sensor picks up the firing density of two non-interacting clusters located close to each other. Let us denote these clusters as cluster a and b, respectively. Accordingly, their firing densities are denoted as p_a and p_b. Let us assume that our sensor measures the compound signal

$$P(t) = c_a p_a(t) + c_b p_b(t) , \qquad (8.63)$$

where the constant coefficients c_a and c_b depend on, e.g., the distance between the sensor and the respective cluster. With one observable, P, one cannot determine two unknown variables, p_a and p_b. That is why, in general, we cannot simply apply an algorithm like that presented in Sect. 8.2.2 to *seperately* extract the modes of the clusters a and b. A related problem is the following: Suppose we measure a firing density. How can we judge whether the signal is generated by only one cluster or by, e.g., two clusters?

This issue is of great neurophysiological importance, for example, concerning the assignment of different functional roles to neurophysiological rhythms, such as the α- and the γ-rhythm. The latter rhythms refer to α-band (8–12 Hz) and γ-band activity (> 30 Hz) measured with, for instance, EEG or

MEG. To shed light upon the role of the γ−rhythm in visual pattern recognition a sort of EEG experiments was performed where different types of visual stimuli are presented to a human subject (see, for instance, Müller et al. 1997). By comparing the stimulation induced α-band and γ-band activity for different stimuli the question is addressed as to whether the γ-rhythm is nothing but a harmonic of the α-rhythm or whether both rhythms have to be considered as separate. Put otherwise: Is the measured electromagnetic field generated by one cluster or do both α-and γ-rhythm originate from separate clusters of synchronously active neurons? In this context it was suggested that a similar time course of α-and γ-rhythm indicates that the γ-rhythm is merely a harmonic of the α-rhythm (Müller et al. 1997).

Our results from Chap. 6 are in contradiction to this criterion: We saw that the time course of different modes of one cluster may be totally different. Let us, for instance, recall a stimulation induced pronounced early response as displayed in Figs. 6.16 and 6.17. Certainly, what is not going on during the early response is that the cluster which was active before the stimulation is temporarily stopped, while a separate cluster with twice the frequency becomes active for a short while. We know that the pronounced early response is a reaction of *one* cluster caused by a stimulation induced excitation of higher order modes. This example illustrates how difficult and dangerous it can be to identify clusters of neurons with Fourier modes of macrovariables like MEG or EEG signals.

8.5 Summary and Discussion

The data analysis methods sketched in this chapter are essentially based on the determination of the instantaneous phases and amplitudes of the modes of a measured signal as explained in Sect. 8.2.2. This approach was used in three different ways:

First, we applied this method to improve the procedure for revealing suitable parameters for a desynchronizing stimulation of a cluster of interacting oscillators. In this case the phase analysis of Sect. 8.2.2 enables us to selectively analyze the phase dynamics of the relevant modes, in particular, the order parameters. In this way we get rid of, e.g., masking contributions of higher order modes which may give rise to a pronounced early response and related transient phenomena (cf. Sect. 6.3).

We have to take into account that the procedure for tracking down the black holes (Sect. 8.3) does only work provided the signal-to-noise ratio of the data is sufficiently large. Obviously, the experimentalist cannot measure the firing density as introduced by (5.95). Rather he measures collective firing activities, e.g., the multi unit activity (MUA), or any other related variable, such as local field potentials (LFP) or MEG and EEG signals which are certainly noisy. If the signal-to-noise ratio is too small, the procedure of Sect. 8.3 has to be modified in a similar way and based on a similar reasoning as in Sect. 8.4.2 for the phase dependent triggered averaging: We want to extract the cluster's stereotyped responses which correspond to the firing density. To this end for the derivation of a phase resetting curve for a certain stimulation duration T and certain intensity parameters $\{I_j\}$ and $\{\gamma_j\}$ it is no longer sufficient to perform a series of stimulation experiments with the initial phase $\varphi_{j,\mathrm{B}}$ equally spaced in $[0, 2\pi]$. As a consequence of the high noise amplitude for each initial phase $\varphi_{j,\mathrm{B}}$ several stimulations have to be carried out and the cluster's responses belonging to each $\varphi_{j,\mathrm{B}}$ have to be averaged seperately. In this way we obtain the averaged response for a particular set of parameters $(\{I_j\}, \{\gamma_j\}, T)$ and for a particular initial phase $\varphi_{j,\mathrm{B}}$. The analysis of the phase dynamics of the averaged responses is then performed by means of the phase resetting curves as explained in Sect. 8.3.

Second, we used the phase and amplitude analysis from Sect. 8.2.2 to perform a phase dependent triggered averaging of, for example, MEG or EEG data. In contrast to conventional triggered averaging techniques this method allows us to analyze whether a stimulus induced response generated by a cluster of synchronously firing neurons depends on the cluster's dynamical state at the beginning of the stimulation.

Third, based on the method of Sect. 8.2.2 new tools were developed for the detection of different types of stimulation induced stereotyped patterns of transient epochs displaying a tightly stimulus locked pattern of the $n : m$ phase difference (Sects. 8.4.3 and 8.4.6). This approach is particularly appropriate for detecting short-term stimulus locked $n : m$ transients and stimulus locked epochs with $n : m$ phase synchronized activity where the latter are special cases of the former. To illustrate the ideas behind these methods and to explain their possible applications we compared these tools with the single run method for detecting $n : m$ phase synchronization in noisy nonstationary data of Tass et al. (1998) which was mainly designed for investigating self-synchronized oscillatory activity.

To study coordinated sequential neuronal processing data analysis tools were presented which make it possible to detect qualitatively different sorts of stimulus locked $n : m$ transients with conduction delay (Sects. 8.4.4 and 8.4.7). Transient phenomena of this kind may occur, e.g., when a sensory stimulus causes a response of a certain brain area. Within this area the sensory input is processed and than sent to another brain area. Due to the conduction delay the responses of both areas may be separated in time. Nevertheless, there may be a stereotyped time course of their time delayed $n : m$ phase difference (8.31).

All these data analysis methods can be applied to the analysis of oscillatory data such as MEG or EEG data. In the context of cerebral synchronization processes we have to take into account that we can decisively improve the spatial resolution of our approach if we apply both data analysis techniques directly to the cerebral current density $\boldsymbol{J}(x, t)$. The latter, more precisely, especially the synaptic current flow generates the electric and the magnetic field of the brain and, thus, reflects the neuronal activity (Hämäläinen et al. 1993). $J(x,t)$ can be assesed, e.g., from the magnetic field measured with the MEG by means of inverse methods like the minimum norm least squares (MNLS) method (Hämäläinen et al. 1993, Wang, Williamson, Kaufman 1995), the cortical current imaging (CCI) (Fuchs et al. 1995), the probabilistic reconstruction of multiple sources (PROMS) (Greenblatt 1993), the low resolution electromagnetic tomography (LORETA) (Pascual-Marqui, Michel, Lehmann 1994), or the magnetic field tomography (MFT) (Clarke and Janday 1989, Clarke, Ioannides, Bolton 1990, Ioannides, Bolton, Clarke 1990). In this way for each cerebral volume element we obtain the corresponding current density.

In summary, we first determine the cerebral current density by applying an appropriate inverse method to the MEG data. Next, we investigate phase synchronization and stimulus locked $n : m$ transients with the methods from Sects. 8.4.3 and 8.4.8. In this way we can study the interactions between different cerebral areas on the one hand and between a cerebral area and the related peripheral activity such as muscular contraction on the other hand:

1. *$n : m$ phase synchronization:* With the single run analysis from Sect. 8.4.8 we can study synchronization processes between different cerebral areas, for example, *cortico-cortical synchronization*, i.e. the synchronization of neuronal activity of different cortical areas. To understand the physiological meaning of such synchronization processes the latter have to be related with peripheral processes like muscular contraction. With this aim in view we additionally apply our phase synchronization analysis to peripheral data such as EMG data.

By measuring the EMG in parallel we obtain information about the synchronized activity of motor units which control the muscle activity. Studying $n : m$ phase synchronization between the EMG and the cortical current density enables us to investigate the *corticomuscular synchronization*. In this way we can address several questions referring to motor control, for instance: Which areas display a current density which is phase synchronized to the EMG and if so which $n : m$ type of synchronization can be observed? How does the strength of the corticomuscular synchronization, its $n : m$ type and the corresponding preferred values of the phase difference change provided a transition between different types of coordinated movements occurs? How do different cortical areas, for example, sensory and motor areas, coordinate their activity in order to integrate their information processing? In a nutshell, studies of this kind enable us to investigate the functional role of cerebral areas noninvasively in humans.

In a similar way we can analyze the *synchronization between cortical activity and other peripheral signals, e.g., a periodic stimulus* by investigating the $n : m$ phase synchronization between the cerebral current density and the stimulus. By varying stimulus parameters like frequency and amplitude we can address, for instance, the question as to which neuronal synchronization processes are related to which psychophysical properties of a stimulus. Such experimental studies can be deepened by theoretical investigations dedicated to synchronization processes between periodic stimuli and clusters of phase oscillators as in Chap. 7. Indeed, the physiological meaning of certain dynamical phenomena can be understood the better the more we comprehend the dynamics of the observed processes.

2. *The cerebral flow of stimulus locked $n : m$ transients and short-term epochs of $n : m$ synchronized activity:* Sensory information processing in humans is typically studied by performing experiments where sensory stimuli of different modalities are administered and the corresponding neuronal activity is registered, for example, with MEG (cf. Hämäläinen et al. 1993, Hari and Salmelin 1997). The results obtained by animal experiments addressing this issue show that for the study of cerebral information processing we have to advance towards dynamical phenomena acting on short time scales, that means we have to trace characteristic epochs of a length of, say, 200 ms and less (for a review see Singer and Gray 1995).

With the single run analysis from Sect. 8.4.8 we can study synchronization processes no matter whether or not they are stimulus locked. The time resolution of this method is restricted because it is a sliding window analysis which needs at least about eight periods of an oscillation per window (Tass et al. 1998). However, this can be sufficient for analyzing oscillatory activity in the γ-band range (> 30 Hz). On the other hand with the methods from Sects. 8.4.3–8.4.7 we can probe stimulus locked $n : m$ transients with an enormous time resolution. The latter is only restricted by the sampling rate of the MEG and, correspondingly, reaches down to the millisecond range. Hence, the different data analysis tools presented in this chapter provide us with complementary information.

Using these methods will enable us to study a cerebral flow consisting of short-term epochs of stimulus locked $n : m$ transients and short-term epochs of $n : m$ synchronized activity noninvasively in humans. This flow would be a cascade of synchronous epochs and $n : m$ transients where the successive epochs and transients may emerge in different groups of brain areas. Such a flow may be especially relevant in the context of cerebral short-term information processing. In particular, we may distinguish between those features of this flow which are stimulus locked and those which are not stimulus locked. Different physiological mechanisms might be related to stimulus locked or to non-locked activity, respectively. Moreover, certain neurological diseases might preferably impair one of these mechanisms.

An important aspect of such an investigation of the phase dynamics of the cerebral current density is that it facilitates a comparison with the other imaging techniques which register signals of totally different nature. The functional magnetic resonance imaging (fMRI) assesses the cerebral activity typically by detecting the blood oxygenation level (Ogawa et al. 1990, for a review see Toga and Mazziotta 1996, Andrä and Nowak 1998) whereas the positron emission tomography (PET) measures the cerebral activity, e.g., in terms of the regional cerebral blood flow (rCBF) or the regional cerebral metabolic rate of oxygen (rCMRO$_2$). A comparison of this kind will certainly reveal more comprehensive and deeper insights into human brain functioning.

9. Modelling Perspectives

9.1 Neural Oscillators

Before we dwell on prospective modelling studies in the context of neuronal phase resetting, let us first recall the starting point of our modelling approach. The repetitive firing of a neuron in the Hodgkin and Huxley (1952) model for the squid axon corresponds to a motion along a limit cycle (cf. Murray 1989). As explained in Sect. 3.4 a limit cycle oscillator can be approximated by means of a phase oscillator (Winfree 1967, Kuramoto 1984), and, thus, a population of repetitively firing neurons can be modeled by a cluster of phase oscillators.

The very assumption behind our phase oscillator model (4.3) is that both the coupling and the stimulus remove the single oscillator from its limit cycle in a way which can be neglected. Therefore, in a first approximation we assume that the stimulus does only act on the phase of each oscillator. Accordingly, delivered to a cluster of this kind, a stimulus may modify the pattern of the oscillators' mutual phase differences which may result in changes of the cluster's macroscopic state of synchronization. From the neurophysiological standpoint the effects of stimulation on synchronization processes are significant for several reasons:

For instance, in the context of neuronal information processing induced by sensory stimulation, synchronization plays an important role (Gray and Singer 1987, 1989, Eckhorn et al. 1988): Within networks of cortical neurons patterns of coincident firing appear to be a means for detecting and filtering out distributed but related information (König, Engel, Singer 1996). On the other hand during, for example, epileptic seizures or epochs of tremor pathological synchronized neuronal activity severely impairs normal brain functioning (cf. Freund 1983, Engel and Pedley 1997). In this context stimulation techniques open up promising therapeutic approaches (Benabid et al. 1987). Accordingly, stimulation induced desynchronization may both disturb physiological processes and annihilate pathological rhythms which typically take place in different brain areas. From a physiological point of view it is, therefore, important to extend the model by considering arrangements of interacting clusters reflecting brain anatomy (Sect. 9.1.2).

Let us, next, turn to a stimulus which seriously affects the dynamics of each individual oscillator, so that the assumption behind the phase oscillator concept is no longer valid. In other words, the stimulus removes the single oscillator from its limit cycle so strongly, that we have to take into account the amplitude dynamics of each oscillator in order to understand the cluster's reaction to the stimulus. In this case, e.g., clusters of limit cycle oscillators (Sect. 9.2) or clusters of chaotic oscillators (Sect. 9.3) may be considered as suitable models.

One may even imagine situations when the amplitude dynamics of the individual oscillator comes into effect in a dramatic way: A suitable stimulus may provoke an amplitude death of each oscillator, i.e., the stimulus may stop the neuronal firing (Sect. 9.4). Here, a collective rhythm is abolished already on a microscopic level by stopping the dynamical units the rhythm consists of.

9.1.1 Time-Delayed Interactions

The model should be improved by taking into account further details concerning the neuronal interactions within a cluster, such as time-delayed couplings, randomly distributed eigenfrequencies and coupling strengths as well as coloured noise. This section will serve to illustrate that our model introduced in Chap. 4 also accounts for transmission delays provided they do not exceed one or a few periods of an oscillation. With this aim in view we first consider a system of N weakly interacting limit cycle oscillators given by

$$\frac{\mathrm{d}\boldsymbol{x}_j(t)}{\mathrm{d}t} = \underbrace{\boldsymbol{f}_j(\boldsymbol{x}_j)}_{A} + \underbrace{\varepsilon \boldsymbol{g}_j[\boldsymbol{x}_1(t),\ldots,\boldsymbol{x}_N(t)]}_{B} , \qquad (9.1)$$

where $j = 1,\ldots,N$. Term A gives rise to a limit cycle oscillation in an m-dimensional space, that means $\boldsymbol{x}_j \in \mathbb{R}^m$. For instance, in the Hodgkin-Huxley membrane model the components of $\boldsymbol{x}_j$ describe the dynamics of different sorts of ion channels, respectively (Hodgkin and Huxley 1952). The interactions of the jth oscillator with all other oscillators are modeled by term B. Since these interactions are assumed to be weak we set $0 < \varepsilon \ll 1$.

To illustrate the main impact of delays it is sufficient to consider a simple case where the oscillators have identical eigenfrequencies, i.e. $\omega_1 = \ldots = \omega_N = \Omega$, and are globally coupled. As explained in Chap. 3 the dynamics of (9.1) can be appoximated by a phase model obeying

$$\frac{\mathrm{d}\psi_j(t)}{\mathrm{d}t} = \Omega + \varepsilon \sum_{k=1}^{N} \{K \sin[\psi_j(t) - \psi_k(t)] + C \cos[\psi_j(t) - \psi_k(t)]\} , \qquad (9.2)$$

where ψ_j denotes the phase of the jth oscillator (Ermentrout 1981, Kuramoto 1984, Hoppensteadt and Izhikevich 1997). To illustrate the effect of delays it is sufficient to restrict ourselves to coupling terms of lowest order in (9.2).

Taking into account a transmission delay τ in the interactions of (9.1) yields

$$\frac{d\boldsymbol{x}_j(t)}{dt} = \boldsymbol{f}_j[\boldsymbol{x}_j(t)] + \varepsilon \boldsymbol{g}_j[\boldsymbol{x}_1(t-\tau),\dots,\boldsymbol{x}_N(t-\tau)] , \quad (9.3)$$

where for the sake of simplicity we assume that the delays between all oscillators are identical. Now we encounter the important issue as to how the corresponding phase model looks like. To answer this question we have to distinguish between short and long transmission delays:

1. *Short transmission delay τ:* If τ is comparable with one or a few periods of an oscillation the phase model takes the simple form

$$\frac{d\psi_j(t)}{dt} = \Omega + \varepsilon \sum_{k=1}^{N} \big\{ K \sin[\psi_j(t) - \psi_k(t) - \theta] \quad (9.4)$$

$$+ C \cos[\psi_j(t) - \psi_k(t) - \theta] \big\} , \quad (9.5)$$

where $\theta = \tau\Omega \bmod 2\pi$ (Ermentrout 1994, Hoppensteadt and Izhikevich 1997, Izhikevich 1998). In other words, the delay merely gives rise to an additional phase shift in the interaction terms. By means of elementary trigonometric formulas we can recast (9.5) into the form of (9.2). With the abbreviations

$$\tilde{K} = K \cos\theta + C \sin\theta , \quad \tilde{C} = C \cos\theta - K \sin\theta \quad (9.6)$$

(9.5), thus, reads

$$\frac{d\psi_j(t)}{dt} = \Omega + \varepsilon \sum_{k=1}^{N} \big\{ \tilde{K} \sin[\psi_j(t) - \psi_k(t)] + \tilde{C} \cos[\psi_j(t) - \psi_k(t)] \big\} . \quad (9.7)$$

Accordingly, the delay τ modifies the coupling parameters, and a comparison of (9.7) with (4.3) and (4.4) immediately shows that our model derived in Chap. 4 is valid in this case, too. Taking into account the results of Sect. 6.4.1 we can, thus, conclude that delays may, for example, modify the velocity of the resynchronization following a stimulation induced desynchronization.

Long transmission delay τ: A different situation occurs provided τ is long compared to the period of an oscillation, that means if τ has order of magnitude of $1/\varepsilon$ periods. In this case the transmission delay persists and one obtains the phase model

$$\frac{d\psi_j(t)}{dt} = \Omega + \varepsilon \sum_{k=1}^{N} \big\{ K \sin[\psi_j(t) - \psi_k(t-\tau)] \quad (9.8)$$

$$+ C \cos[\psi_j(t) - \psi_k(t-\tau)] \big\} \quad (9.9)$$

(Izhikevich 1998). This means that as a consequence of the delay in this case the cluster's phase dynamics may become richer and more complex (Schuster and Wagner 1989).

As explained in Sect. 3.4.1 both spiking as well as bursting neurons are modeled by means of phase oscillators with time-dependent phases. Whenever

$\psi_j(t)$ equals a fixed value ρ mod 2π the jth model neuron generates a spike or a burst depending on whether it is active in a spiking or bursting mode. From the neurophysiological point of view it is important to assess whether the cerebral transmission delays have to be considered as short or long compared to the neurons' frequencies. In the brain both the spiking frequencies as well as the frequencies of the neuronal dynamics generating the neurons' bursts range from only a few Hz up to more than 40 Hz (cf. Creutzfeldt 1983). Transmission delays in the brain typically do not exceed, e.g., 100 ms. Hence, for neurons with frequencies of at least up to 10 or even 20 Hz a delay simply induces a phase shift according to (9.5). For such clusters of neurons our model presented in this book is valid. Especially, this holds for the Parkinsonian resting tremor discussed in detail in Chap. 10.

Nevertheless, it is also important to study the dynamical phenomena emerging provided the coupling strength is remarkably enlarged for a short delay or provided one encounters a long transmission delay.

9.1.2 Anatomy of Interacting Clusters

Additionally it is necessary to study the impact of a stimulus on a *group* of interacting clusters. Such groups abound in the brain, where one encounters both mutually interacting clusters and unidirectionally interacting clusters arranged in loops (cf. Creutzfeldt 1983, Steriade, Jones, Llinás 1990, Nieuwenhuys, Voogd, van Huijzen 1991). However, the dynamical synchronization patterns emerging in groups of coupled clusters are not yet understood in a sufficient way. Therefore, similar to the analysis in Chap. 5 the dynamics of interacting clusters of phase oscillators subjected to noise should be analyzed.

In this way suitably prepared, one could apply oneself to study how the interacting clusters are effected by a stimulus. The latter might act on each cluster in a different way. In particular, it might directly act on one of the clusters, while the impact of the stimulation is distributed among all clusters via the oscillators' interactions. For example, a stimulus administered to one cluster within a loop of clusters might induce dynamical features propagating in a wave-like manner through this particular loop.

9.2 Limit Cycle Oscillators

As already discussed in Sect. 3.5 the dynamics of a cluster of weakly interacting limit cycle oscillators can be approximated by means of the dynamics of a population of phase oscillators (Winfree 1967, Ermentrout and Rinzel 1981, Kuramoto 1984). The assumption behind this approximation is that the coupling predominately affects the motion of each oscillator around its limit cycle. As far as the synchronization behavior of several clusters of limit cycle oscillators is concerned this approach turned out to be valid in the case of strong coupling, too (Tass and Haken 1996).

However, a strong stimulus applied to a limit cycle oscillator might remove the oscillator from its limit cycle in a way which cannot be neglected. In this case the assumption behind the phase oscillator concept would be violated. Correspondingly, the cluster's reaction to the stimulus might relevantly depend on effects which are due to the oscillators' amplitudes. Hence, one should analyze the stimulation induced desynchronization in clusters of limit cycle oscillators, e.g., Van der Pol oscillators or oscillators similar to those defined by (2.9).

9.3 Chaotic Oscillators

In a first approximation the dynamics of the phase of an autonomous chaotic oscillator is governed by

$$\dot{\psi} = \Omega + F(A) , \tag{9.10}$$

where ψ and ω denote the phase and the mean frequency of the oscillation, and $F(A)$ models the influence of the chaotic amplitude $A(t)$ on the momentary frequency (Farmer 1981, Pikovsky 1985). Correspondingly, two interacting chaotic oscillators with phases ψ_1 and ψ_2 obey the phase dynamics given by

$$\dot{\psi}_j = \Omega_j + F_j(A_j) + \varepsilon G(\psi_j, \psi_k) , \tag{9.11}$$

where $j \neq k$. The coupling G is 2π periodic in each argument function, and in the simplest case one is allowed to assume that $G(\psi_1, \psi_2) = \sin(\psi_2 - \psi_1)$ holds (Rosenblum, Pikovsky, Kurths 1996). ε is a small quantity fulfilling $0 < \varepsilon \ll 1$. Introducing the phase difference $\phi(t)$ and the frequency detuning $\Delta\Omega$ by

$$\phi(t) = \psi_1(t) - \psi_2(t) , \quad \Delta\Omega = \Omega_1 - \Omega_2 , \tag{9.12}$$

the evolution equation of the phase difference reads

$$\dot{\phi} = \Delta\Omega - 2\epsilon\sin(\phi) + F_1(A_1) - F_2(A_2) . \tag{9.13}$$

Replacing $F_1(A_1) - F_2(A_2)$ in (9.13) by an external random force, one obtains an equation governing the synchronization of two phase oscillators in the presence of noise (Stratonovich 1963, Haken 1983). Rosenblum, Pikovsky, Kurths (1996) have shown that weakly coupled self-sustained chaotic oscillators can synchronize their phases, while their amplitudes remain chaotic and are practically uncorrelated. This dynamical principle, the so-called *phase synchronization of chaotic oscillators*, also governs the synchronization behavior of a cluster of globally coupled chaotic oscillators (Pikovsky, Rosenblum, Kurths 1996). In other words, the synchronization behavior of weakly interacting chaotic oscillators is similar to that of limit cycle oscillators. Remarkably, in this case the chaotic amplitudes act qualitatively in the same way as external noise (Rosenblum, Pikovsky, Kurths 1996, Pikovsky, Rosenblum, Kurths 1996).

Accordingly, the question suggests itself as to whether the stimulation induced desynchronization processes in a cluster of chaotic oscillators are similar to those in a cluster of limit cycle or phase oscillators. The chaotic amplitudes might, for instance, give rise to additional dynamical features in the course of transient desynchronization processes. For this reason one should investigate the impact of different types of stimuli, such as pulsatile or periodic stimuli, on a cluster of chaotic oscillators in the presence of noise and without noise.

9.4 Macroscopic Versus Microscopic

Let us consider a cluster of oscillators, where the amplitude dynamics of each oscillator has a potential as shown in the lower plot of Fig. 6.39. This means that for each oscillator there are two stable dynamical regimes: a limit cycle oscillation and a fixed point with vanishing amplitude, i.e. a black hole (cf. Sect. 6.6). Correspondingly, a neuron modeled by such an oscillator fires or bursts periodically or its firing ceases, where a sufficiently strong stimulus may induce a switching between both states. In particular, a well-timed stimulus may stop the firing of a single neuron. This phenomenon was theoretically and experimentally shown for oscillations in the space-clamped membrane of the squid giant axon: The Hodgkin and Huxley (1952) model describing the neuronal firing of the squid axon has a stable limit cycle regime. Best's (1979) numerical analysis of this model revealed type 0 and type 1 resetting and, consequently, a black hole behavior. Guttman, Lewis, Rinzel (1980) verified the black hole experimentally: The repetitive firing of an axon is stopped by a well-timed stimulus of the right size.

To annihilate synchronized activity of such a cluster a stimulus may act in two different ways:

1. On a *macroscopic scale* the stimulus may desynchronize the cluster as studied in the previous chapters. Incoherency occurs provided the stimulus induces a uniform desynchronization (cf. Sect. 6.2.2). In this case all model neurons remain active and fire incoherently, and the neurons' interactions determine whether or not the cluster resynchronizes after the stimulation.

2. In contrast, the stimulus may also affect the oscillators on a *microscopic scale* by kicking each oscillator into its black hole. In other words, the stimulus provokes an amplitude death of each oscillator, and, corrsepondingly, the neurons' firing stops.

Obviously, there may be combinations of both effects, for instance, only a part of the neurons may be stopped, whereas the other part may be desynchronized. Thus, the cluster's reaction to the stimulus may be rather complex. Nevertheless, from a physiological point of view it is important to analyze such processes.

10. Neurological Perspectives

In the nervous system, in particular, in the brain innumerable clusters of neurons are densely interconnected in a complex way forming groups and loops (cf. Creutzfeldt 1983, Zilles and Rehkämper 1988, Steriade, Jones, Llinás 1990). The functional interactions of these clusters are no less complex and, as yet, they are not sufficiently understood. Different groups of clusters participate in a variety of rhythms, respectively (Niedermeyer and Lopes da Silva 1987, Steriade, Jones, Llinás 1990, Cohen, Rossignol, Grillner 1988, Hari and Salmelin 1997).

Clinical and experimental observations have shown that certain types of stimuli administered to particular neuronal clusters may serve as a treatment for neurological diseases, especially for movement disorders. Stimulation may both induce physiological rhythms and suppress pathological rhythms.

10.1 Therapeutic Stimulation Techniques

There are several types of therapeutic stimulation techniques which are applied to different parts of the nervous system (for a review see Eccles and Dimitrijevic 1985):

1. *Electrical peripheral nerve stimulation:* By administering electrical pulses to peripheral nerve fibers functional movements of an abnormal neuromuscular system are improved or induced, for example, in paraplegic patients (for a review see Vodovnik, Kralj, Bajed 1985).
2. *Spinal cord stimulation:* By means of one or more electrodes the spinal cord is stimulated at the cervical, mid- or upper thoracic level. This type of stimulation is applied to chronic pain syndromes or to movement disorders, for instance, to achieve a modification of dystonic and hyperkinetic conditions (Cook and Weinstein 1973, for a review see Gybels and Van Roost 1985).
3. *Deep brain stimulation:* Chronic therapeutic brain stimulation was first performed by Cooper (1973) with surface electrodes and by Mundinger (1977a) with electrodes implanted in several parts of the brain such as thalamus, pulvinar, and dentate nucleus (for a review see Siegfried and Hood 1985). Initially, deep brain stimulation was used for the treatment

of spasmodic torticollis (Mundinger 1977a), which is a particular sort of movement disorder, and chronic pain syndromes (Mundinger 1977b). In the meantime deep brain stimulation turned out to be a promising therapy, for instance, for parkinsonian tremor (Benabid et al. 1987, 1991, Blond et al. 1992, for a review see Volkmann and Sturm 1998, Volkmann, Sturm, Freund 1998).

All of the above mentioned stimulation techniques were developed from clinical observations and are empirically based. A better theoretical understanding of the mechanisms of stimulation may help to improve these techniques.

10.2 Parkinsonian Resting Tremor

This section is devoted to the parkinsonian resting tremor. First, it will be sketched how pathological rhythmic activity in different brain areas acts in combination to produce the tremor. This will prepare us to understand the impact of irreversible or temporary modifications of the tremor generating areas by means of stereotactic operations or deep brain stimulation. Finally, it will be outlined why and how deep brain single pulse stimulation should be effective in patients suffering from parkinsonian resting tremor.

10.2.1 Disease Mechanism

The genesis of physiological and pathological tremors is not yet sufficiently understood (Freund 1983, Elble and Koller 1990). Resting tremor in Parkinson's disease (PD) is an involuntary shaking with a frequency around 3 Hz to 6 Hz which predominantly affects the distal portion of the upper limb and classically decreases or vanishes during voluntary action (Freund 1983).

Several studies clearly indicate that parkinsonian resting tremor is caused by clusters of neurons synchronously firing at a frequency similar to that of the tremor (Llinás and Jahnsen 1982, Pare, Curro'Dossi, Steriade 1990): In the anterior nucleus of the ventrolateral thalamus (VLa) there are the so-called no-response cells which are neither modulated by somatosensory stimuli nor by active or passive movements (Lenz et al. 1994). The synchronized output of these no-response cells acts on the periphery via the motor cortex. This follows from recordings of tremor locked activity from the sensorimotor cortex during neurosurgery in PD patients (Alberts, Wright, Feinstein 1969) and in monkeys with experimentally induced Parkinson-like tremor (Lamarre and Joffroy 1979). Apart from the thalamic cluster there are also clusters of synchronously active neurons in the basal ganglia (pallidum and subthalamic nucleus) which contribute to the tremor generation (cf. Bergman et al. 1994, 1998).

Along the lines of a first qualitative model Volkmann et al. (1996) suggested that the synchronized oscillatory activity of VLa feeds into two loops of rhythmically active neuronal clusters (Fig. 10.1): The thalamus (VLa) drives the supplementary motor area (SMA) and the premotor cortex (PMC), where both SMA and PMC drive the primary motor cortex (M1). The intrinsic loop is closed as SMA, PMC and M1 feed back into VLa via the basal ganglia. On the other hand peripheral feedback from muscle spindles and joint receptors reaches the motor cortex via the thalamus, in this way forming an extrinsic loop.

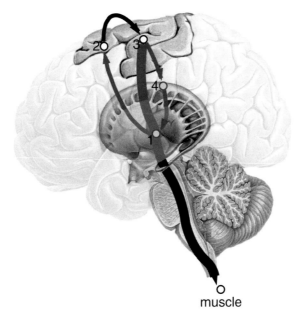

muscle

Fig. 10.1. *Schematic illustration of the disease mechanism of the parkinsonian resting tremor* according to Volkmann et al. (1996): Synchronized neuronal activity in the thalamus (VLa, denoted by "1") drives premotor areas (SMA and PMC, "2"). The latter drive the primary motor cortex (M1, "3") which activates the muscles via the pyramidal tract (thick arrow). M1 ("3") feeds back into the thalamus ("1") via the basal ganglia ("4") in this way closing the intrinsic loop. Additionally there is an extrinsic loop (not marked in the figure) because the peripheral feedback from muscle spindles and joint receptors reaches the motor cortex via the cerebellum and thalamus. (Courtesy of J. Volkmann)

Concerning the interaction of the different neuronal clusters, in an MEG study it was shown that during epochs of tremor activity one observes a pronounced corticomuscular synchronization (CMS) as well as a pronounced cortico-cortical synchronization (CCS) (Tass et al. 1998). In that study $n : m$ phase synchronization between MEG and EMG data (Fig. 10.2) was analyzed as explained in Sect. 8.4.8. Figure 10.3 shows the different synchronization processes observed in a patient who had a tremor of the right hand and fore-

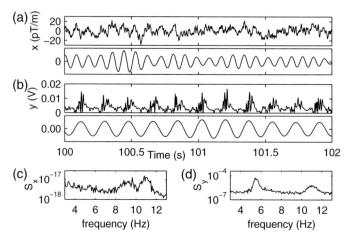

Fig. 10.2a–d. *MEG and EMG signals and their power spectra:* An original and the correspoding bandpass filtered (10-14 Hz) MEG signal from a channel over the left sensorimotor cortex (**a**) and the power spectrum of the original MEG signal (**c**). The original and the bandpass filtered (5-7 Hz) EMG signal of the right flexor digitorum superficialis muscle (**b**) and the power spectrum of the original EMG signal (**d**). (From Tass et al. (1998)).

arm with a principal frequency component between 5 and 7 Hz. Pronounced tremor activity starts after approx. 50 s (Fig. 10.3a). During this epoch we observe a peripheral coordination in terms of a 1:1 antiphase locking of the EMG of two antagonistic muscles, namely, the right flexor digitorum superficialis and the right extensor indicis muscle (Fig. 10.3b). Moreover, during that period of tremor activity we observe a CMS as well as a CCS. The activity of both sensorimotor and premotor areas is 1:2 phase locked with the EMG activity of flexor as well as extensor muscles (Figs. 10.3c and 10.4), whereas the activities of these two brain areas are 1:1 locked (Fig. 10.3d). In this way it turned out that the peripheral coordination of the resting tremor, i.e. the strength of the anti-phase synchronization of flexor and extensor muscle, reflects the strength of the CCS.

The window length T and the filter parameters may influence the $n : m$ synchronization analysis discussed in Sect. 8.4.8. Hence, the impact of variations of these parameters on the results has to checked. Tass et al. (1998) used bandpass FIRCLS filters (Selesnick, Lang, Burrus 1996). The filters were tested by varying the band edges in equidistant steps between 3.5–7 Hz and 5–6.5 Hz for the lower band and between 7–14 Hz and 10–13 Hz for the higher band. The window length T was varied between 2 and 20 seconds in equidistant steps. The results were robust with respect to variations of T and the band edges of all filters used. Figures 10.3 and 10.4 show the findings obtained for the following parameters: bandpass of EMG signals: 5–7 Hz, bandpass of MEG signals: 5–7 and 10–14 Hz (for the quantification of 1:1

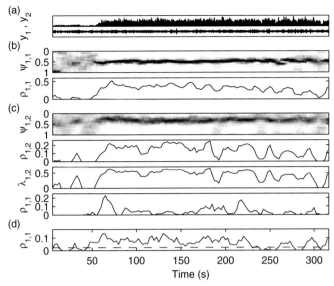

Fig. 10.3a–d. *Related synchronization processes:* (**a**) EMG of the right flexor digitorum superficialis muscle (RFM, *upper trace*) and an MEG channel over the left sensorimotor cortex (LSC, *lower trace*). (**b**) 1:1 synchronization between RFM and right extensor indicis muscle: The distribution of the cyclic phase difference $\theta_{1,1} = \psi_{1,1}/(2\pi)$ (cf. (8.57)) calculated in the running window $[t - 5s, t + 5s]$ is displayed as a gray-scale plot (*upper plot*). White and black correspond to minimal and maximal values, respectively. The lower plot shows the corresponding significant synchronization index $\rho_{1,1}$ (cf. (8.61)). (**c**) 1:2 corticomuscular synchronization: Time course of the distribution of the cyclic phase difference $\theta_{1,1} = \psi_{1,2}/(2\pi)$ between an MEG signal from the LSC and the EMG of the RFM (*uppermost plot*). The corresponding indices $\rho_{1,2}$ and $\lambda_{1,2}$ are plotted below and can be compared with the 1:1 synchronization index $\rho_{1,1}$ between LSC and RFM (*lowest plot*). (**d**) 1:1 cortico-cortical synchronization between LSC and a premotor MEG channel. The value of $\rho_{1,1}$ belonging to the 99.9th percentile of the surrogates is indicated by the dashed line. Significance levels are $\rho^s_{1,2} = 0.03$, $\lambda^s_{1,2} = 0.26$, $\rho^s_{1,1} = 0.07$ [(**b**) and (**c**)], $\rho^s_{1,1} = 0.03$ (**d**). (From Tass et al. (1998)).

and 1:2 locking, respectively). The comparison between 1:2 and 1:1 CMS was additionally confirmed by analyzing the data filtered with two-band filters (e.g., 5–7 Hz plus 10–14 Hz).

This survey suggests that the rhythmic activity in the thalamus and the basal ganglia dominating the tremor generation may be considered as a therapeutic target, so to speak an Achilles heel of the disease mechanism. Suppressing this particular rhythm might in a sense rob the complex tremor inducing neuronal dynamics of its driving force. Stereotactic lesioning and deep brain stimulation are two surgical approaches aiming at selectively suppressing the effects of the pathological activity in the thalamus and the basal ganglia.

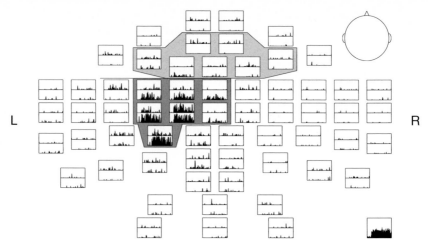

Fig. 10.4. *1:2 phase synchronization between cerebral and tremor activity:* Plot shows the time dependence of the significant synchronization index $\rho_{1,2}$ from (8.61) which characterizes the strength of the 1:2 locking between the EMG of the right flexor digitorum superficialis muscle (reference channel, plotted in the lower right corner) and all MEG channels. Each rectangle corresponds to an MEG sensor, i.e. a SQUID. The time axis runs from 0 to 310 s. The y-axis scales from 0 to 0.25. The head is viewed from above, where "L" and "R" correspond to the left and right side as indicated by the "head" in the upper right corner. The upper and lower gray regions correspond to premotor areas (SMA and PMC) and contralateral sensorimotor areas (M1), respectively. The results for the extensor muscle are similar. Window length $T = 10$ s and significance level $\rho_{1,2}^{s} = 0.03$. Same data as in Fig. 10.3. (From Tass et al. (1998)).

10.2.2 Stereotactic Treatment

Brain surgery is applied for the treatment of various types of movement disorders (for a review see Gildenberg 1985). The modern surgical treatment of tremor was introduced by Cooper (1953, 1956) and Cooper and Bravo (1958) who accidentally observed that due to a destruction of portions of the thalamus the tremor in a PD patient vanished. Thereupon a series of studies was dedicated to reveal the most effective target area in the brain which at that time was considered to be the so-called ventralis intermedius nucleus (VIM) of the thalamus (Ohye and Narabayashi 1979). The corresponding tremor suppressing operation is called VIM thalamotomy.

With the introduction of effective drug therapy for PD (L-DOPA) interest in the stereotactic treatment of parkinsonian tremor declined dramatically, so that for a long time the surgical treatment of PD was practically fallen into oblivion (Tasker et al. 1983). After six years of L-DOPA therapy, however, half the patients suffer from side effects such as dyskinesias (uncontrolled excessive movements), so-called motor fluctuations or psychosis (Poewe, Lees, Stern 1986). Accordingly, the disadvantages and complications of the late stage L-DOPA therapy revived a need for surgical treatment (cf. Benabid et al. 1996).

Nowadays, conventional drug therapy is still the treatment of choice for the early years of PD. Nevertheless, the surgical treatment is meanwhile established since it opened up important therapeutic alternatives. Besides VIM thalamotomy another stereotactic operation, the so-called pallidotomy, was developed (Laitinen, Bergenheim, Hariz 1992, for a review see Volkmann and Sturm 1998, Volkmann, Sturm, Freund 1998). During pallidotomy lesions are stereotactically placed in a particular target area of the basal ganglia, namely in the posteroventral aspect of the internal pallidum (Laitinen, Bergenheim, Hariz 1992). This stereotactical operation turned out to be an effective therapy for the main symptoms of PD, i.e., the tremor, the rigidity and the akinesia. Additionally, it reduces the drug-induced late stage dyskinesias.

During stereotactic tremor surgery small circumscribed lesions may be placed within the brain with millimeter precision by using a stereotactic frame system attached to the patient's head. Hereby the most effective target area has to be localized in order to avoid complications. The localization of the target area is achieved by means of deep brain recording or electrical stimulation (cf. Benabid et al. 1996). During procedures of this kind it was observed that parkinsonian tremor was suppressed acutely with electrical stimulation (for a review see Benabid et al. 1996). Consequently, Benabid et al. (1987) proposed chronic high-frequency deep brain stimulation instead of thalamic lesioning in PD for the treatment of tremor. For this a stimulating electrode is stereotactically implanted into the thalamic target area and connected to a subcutaneous current generator.

Today deep brain stimulation with chronically implanted electrodes is a new and promising therapeutical alternative to the established surgical therapies (Benabid et al. 1991, Blond et al. 1992, for a review see Volkmann and Sturm 1998, Volkmann, Sturm, Freund 1998). Indeed, the stimulation appears to have several advantages. For instance, deep brain stimulation seems to be a more gentle procedure with a lower risk of permanent neurological deficits. Moreover, the stimulation parameters can individually be adjusted to the patient's conditions during the course of the disease. Apart from the thalamus (Benabid et al. 1991, Blond et al. 1992) other target areas in the basal ganglias like the so-called subthalamic nucleus (Limousin et al. 1995) and the pallidum (Siegfried and Lippitz 1994, Volkmann et al. 1998) may be used for the deep brain stimulation. Future studies have to reveal the most effective electrode localization.

Since deep brain stimulation advanced to an important therapeutic means, it is necessary to understand the stimulation mechanism thoroughly, e.g., in order to improve this procedure. Empirically it is known that only high-frequency stimulation (> 100 Hz) causes a reversible blockade of the stimulated area. However, as yet, it is unclear how thalamic stimulation suppresses parkinsonian tremor (cf. Andy 1983, Benabid et al. 1991, 1993, Blond et al. 1992, Caparros-Lefebvre et al. 1994, Strafella et al. 1997).

10.2.3 Resetting the Tremor Rhythm

The different types of stimulation techniques theoretically studied in this book may be applied in different neurological diseases associated with pathological synchronized neuronal activity. Besides other movement disorders parkinsonian resting tremor is particularly suitable for studying the effects of single pulse stimulation for several reasons:

1. Circumscribed clusters of synchronously active neurons in the thalamus and the basal ganglia act on cortical motor areas in this way giving rise to the parkinsonian resting tremor (cf. Bergman et al. 1994, 1998, Volkmann et al. 1996).

2. By means of deep brain electrodes in PD patients it is possible to record the pathological activity of these particular clusters as well as to stimulate them. This involves a remarkable chance for administering stimuli to the very neuronal clusters responible for the tremor activity and to measure the effects of stimulation directly. A stimulation of this kind is superior to peripheral nerve stimulation. In the latter case, for instance, peripheral transfer effects and contributions of reflex arcs restrict the possibilities of performing an effective stimulation.

3. Suitable parameters for a desynchronizing single pulse stimulation should be revealed as explained in Sect. 8.3. If it is possible to induce a transient desynchronization of the tremor activity in the thalamus or the basal ganglia we are able to study whether and if so how this activity recovers and how the transient desynchronization influences other neuronal clusters like those of the intrinsic and extrinsic loops mentioned above.

4. It is not necessary to use high intensities for carrying out deep brain single pulse stimulation. Rather one may profit from the subcritical long pulses with their gentle stimulation intensities and their advantageous dependence of the stimulation outcome on the initial phase (Sect. 6.7).

5. Our single pulse approach can be based on a broad foundation of experimental facts, which was revealed with different stimulation techniques:

Periodic pulsatile deep brain stimulation in PD patients:

Although appropriate data analysis tools were not available in the late 1950s the results of Hassler et al. (1960) clearly indicate that, for example, periodic pulsatile stimulation of the pallidum (a part of the basal ganglia) at rates similar to the tremor frequency induces a 1:1 entrainment of the tremor combined with an increase of the tremor amplitude. Periodic pulsatile stimulation with higher frequencies (25–100 Hz) mostly dampens the tremor amplitude and may even lead to a complete tremor arrest (Hassler et al. 1960).

Benabid et al. (1991) studied the effects of periodic deep brain stimulation of the thalamus (VIM) with low-intensity current pulses (0.2–2.0 mA) in PD patients in detail: The tremor suppressing effect started at pulse frequencies higher than 60 Hz and was most pronounced for frequencies between 100 and 250 Hz. Pulse width was varied from 60 to 210 μsec and turned out not to be a critical parameter for the stimulation outcome. Hence, to spare the battery, for the chronic therapeutic stimulation short pulses (60 μsec) were selected.

Peripheral single pulse and sinusoidal stimulation:
A large number of investigations was devoted to the impact of peripheral stimulation on different sorts of tremor, thereby addressing various aspects of stimulation induced neural dynamics (for a review see Elble and Koller 1990). To this end the peripheral nervous system was stimulated with, for instance, electrical or mechanical stimuli such as torque-pulse perturbation to the wrists of patients (Lee and Stein 1981, Elble, Higgins, Moody 1987) or electrical stimulation of peripheral nerves (Bathien, Rondot, Toma 1980, Deuschl, Lucking, Schenk 1987, Britton et al. 1993a). The stimulation induced reponse in the peripheral nervous system is forwarded to the central nervous system, where it finally influences the tremor generating areas.

The majority of these resetting studies was devoted to examining another type of tremor, namely the essential tremor. The disease mechanism of the latter is different compared to the parkinsonian resting tremor. It was suggested that essential tremor is due to synchronized neuronal activity emerging not in the thalamus but in the olivocerebellar system (Lamarre et al. 1971, Lamarre et al. 1975, for a review see Freund 1983, Elble and Koller 1990). Nevertheless, also in the present context it is remarkable that Elble, Higgins, Hughes (1992) were able to induce 1:1 and 2:1 entrainment between a sinusoidal force delivered by a computer-controlled torque motor and the essential tremor in all 10 investigated patients. Performing single pulse perturbation with the same torque motor type 0 resetting was observed only in 5 patients, whereas in the other 5 patients phase resetting curves could not be classified reliably. In this case the data analysis method explained in Sects. 8.2.2 and 8.3 might improve the evaluation of the stimulation induced new phase. However, since type 1 resetting is trivially achieved, e.g., by negligible stimuli, the observation of type 0 resetting implies the possibility of a desynchronizing peripheral stimulation (cf. Sect. 6.2.3).

Tremor stimulation experiments were also performed with transcranial magnetic stimulation (cf. Britton et al. 1993b, Pascual-Leone et al. 1994), where magnetic stimulators noninvasively act on brain tissue (Barker, Jalinous, Freeston 1986). Although these studies were not addressing phase resetting in the way described in Sect. 8.3, such experiments are certainly worth performing. However, transcranial magnetic stimulation can only influence cortical neurons and its effects on rhythmical brain activity are difficult to monitor. Up to now a direct stimulation and registration of thalamic activity can only be achieved by means of deep brain stimulation.

6. Deep brain single pulse stimulation might contribute to an improvement of the therapy of the cardinal features of PD (tremor, rigidity and akinesia)

on the one hand and drug-induced dyskinesias on the other hand. One might imagine a stimulation technique consisting of a calibration and a stimulation with feedback control:

Calibration: As explained in Sect. 8.3 appropriate parameters for a desynchronizing stimulation have to be determined. The strength of the neuronal interactions and other related physiological parameters may undergo variations, for instance, of a circadian type. Correspondingly, the optimal stimulation parameters may vary, too. For this reason it may possibly be necessary to perform the calibration several times a day.

Stimulation with feedback control: To carry out a stimulation as gentle as possible and only during periods of tremor activity, one might repetitively administer a subcritical long pulse to the thalamic target area: The strength of the particular pathologically active cluster is, e.g., assessed by means of measuring the cluster's LFP via the chronically implanted electrode. Whenever this cluster resynchronizes, that means whenever a threshold of the LFP is exceeded, a well-timed subcritical long pulse is again administered via that electrode in order to desynchronize afresh. This iterative stimulation mode would reduce stimulation induced side effects due to current spread and it would prevent from possible tissue damage. Moreover it would save the stimulator's battery since the stimulation intensity required for this sort of stimulation is low compared to the strength of the neurons' mutual interactions (cf. Sect. 6.7).

11. Epilogue

The intention behind this book is to promote the investigation of phase resetting and stimulation in populations of neuronal oscillators. To this end it is necessary to bring together scientists who belong to different communities: the biologists, neurologists, neuroscientists, neurosurgeons and physiologists on the one hand and the computer scientists, engineers, mathematicians and physicists on the other hand. Their particular knowledge and intuition may contribute to an interdisciplinary enterprise truely devoted to neurophysiological issues.

This need not be a difficult undertaking. Without doubt, the interests of scientists from different disciplines are often diametrically opposed, and to compound matters, there are the communities' different languages, the different types of thinking and argueing. Anyhow, with good reason we may expect common efforts of this kind to get going: The study of phase resetting already turned out be fruitful in other branches of biology and physiology (Winfree 1980, Glass and Mackey 1988). Moreover, the approach presented in this book is based on Haken's (1977, 1983) synergetics, a theory of self-organization which smoothed the way for an interdisciplinary study of brain functioning (Haken 1996). Correspondingly, this book may be considered as an attempt at giving an impression of both the clinical importance as well as the imperative scientific aesthetics of phase resetting in populations of neuronal oscillators.

The success of future studies, although not predictable, will certainly be the larger the more these studies are guided by important and promising neurophysiological questions. Hence, this approach may centre upon three main themes: natural sensory stimulation, experimental electrical and magnetic stimulation, and therapeutic stimulation.

11.1 Natural Sensory Stimulation

A huge flow of information permanently streams into the brain via the sensory system. How is the relevant information extracted? How does the brain, for instance, manage to recognize single objects of the visual world? The local features of an object are encoded by the firing of the corresponding clusters of neurons, respectively (Hubel and Wiesel 1959). To recognize a single object as a whole it is necessary to separate different objects from each other and from the background (cf. Marr 1976, Treisman 1980, 1986, Singer 1989, Julesz 1991). In other words, those clusters of neurons which belong to the same object have to somehow combine themselves to form a functional group even if they are located in remote brain areas. This issue, the so-called binding problem, is not yet entirely resolved.

It was suggested that this binding process is realized by means of a temporal neuronal firing code (cf. Hebb 1949, Milner 1974, Abeles 1982, Crick 1984, von der Malsburg and Schneider 1986, Shimizu, Yamaguchi, Tsuda, Yano 1985, Singer 1989), in particular, an in-phase synchronization (Milner 1974, von der Malsburg and Schneider 1986). Accordingly, the recognition of an object requires a synchronous firing of all neurons that encode local features belonging to this particular object. This hypothesis was supported by numerous experimental results in rabbit (Freeman 1975), cat (Gray and Singer 1987, Eckhorn et al. 1988), pigeon (Neuenschwander and Varela 1993) and monkey (Livingstone 1991, Kreiter and Singer 1992, Murthy and Fetz 1992, Eckhorn et al. 1993).

Binding processes and, more general, the interactions between different brain areas can noninvasively be studied in humans in the following way: Synchronously firing neurons generate a magnetic field which can be registered outside the head with multichannel magnetoencephalography (MEG) (Cohen 1972, cf. Ahonen et al. 1991, Hämäläinen et al. 1993). In order to directly study the neuronal activity, e.g., of cortical areas we have to assess the cerebral current density from the magnetic field measured with the MEG by means of a suitable inverse method such as the minimum norm least squares (MNLS) method (Hämäläinen et al. 1993, Wang, Williamson, Kaufman 1995), the cortical current imaging (CCI) (Fuchs et al. 1995), the probabilistic reconstruction of multiple sources (PROMS) (Greenblatt 1993), the low resolution electromagnetic tomography (LORETA) (Pascual-Marqui, Michel, Lehmann 1994), or the magnetic field tomography (MFT) (Clarke and Janday 1989, Clarke, Ioannides, Bolton 1990, Ioannides, Bolton, Clarke 1990). In this way we can estimate the current density belonging to each cerebral volume element, respectively. The interactions between different areas of the brain can then be investigated by analyzing synchronization processes between the cerebral current density of the different areas.

The function of different brain areas in sensory information processing in humans is typically explored by performing experiments where stimuli of different modalities like tones or flashlights are administered and the corresponding neuronal activity is measured, e.g., with MEG (cf. Hämäläinen et al. 1993). Different features of synchronization processes of the cerebral current density during stimulation experiments of this kind can be analyzed by means of two complementary approaches:

The first technique is a single run method for the study of $n : m$ phase synchronization in noisy nonstationary data (Tass et al. 1998). As explained in Sect. 8.4.8 in this case synchronization is understood as a process in the course of which two oscillators tend to adjust their rhythms despite of noise. Accordingly, $n : m$ phase synchronization is understood in a statistical sense as an existence of preferred values of the oscillators' $n : m$ phase difference (modulo 2π). In order to detect short-term epochs of $n : m$ phase locked current density a sliding window analysis is performed: Within each window the strength of the synchronization is quantified by comparing the actual distribution of the $n : m$ phase difference with the uniform distribution. Thus, the time resolution of this method is restricted since one needs a window length which corresponds to at least about eight periods of an oscillation. As a consequence of this type of analysis epochs of $n : m$ phase locked current density are detected no matter whether or not these epochs are stimulus locked.

The second approach was presented in Sects. 8.4.3–8.4.7. The spirit of this approach is totally different compared to that of the first one. In Chap. 6 we saw that a stimulus which is applied to a cluster of synchronized oscillators typically induces a transient response. Thus, we may expect that the stimulation induced responses of interacting neuronal clusters located in different brain areas may be related in some way. More precisely, the stimulus may give rise to short-term epochs displaying a stereotyped time course of the $n : m$ phase difference of the clusters' current densities. It is important to take into account that the $n : m$ phase difference of such responses can be stimulus locked although the responses themselves are not stimulus locked. Hence, we can easily miss cerebral interaction patterns of this kind if we only use standard averaging techniques which only detect stimulus locked responses.

For this reason it was necessary to design new methods for the detection of different sorts of stimulus locked $n : m$ transients, that means stereotyped epochs of the $n : m$ phase difference, $n\varphi_1(t) - m\varphi_2(t)$, which follow the stimulus with a fixed time delay, where φ_1, φ_2 denote the phases of the signals belonging to both brain areas, respectively. Such epochs may but need not display $n : m$ phase locking which corresponds to a plateau of the $n : m$ phase difference. The time resolution of these methods is only restricted by the sampling frequency of the MEG and, thus, reaches down to the millisecond range. Therefore these techniques are particularly appropriate for detecting short-term eruptions of stimulation induced patterns of the $n : m$ phase differences of the current densities of different brain areas. Such transient epochs may be very short, for instance, they may correspond to less than two periods of an oscillation.

Another important target of such an analysis are stimulus locked $n : m$ transients with conduction delay. In this case, for example, a sensory stimulus induces a response in a certain brain area, within which this sensory input is processed and than forwarded to another brain area. Due to the conduction delay the responses of both areas do not need to overlap in time. Nevertheless, the time course of their time delayed $n : m$ phase difference, $n\varphi_1(t - \tau) - m\varphi_2(t)$, may be stereotyped where τ denotes the conduction delay, and φ_1, φ_2 are the phases of the signals belonging to both brain areas.

Using the data analysis methods of Chap. 8 enables us to probe the coordinated sequential and spatially distributed neuronal processing. In particular, we can study whether a sensory stimulus induces a flow of $n : m$ synchronized epochs and stimulus locked $n : m$ transients. Such a flow would be a cascade consisting of both short-term epochs of fully developed $n : m$ synchronized current density on the one hand and short-term epochs of stimulus locked $n : m$ transients of the current density on the other hand. According to the particular function of the brain areas the successive epochs of this flow may emerge in different groups of brain areas. In this way a flow of this kind can be the correlate of a neuronal information processing caused by the sensory stimulus.

It is important to distinguish between those features of this flow of synchronized and transient cerebral activity which are stimulus locked and those which are not stimulus locked. Both types of dynamical phenomena may belong to submechanisms serving different physiological purposes. For example, one submechanism may correspond to a flow of tightly stimulus locked synchronous epochs and $n : m$ transients which follow every single stimulus in a stereotyped way. The other submechanism may consist of a flow of synchronous epochs which are not stimulus locked and which may, e.g., reflect an intrinsic variablity which is necessary for adaptation and learning. Moreover, under certain pathological conditions submechanisms of this kind may undergo characteristic modifications.

Additionally, we have to be aware of the fact that the stimulus locked skeleton of this flow need not go off in a rigid way like a reflex. Rather we might expect that there may be a certain variability of the stimulus locked parts of the synchronous and transient activity. This variability may depend on the state of one or more relevant physiological rhythms at the beginning of the sensory stimulation. The time course of the successively emerging epochs as well as the involved brain areas may decisively depend on the rhythms' initial phases at which the stimulus is administered. This variablility of stimulus locked reactions might be the physiological basis of an internal receptivity and susceptibility to sensory stimuli. On the other hand, pattern recognition processes may give rise to variable coordinated neuronal responses in terms of multiple stimulus locked $n : m$ transients (Sect. 8.4.5). The different branches of multiple stimulus locked $n : m$ transients might be related to qualitatively different perceptions or sensations. The data analysis tools presented in Sects. 8.4.4–8.4.7 enable us to study processes of this kind.

11.2 Experimental Electrical and Magnetic Stimulation

Neuronal synchronization is not only important in the context of pattern recognition. Rather under physiological as well as pathological conditions synchronization appears to be a basic mechanism concerning the neuronal interactions within and between brain areas. On the one hand many animal experiments clearly indicate that the control of coordinated movements is based on a synchronization of the firing of clusters of neurons in different motor areas (Roelfsema et al. 1997, for a review see MacKay 1997). On the other hand neuronal synchronization is observed during pathological processes like epochs of pathological tremor (Freund 1983, Elble and Koller 1990) or epileptic seizures (Engel and Pedley 1997). In this context it is particularly important to understand how structurally intact brain areas become functionally impaired by, for instance, entrainment processes caused by neuronal populations generating pathological rhythmic activity.

Experimental electrical and magnetic stimulation is an indispensable tool for investigating the repertoire of spatiotemporal cerebral synchronization processes. A periodic stimulus, for example, may mimic the driving exerted by a neuronal population displaying pathological rhythmic activity. In contrast to that, pulsatile stimuli administered in different brain areas may be used to analyze characteristic patterns of evoked changes of the brain activity in different areas and the corresponding behavioral changes. Such examinations can be performed in animal experiments (Fritsch and Hitzig 1870, for a review see Creutzfeldt 1983, Steriade, Jones, Llinás 1990) and in patients during neurosurgery (Foerster 1936, Penfield, Boldrey 1937, for a review see Freund 1987). Additionally, in neurology stimulation is used for diagnostic purposes (cf. Barker, Jalinous, Freeston 1986, Meyer 1992, Schnitzler and Benecke 1994, Classen et al. 1997).

By means of the data analysis tools of Chap. 8 we can assses the relevant variables out of the data obtained during stimulation experiments of this kind. If we, for instance, perform a repetitive pulsatile stimulation and register the local field potential with several electrodes placed in different parts of the animal's brain we can use the data analysis techniques of Sects. 8.4.3–8.4.8 which were already discussed above. In this way we can study the flow of $n : m$ synchronized epochs and stimulus locked $n : m$ transients. On the other hand as far as periodic stimuli are concerned the single run $n : m$ synchronization analysis by Tass et al. (1998) is just made for investigating cerebral entrainment processes.

Using the data analysis methods of Chap. 8 it, thus, becomes possible to compare experimentally observed synchronization and desynchronization processes with the multitude of theoretically revealed stimulation induced dynamical phenomena which were presented in Chaps. 6 and 7.

11.3 Therapeutic Stimulation

It is the physician's pressing wish to use the insights into stimulation induced neuronal dynamics for therapeutic purposes. For instance, several movement disorders such as Parkinson's disease are suitable for this. In Parkinson's disease pathological rhythmic activity feeds into loops of neuronal clusters in this way driving the involved clusters (cf. Llinás and Jahnsen 1982, Pare, Curro'Dossi, Steriade 1990, Volkmann et al. 1996). Such diseases can be treated by suppressing pathological neuronal activity which is realized by means of high-frequency deep brain stimulation (Benabid et al. 1991, Blond et al. 1992). To this end electrodes are chronically implanted within the brain with millimeter precision within a particular neuronal cluster. Of course, this kind of treatment is only done in patients that do not benefit from drug therapy any more (for a review see Volkmann and Sturm 1998).

Remarkably, observations during stereotactic surgery were the decisive factor for developing the high-frequency deep brain stimulation of prakinsonian resting tremor (cf. Benabid et al. 1996). As yet, it is not understood how this type of stimulation suppresses the tremor generating rhythmic neuronal activity (cf. Andy 1983, Benabid et al. 1991, 1993, Blond et al. 1992, Caparros-Lefebvre et al. 1994, Strafella et al. 1997). For this reason we need a profound theoretical basis to understand the experimental observations and to improve the stimulation techniques.

Based especially on the theoretical results of Sect. 6.7 a new stimulation mode, namely single pulse deep brain stimulation with feedback control, was suggested in Sect. 10.2.3 for the treatment of the parkinsonian resting tremor in patients with chronically implanted electrodes. The very idea behind this type of stimulation is to desynchronize a cluster of neurons that generates pathological activity which feeds into loops of neuronal populations. The desynchronization is achieved with a well-timed pulsatile stimulus administered via a chronically implanted electrode. As soon as the cluster resynchronizes, and the extent of the cluster's synchronization exceeds a certain threshold, the next well-timed pulsatile stimulus is administered in order to desynchronize the cluster again. According to our theoretical investigations such a stimulation mode can be carried out with pulses of low intensity. Hence, this treatment would be gentle but, nevertheless, effective: Gentle pulses are used to trip the temor rhythm up whenever it gets going.

A stimulation method of this kind may be a starting point for developing further techniques which aim at influencing the state of synchronization by means of flexible and feedback based approaches. With this aim in view depending on the clusters' dynamics we might use single pulses, pulse trains, periodic and chaotic signals or even an appropriately modified feedback signal like a time-delayed local field potential as stimuli.

No matter which topic we want to turn our attention to – the natural sensory, the experimental or the therapeutic stimulation – we may expect to obtain more and deeper insights if experiments, clinical studies, data analysis and theoretical modelling are well-suited to each other. To compare the experimental data with the variety of theoretically revealed stimulation induced dynamical phenomena presented in Chaps. 6 and 7, the data analysis tools of Chap. 8 are available.

I hope that this book will form the foundations for a sound understanding of stimulation induced processes in clusters of neuronal oscillators subjected to noise. Moreover, it is my sincere desire that one day patients will benefit from these efforts.

Appendices

A. Numerical Analysis
of the Partial Differential Equations

The numerical analysis of the Fokker–Planck equation (2.30) presented in Chap. 2 was performed by integrating the Fourier transformed Fokker–Planck equation (2.47). The latter is an infinite-dimensional system of ordinary differential equations. Accordingly, (2.47) was approximated by an M-dimensional system of ordinary differential equations of the Fourier modes with wave vectors $k = 1, \ldots, M$ in the following way: Fourier modes with wave number k fulfilling $|k| > M$ were set equal to zero. $\hat{f}(0, t)$ is given by (2.46). Fourier modes with negative wave vectors (i.e. $k = -1, \ldots, -M$) were determined according to (2.45).

For the numerical integration, for instance, by means of a 4th order Runge-Kutta algorithm the time step Δt and the number of Fourier modes M were appropriately chosen so that a sufficient accuracy was achieved. The latter was tested by comparing the numerically determined stationary solution with the analytically derived stationary solution f_{st} from (2.39). The numerical accuracy was estimated by calculating the relative error of the distribution

$$E_f = \frac{\|f_{st} - f_{num}\|}{\|f_{st}\|}, \tag{A.1}$$

where f_{num} denotes the numerical solution to which the M-dimensional system converges. The norm is defined by $\|f\| = \sqrt{\int_0^{2\pi} f^2(\xi)\, d\xi}$. Additionally the relative error of the ensemble variable

$$E_Z = \frac{|Z_{st} - Z_{num}|}{|Z_{st}|} \tag{A.2}$$

was determined, where Z_{st} and Z_{num} were determined by inserting f_{st} and f_{num} into (2.33), respectively.

With increasing noise level the shape of f becomes smoother. Hence it was possible to reduce M for higher noise level. Accordingly for the numerical integrations of Chap. 2 the parameters $\Delta t = 0.00025$ and $M = 130$ (200) were chosen for $Q = 0.4$ (0.2). In all simulations the relative error of the distribution E_f was ≈ 0.005, whereas the relative error of the ensemble variable E_Z was ≈ 0.002. As we are only interested in qualitative results the numerical investigation can, thus, be considered to be sufficiently accurate.

The numerical investigation of the evolution equation of the average number density (4.21) in Chaps. 5–7 was carried out by integrating the Fourier transformed equation of the average number density (4.29). To this end in (4.29) M Fourier modes were taken into account as explained above. $M = 200$ and $\Delta t = 0.0001$ turned out to be appropriate parameters which was tested by varying M and Δt systematically. Comparing, e.g., simulations for $M = 200$ and $\Delta t = 0.0001$ with those for $\Delta t < 0.0001$ and $M = 200$ or those for $M > 200$ and $\Delta t < 0.0001$ revealed relative errors of f and Z between both simulations which did not exceed ≈ 0.005.

B. Phase and Frequency Shifts Occurring in Chap. 3

B.1 Two Clusters

Frequency shifts of higher order are given by

$$\delta_{(3)} = d_1 V + d_2(n_1^2 V_1 + n_2^2 V_2) + d_3(n_2 - n_1)n_1 \Lambda_1 \ , \tag{B.1}$$

$$\delta_{(3)}^{\dagger} = d_4(n_1 - n_2)^2 n_1 n_2 \ , \tag{B.2}$$

where V, V_j, and Λ_1 are defined in Sect.3.5, and

$$d_1 = \frac{C_1 - 8C_2 + 9C_3 - 32C_4}{8\varepsilon^2(K_2 + 2K_4)^2} \ , \qquad d_2 = -\frac{3(C_1 + 9C_3)}{8\varepsilon^2(K_2 + 2K_4)^2} \ , \tag{B.3}$$

$$d_3 = \frac{(C_1 + C_3)(C_1 - 8C_2 + 9C_3 - 32C_4)}{2\varepsilon(K_2 + 2K_4)^2} \ , \tag{B.4}$$

$$d_4 = \frac{(C_1 + C_3)^2(C_1 - 8C_2 + 9C_3 - 32C_4)}{2(K_2 + 2K_4)^2} \ . \tag{B.5}$$

Higher order terms of the phase shift in the synchronized state read

$$\begin{aligned}
\chi_j &= \frac{1}{2K_2 + 4K_4}\eta_j + \varepsilon\frac{K_1 + 3K_3}{4(K_2 + 2K_4)^2} \\
&\quad \times [(n_\nu - n_\mu)\eta_j + n_\nu \Lambda_\nu - n_\mu \Lambda_\mu] \ , \tag{B.6}
\end{aligned}$$

$$\begin{aligned}
\chi_j^{\dagger} &= \varepsilon\frac{C_1 + C_3}{2K_2 + 4K_4}\left[n_\mu - n_\nu + (n_\nu - n_\mu)^2\right] \\
&\quad + \varepsilon^2\frac{(C_1 + C_3)(K_1 + 3K_3)}{2(K_2 + 2K_4)^2} \\
&\quad \times \left[n_\nu(n_\mu - n_\nu) + (n_\nu - n_\mu)^3\right] + O(\|x_c\|^3) \tag{B.7}
\end{aligned}$$

for the jth oscillator in the νth cluster, $\mu \neq \nu$ (i.e. $\nu = 1, \mu = 2$ or $\nu = 2, \mu = 1$), and Λ_ν as introduced in Sect. 3.5.

B.2 Three Clusters

H_j and $H_j^\dagger$ are higher order phase shifts. For the jth oscillator in the νth cluster we obtain

$$H_j(\boldsymbol{x}_c) = \frac{\delta_{(2)}\varepsilon^2}{3K_3} + \left[f_j(n_{\nu+} + n_{\nu-} - 2n_\nu) - g_j(n_{\nu+} - n_{\nu-})\right]\varepsilon\eta_j$$
$$+ f_j\varepsilon(2n_\nu\Lambda_\nu - n_{\nu+}\Lambda_{\nu+} - n_{\nu-}\Lambda_{\nu-})$$
$$+ g_j\varepsilon(n_{\nu+}\Lambda_{\nu+} - n_{\nu-}\Lambda_{\nu-}) + O(\|\boldsymbol{x}_c\|^3) , \tag{B.8}$$

$$H_j^\dagger(\boldsymbol{x}_c) = \frac{\delta_{(2)}^\dagger\varepsilon^2}{3K_3} + \left[f_j(n_{\nu+} + n_{\nu-} - 2n_\nu) - g_j(n_{\nu+} - n_{\nu-})\right]\varepsilon\eta_j^\dagger$$
$$+ f_j\varepsilon(2n_\nu\eta_{k(\nu)}^\dagger - n_{\nu+}\eta_{k(\nu+)}^\dagger - n_{\nu-}\eta_{k(\nu-)}^\dagger)$$
$$+ g_j\varepsilon(n_{\nu+}\eta_{k(\nu+)}^\dagger - n_{\nu-}\eta_{k(\nu-)}^\dagger) + O(\|\boldsymbol{x}_c\|^3) , \tag{B.9}$$

$f_j = (K_1 + 2K_2 + 4K_4)/(18K_3^2)$, $g_j = (C_1 - 2C_2 + 4C_4)/(6\sqrt{3}K_3^2)$, and $n_{\nu+} = n_{[\nu+1]_3}$, $n_{\nu-} = n_{[\nu-1]_3}$, where $[k]_m = [k]_{\mathrm{mod}\,m}$ (e.g. $[5]_3 = 2$). By $\eta_{k(\nu)}^\dagger$, $\eta_{k(\nu+)}^\dagger$, $\eta_{k(\nu-)}^\dagger$ we denote $\eta_k^\dagger$, where the kth oscillator is in cluster ν, $[\nu+1]_3$, $[\nu-1]_3$. With these notations the shifts of the eigenfrequencies read

$$\eta_j^\dagger = \frac{\varepsilon\sqrt{3}}{2}\left(n_{[\nu+1]_3} - n_{[\nu-1]_3}\right)(K_1 - K_2 + K_4) + \varepsilon(C_1 + C_2 + C_4)$$
$$\times \frac{1}{2}\left[\sum_{k=1}^3 \left(n_k - n_{[k+1]_3}\right)^2 + n_{[k-1]_3} + n_{[k+1]_3} - 2n_k\right] \tag{B.10}$$

for the jth oscillator in the νth cluster. Terms of second order of the synchronization frequency are

$$\delta_{(2)} = \frac{C_1 - 2C_2 + 4C_4}{\sqrt{3}\varepsilon K_3} \sum_{k=1}^3 \left(n_{[k+1]_3} - n_{[k-1]_3}\right)n_k\Lambda_k \tag{B.11}$$

with Λ_k from Sect. 3.5, and

$$\delta_{(2)}^\dagger = \frac{C_1 - 2C_2 + 4C_4}{K_3}\left[\frac{1}{2}(K_1 - K_2 + K_4)\left(\sum_{k=1}^3 n_k n_{[k+1]_3} - 9n_1n_2n_3\right)\right.$$
$$\left. + \frac{1}{2\sqrt{3}}(C_1 + C_2 + C_4)\sum_{k=1}^3 (n_k - n_{[k+1]_3})^2\right] . \tag{B.12}$$

B.3 Four Clusters

Configuration dependent shifts of the eigenfrequencies are given by

$$\eta_j^\dagger = \varepsilon(K_1 - K_3)\left(n_{[\nu+1]_4} - n_{[\nu-1]_4}\right)$$
$$+ \varepsilon C_2\left\{(n_1 + n_3 - n_2 - n_4)^2 + n_{[\nu+1]_4} + n_{[\nu-1]_4} - n_\nu - n_{[\nu+2]_4}\right\}$$
$$+\varepsilon(C_1 + C_3)\left[(n_1 - n_3)^2 + (n_2 - n_4)^2 + n_{[\nu+2]_4} - n_\nu\right] \tag{B.13}$$

for the jth oscillator in the νth cluster. Higher order frequency shifts read

$$\delta_{(2)} = \frac{C_1 - 3C_3}{2\varepsilon K_4}\left[(n_2 - n_4)(n_1\Lambda_1 - n_3\Lambda_3) + (n_3 - n_1)(n_2\Lambda_2 - n_4\Lambda_4)\right] \tag{B.14}$$

with Λ_k as introduced in Sect. 3.5, and

$$\delta_{(2)}^\dagger = \frac{C_1 - 3C_3}{2K_4}(D_1 + D_2) \ , \quad \text{where} \tag{B.15}$$

$$D_1 = (K_1 - K_3)\left[(n_1 + n_3)(n_2 - n_4)^2 + (n_2 + n_4)(n_1 - n_3)^2\right] \tag{B.16}$$
$$D_2 = (C_1 + 2C_2 + C_3)(n_2 + n_4 - n_1 - n_3)(n_1 - n_3)(n_2 - n_4) \ . \tag{B.17}$$

For the jth oscillator of the νth cluster we obtain the higher order phase shifts

$$H_j(\boldsymbol{x}_c) = \frac{\delta_{(2)}\varepsilon^2}{4K_4} + \left[\frac{K_1 + 3K_3}{64K_4^2}(n_{\nu++} - n_\nu)\right.$$
$$+ \frac{K_2}{8K_4^2}(n_{\nu+} + n_{\nu-} - n_\nu - n_{\nu++}) - \frac{C_1 - 3C_3}{64K_4^2}$$
$$\left.\times(n_{\nu+} - n_{\nu-})\right]\varepsilon\eta_j + \frac{K_1 + 3K_3}{64K_4^2}\varepsilon(n_\nu\Lambda_\nu + n_{\nu++}\Lambda_{\nu++})$$
$$+ \frac{K_2}{8K_4^2}\varepsilon\left(n_{\nu++}\Lambda_{\nu++} - \sum_{\substack{\mu \\ \mu \neq \nu++}} n_\mu\Lambda_\mu\right)$$
$$+ \frac{C_1 - 3C_3}{64K_4^2}\varepsilon(n_{\nu+}\Lambda_{\nu+} - n_{\nu-}\Lambda_{\nu-}) + O(\|\boldsymbol{x}_c\|^3) \ , \tag{B.18}$$

$$H_j^\dagger(\boldsymbol{x}_c) = \frac{\delta_{(2)}^\dagger\varepsilon^2}{4K_4} + \left[\frac{K_1 + 3K_3}{64K_4^2}(n_{\nu++} - n_\nu)\right.$$
$$+ \frac{K_2}{8K_4^2}(n_{\nu+} + n_{\nu-} - n_\nu - n_{\nu++}) - \frac{C_1 - 3C_3}{64K_4^2}$$
$$\left.\times(n_{\nu+} - n_{\nu-})\right]\varepsilon\eta_j^\dagger + \frac{K_1 + 3K_3}{64K_4^2}\varepsilon(n_\nu\eta_{k(\nu)}^\dagger + n_{\nu++}\eta_{k(\nu++)}^\dagger)$$
$$+ \frac{K_2}{8K_4^2}\varepsilon\left(n_{\nu++}\eta_{k(\nu++)}^\dagger - \sum_{\substack{\mu \\ \mu \neq \nu++}} n_\mu\eta_{k(\mu)}^\dagger\right)$$
$$+ \frac{C_1 - 3C_3}{64K_4^2}\varepsilon(n_{\nu+}\eta_{k(\nu+)}^\dagger - n_{\nu-}\eta_{k(\nu-)}^\dagger) + O(\|\boldsymbol{x}_c\|^3) \ , \tag{B.19}$$

and $n_{\nu+} = n_{[\nu+1]_4}$, $n_{\nu++} = n_{[\nu+2]_4}$, $n_{\nu-} = n_{[\nu-1]_4}$. Analogously, $\eta^{\dagger}_{k(\nu)}$, $\eta^{\dagger}_{k(\nu+)}$, $\eta^{\dagger}_{k(\nu++)}$, $\eta^{\dagger}_{k(\nu-)}$ denote $\eta^{\dagger}_k$, with k in cluster ν, $[\nu+1]_4$, $[\nu+2]_4$, $[\nu-1]_4$.

C. Single-Mode Instability

The subsequent sections refer to the order parameter equations of the single-mode instabilities occurring in Sect. 5.4.

C.1 First-Mode Instability

The coefficients of (5.23) and (5.24) read

$$\zeta = 2\pi \frac{Q(2Q - K_2) + C_1(C_1 - C_2) + i[C_1(2Q - K_2) + Q(C_2 - C_1)]}{(2Q - K_2)^2 + (C_1 - C_2)^2},$$

$$(C.1)$$

$$\alpha_1 = 2\pi^2 \left[(2Q - K_2)^2 + (C_1 - C_2)^2\right]^{-1} \Big\{ (Q - K_2)[Q(2Q - K_2)$$

$$+ C_1(C_1 - C_2)] + (C_1 + C_2)[C_1(2Q - K_2) + Q(C_2 - C_1)] \Big\}, \quad (C.2)$$

$$\beta_1 = 2\pi^2 \left[(2Q - K_2)^2 + (C_1 - C_2)^2\right]^{-1} \Big\{ (Q - K_2)[C_1(2Q - K_2)$$

$$+ Q(C_2 - C_1)] - (C_1 + C_2)[Q(2Q - K_2) + C_1(C_1 - C_2)] \Big\}. \quad (C.3)$$

C.2 Second-Mode Instability

The coefficients of order parameter equation (5.44) are given by

$$\alpha_2 = 4\pi^2 \left[(4Q - K_4)^2 + (C_2 - C_4)^2\right]^{-1} \Big\{ (2Q - K_4)[2Q(4Q - K_4)$$

$$+ C_2(C_2 - C_4)] + (C_2 + C_4)[C_2(4Q - K_4) + 2Q(C_4 - C_2)] \Big\}, \quad (C.4)$$

$$\beta_2 = 4\pi^2 \left[(4Q - K_4)^2 + (C_2 - C_4)^2\right]^{-1} \Big\{ (2Q - K_4)[C_2(4Q - K_4)$$

$$- 2Q(C_2 - C_4)] + (C_2 + C_4)[2Q(K_4 - 4Q) + C_2(C_4 - C_2)] \Big\}. \quad (C.5)$$

C.3 Third-Mode Instability

The coefficients of (5.52) are

$$\alpha_3 = 18\pi^2 \frac{Q(18Q^2 + 2C_3^2)}{36Q^2 + C_3^2} \quad , \quad \beta_3 = -6\pi^2 \frac{C_3(9Q^2 + C_3^2)}{36Q^2 + C_3^2} \quad . \tag{C.6}$$

C.4 Fourth-Mode Instability

Coefficients occurring in (5.60) are

$$\alpha_4 = 32\pi^2 \frac{Q(32Q^2 + 2C_4^2)}{64Q^2 + C_4^2} \quad , \quad \beta_4 = -8\pi^2 \frac{C_4(16Q^2 + C_4^2)}{64Q^2 + C_4^2} \quad . \tag{C.7}$$

D. Two-Mode Instability

This appendix is devoted to analyzing bifurcating solutions of the two-mode instability investigated in Sect.5.4.6.

D.1 Center Manifold

As outlined in Sect.3.6.1 on the center manifold the stable modes are given by the center modes according to (5.71). By means of (3.19) and (3.20) third and fourth mode are determined:

$$h_3(\boldsymbol{x}_c) = e_1 \hat{n}(1,t) \hat{n}(2,t) + O(\|\boldsymbol{x}_c\|^3) , \qquad (D.1)$$

where

$$e_1 = \frac{6\pi[3Q + i(C_1 + C_2)]}{3(3Q - K_3) + i(C_1 + 2C_2 - 3C_3)} \qquad (D.2)$$

and

$$h_4(\boldsymbol{x}_c) = e_2 \hat{n}(2,t)^2 + O(\|\boldsymbol{x}_c\|^3) , \qquad (D.3)$$

where

$$e_2 = \frac{4\pi(2Q + iC_2)}{2(4Q - K_4) + i2(C_2 - C_4)} . \qquad (D.4)$$

D.2 Order Parameter Equation

The coefficients of the order parameter equations read

$$a_1 = \pi[Q + (\mu - 2)\varepsilon + i(C_1 + C_2)] , \quad a_2 = \pi[K_3 - 2Q + i(C_2 + C_3)]e_1 , \quad (D.5)$$

$$b_1 = 2\pi(Q + 2\varepsilon + iC_1) , \quad b_2 = 2\pi[K_3 - Q + i(C_1 + C_3)]e_1 , \qquad (D.6)$$

$$b_3 = 2\pi[K_4 - 2Q + i(C_2 + C_4)]e_2 . \qquad (D.7)$$

D.3 Linear Problem (Type I)

To judge whether the fixed points are linearly stable one has to linearize (5.83) to (5.85) around the stationary states fulfilling $\dot{R}_j = 0$ and $\dot{\phi} = 0$. This is achieved with the transformation

$$A_j = A_j^{I,II} + \xi_j \ (j = 1,2) \ , \quad \phi = \phi^{I,II} + \psi \tag{D.8}$$

for the fixed points of type I and II respectively. In this way linearizing around type I fixed points yields

$$\dot{\xi}_j = \lambda_j^I \xi_j \ (j = 1,2) \ , \dot{\psi} = \lambda_3^I \psi \tag{D.9}$$

with eigenvalues

$$\lambda_1^I = (1 - \mu)\varepsilon + \frac{1}{2}\sqrt{\mu\varepsilon Q} + \varepsilon\frac{\mu - 2}{2}\sqrt{\frac{\mu\varepsilon}{Q}} \ , \tag{D.10}$$

$$\lambda_2^I = -2\mu\varepsilon \ , \quad \lambda_3^I = -2\pi[Q + (\mu - 2)\varepsilon]A_2^I \cos\phi^I \ . \tag{D.11}$$

Type I fixed points are linearly stable provided the eigenvalues $\lambda_1^I, \ldots, \lambda_3^I$ are negative. $\lambda_3^I < 0$ requires that both A_2^I and $\cos\phi^I$ have the same sign, i.e. $\mathrm{sign}(A_2^I \cos\phi^I) > 0$ has to be fulfilled. Actually, this is no restriction. As ε and μ are positive $\lambda_2^I < 0$ is negative. Hence, according to (D.9) the sign of λ_1^I determines whether the fixed point is stable ($\lambda_1^I < 0$) or unstable ($\lambda_1^I > 0$).

D.4 Linear Problem (Type II)

From (5.83) one derives

$$A_2^{II} = \frac{[Q + (\mu - 2)\varepsilon]\cos\phi^{II} \pm \sqrt{[Q + (\mu - 2)\varepsilon]^2 \cos^2\phi + 16Q\varepsilon}}{8\pi Q} \ , \tag{D.12}$$

where ϕ^{II} denotes the stationary relative phase. With (5.84) and (D.12) one gets

$$(A_1^{II})^2 = \frac{A_2^{II}\left[\mu\varepsilon - 4\pi^2 Q(A_2^{II})^2\right]}{2\pi[2\pi Q A_2^{II} - (Q + 2\varepsilon)\cos\phi^{II}]} \ . \tag{D.13}$$

Linearization of (5.83) to (5.85) around type II fixed points by means of transformation (D.8) leads to

$$\dot{\xi}_1 = w_1\xi_2 \ , \quad \dot{\xi}_2 = w_2\xi_1 + w_3\xi_2 \ , \quad \dot{\psi} = \lambda_3^{II}\psi \tag{D.14}$$

with

$$w_1 = \pi A_1^{II}\left\{[Q + (\mu - 2)\varepsilon]\cos\phi^{II} - 8\pi Q A_2^{II}\right\} \ , \tag{D.15}$$

$$w_2 = 4\pi A_1^{II}\left[(Q + 2\varepsilon)\cos\phi^{II} - 2\pi Q A_2^{II}\right] \ , \tag{D.16}$$

$$w_3 = \mu\varepsilon - 4\pi^2 Q \left[\left(A_1^{II} \right)^2 + 3 \left(A_2^{II} \right)^2 \right] , \tag{D.17}$$

$$\lambda_3^{II} = -2\pi \left\{ [Q + (\mu - 2)\varepsilon] A_2^{II} + (Q + 2\varepsilon) \frac{(A_1^{II})^2}{A_2^{II}} \right\} (\cos\phi^{II}) . \tag{D.18}$$

As a consequence of (D.18) linear stability requires that $\text{sign}(A_2^{II}\cos\phi^{II})$ is positive. Thus, in (5.89) one has to choose the upper (lower) sign iff $\cos\phi^{II}$ is positive (negative), i.e. iff $\phi^{II} = 0$ ($\phi^{II} = \pi$). Taking into account (5.77) and (5.82) one immediately shows that, actually, there are only two different type II fixed points. In both cases the relative phase reads

$$\phi^{II} = 0 , \tag{D.19}$$

whereas the amplitudes are given by $(\pm A_1^{II}, A_2^{II})$, where

$$A_2^{II} = \frac{Q + (\mu - 2)\varepsilon + \sqrt{[Q + (\mu - 2)\varepsilon]^2 + 16Q\varepsilon}}{8\pi Q} , \tag{D.20}$$

$$A_1^{II} = \sqrt{\frac{A_2^{II} \left[\mu\varepsilon - 4\pi^2 (A_2^{II})^2 \right]}{2\pi \left[2\pi Q A_2^{II} - (Q + 2\varepsilon) \right]}} . \tag{D.21}$$

A_1^{II} is a real quantity and, thus, the square root's argument in (D.21) has to be nonvanishing. For this reason the fixed point as a whole only exists for certain values of Q, ε and μ as discussed in Sect. 5.4.6. With a little calculation one obtains the eigenvalues

$$\lambda_{1,2}^{II} = \frac{w_3 \pm \sqrt{w_3^2 + 4w_1 w_2}}{2} \tag{D.22}$$

belonging to the dynamics of ξ_1 and ξ_2 governed by (D.14). Note that according to (D.15) and (D.16) for vanishing A_1^{II} both w_1 and w_2 vanish. From (D.22) it follows that in this case one of the eigenvalues $\lambda_{1,2}^{II}$ vanishes, too. It is important to stress that as a consequence of (D.15), (D.16) and (D.22) both type II fixed points, (A_1^{II}, A_2^{II}) and $(-A_1^{II}, A_2^{II})$, have the same eigenvalues.

D.5 Singularities

The amplitude of the first mode of the type II fixed point analyzed in Sect.5.4.6 is given by $A_1^{II} = \sqrt{D_1^{II}}$. This Sect.is devoted to the singularities which occur provided the denominator of D_1^{II} vanishes (cf. (5.90)). Two combinations occur:

1. *Both denominator and numerator of D_1^{II} vanish:* A straightforward calculation shows that denominator and numerator of D_1^{II} vanish if and only if

$$\mu = \frac{Q}{\varepsilon} + 4 + 4\frac{\varepsilon}{Q} \tag{D.23}$$

is fulfilled. From (D.23) one immediately derives that there is only one minimum of μ which is located in $\varepsilon_{min} = Q/2$. As $\mu(\varepsilon_{min}) = 8$ one obtains $\mu(\varepsilon_{min})\varepsilon_{min} = 4Q$. $\mu\varepsilon$ is the bifurcation parameter of the second mode (cf. (5.84)). So, whenever both denominator and numerator of D_1^{II} vanish the parameter $\mu\varepsilon$ is not small. On the other hand the center manifold theorem is only valid for small values of the bifurcation parameters ε and $\varepsilon\mu$ (cf. Pliss 1964, Kelley 1967). Hence, in the parameter range where the order parameter equation is valid denominator and nominator of D_1^{II} do not vanish simultaneously. By the way, from a pragmatic point of view one might expect to get rid of the vanishing denominator by taking into account terms of fourth and higher order of the center manifold.

2. *The denominator of D_1^{II} vanishes, whereas its numerator does not vanish:* In this case it was not possible to derive a simple expression like (D.23). For this reason a numerical analysis was performed. It turned out that a vanishing denominator of D_1^{II} is associated with rather large values of $\mu\varepsilon$, e.g., $\mu = 6.1$, $\varepsilon = 0.2$ or $\mu = 7$, $\varepsilon = 0.05$. For these parameter values a numerical integration of the order parameter equations reveals singular, i.e. not damped, behavior of the mode amplitudes. On the contrary, the numerical integration of the total system (5.4) and (5.5) does not display any singular behavior. Note that the total system was numerically approximated by taking into account Fourier modes with wave numbers $k = 0, \pm 1, \pm 2, \ldots, \pm 200$. Clearly this indicates that for larger values of the bifurcation parameters on the center manifold one has to take into account terms of fourth and higher order, too.

References

1. Introduction

Adam, D.R., Smith, J.M., Akselrod, S., Nyberg, S., Powell, A.O., Cohen, R.J. (1984): Fluctuations in T-wave morphology and susceptibility to ventricular fibrillation, J. Electrocardiol. **17**, 209–218

Allessie, M.A., Lammers, W.J.E.P., Bonke, F.I.M., Hollen, J. (1985): Experimental evaluation of Moe's multiple wavelet hypothesis of atrial fibrillation, In: *Cardiac Electrophysiology and Arrhythmias*, Zipes, D.P., Jalife, J. (eds.), Grune and Stratton, Orlando

Andy, O.J. (1983): Thalamic stimulation for control of movement disorders, Appl. Neurophysiol. **46**, 107–113

Arnold, V.I. (1983): *Geometrical methods in the theory of ordinary differential equations*, Springer, Heidelberg

Basar, E. (1998a): *Brain Oscillations*, Springer, Berlin; (1998b): *Integrative Brain Function*, Springer, Berlin

Benabid, A.L., Pollak, P., Gervason, C., Hoffmann, D., Gao, D.M., Hommel, M., Perret, J.E., De Rougemont, J. (1991): Long-term suppression of tremor by chronic stimulation of the ventral intermediate thalamic nucleus, The Lancet **337**, 403–406

Benabid, A.L., Pollak, P., Seigneuret, E., Hoffmann, D., Gay, E., Perret, J. (1993): Chronic VIM thalamic stimulation in Parkinson's disease, essential tremor and extra-pyramidal dyskinesia, Acta Neurochir. Suppl. (Wien) **58**, 39–44

Benabid, A.L., Pollak, P., Gao, D., Hoffmann, D., Limousin, P., Gay, E., Payen, I., Benazzouz, A. (1996): Chronic electrical stimulation of the ventralis intermedius nucleus of the thalamus as a treatment of movement disorders, J. Neurosurg. **84**, 203–214

Blond, S., Caparros-Lefebvre, D., Parker, F., Assaker, R., Petit, H., Guieu, J.-D., Christiaens, J.-L. (1992): Control of tremor and involuntary movement disorders by chronic stereotactic stimulation of the ventral intermediate thalamic nucleus, J. Neurosurg. **77**, 62–68

Caparros-Lefebvre, D., Ruchoux, M.M., Blond, S., Petit, H., Percheron, G. (1994): Long term thalamic stimulation in Parkinson's disease, Neurology **44**, 1856–1860

Cohen, A.H., Rossignol, S., Grillner, S. (eds.) (1988): *Neural Control of Rhythmic Movements in Vertebrates*, John Wiley & Sons, New York

Eckhorn, R., Bauer, R., Jordan, W., Brosch, M., Kruse, W., Munk, M., Reitboeck, H.J. (1988): Coherent oscillations: a mechanism of feature linking in the visual cortex?, Biol. Cybern. **60**, 121–130

Elble, R.J., Koller, W.C. (1990): *Tremor*, The Johns Hopkins University Press, Baltimore

Engel, J., Pedley, T. A. (eds.) (1997): *Epilepsy : A comprehensive textbook*, Lippincott-Raven, Philadelphia

Freeman, W.J. (1975): *Mass action in the nervous system*, Academic Press, New York

Freund, H.-J. (1983): Motor unit and muscle activity in voluntary motor control, Physiological Reviews **63**, 387–436; (1987): Abnormalities of motor behavior after cortical lesions in man, In: *The Nervous System: Higher Functions of the Brain*, Mountcastle, V.B. (Section ed.), Plum, F. (Vol. ed.), Sect. 1, vol. 5 of Handbook of Physiology, Williams and Wilkins, Baltimore 763–810

Glass, L., Mackey, M.C. (1988): *From Clocks to Chaos. The Rhythms of Life*, Princeton University Press

Golomb, D., Hansel, D., Shraiman, B., Sompolinsky, H. (1992): Clustering in globally coupled phase oscillators, Phys. Rev. A **45**, 3516–3530

Gray, C.M., Singer, W. (1987): Stimulus specific neuronal oscillations in the cat visual cortex: a cortical function unit, Soc. Neurosci. **404**, 3; (1989): Stimulus-specific neuronal oscillations in orientation columns of cat visual cortex, Proc. Natl. Acad. Sci. USA **86**, 1698–1702

Guckenheimer, J., Holmes, P. (1990): *Nonlinear Oscillations, Dynamical Systems, and Bifurcations of Vector Fields*, Berlin, Heidelberg

Hämäläinen, M., Hari, R., Ilmoniemi, R.J.,Knuutila, J., Lounasmaa, O.V. (1993): Magnetoencephalography – theory, instrumentation, and applications to non-invasive studies of the working human brain, Rev. Mod. Phys. **65**, 413–497

Haken, H. (1970): *Laser Theory*, Springer, Berlin; (1977): *Synergetics, An Introduction*, Springer, Berlin; (1983a): *Advanced Synergetics*, Springer, Berlin; (1983b): Synopsis and Introduction, In: *Synergetics of the brain*, Basar, E., Flohr, H., Haken, H. (eds.), Springer, Berlin; (1996): *Principles of Brain Functioning, A Synergetic Approach to Brain Activity, Behavior and Cognition*, Springer, Berlin

Haken, H., Koepchen, H.P. (eds.) (1991): Rhythms in Physiological Systems, Springer, Berlin

Hakim, V., Rappel, W. (1992): Dynamics of the globally coupled complex Ginzburg–Landau equation, Phys. Rev. A **46**, R7347–R7350

Hansel, D., Mato, G., Meunier, C. (1993): Clustering and slow switching in globally coupled phase oscillators, Phys. Rev. E **48**, 3470–3477

Hari, R., Salmelin, R. (1997): Human cortical oscillations: a neuromagnetic view through the skull, TINS **20**, 44–49

Hildebrandt, G. (1982): The time structure of autonomous processes, In: *Biological Adaptation*, Hildebrandt, G., Hensel, H. (eds.), Georg Thieme, Stuttgart (1987): The autonomous time structure and its reactive modifications in the human organism, In: *Temporal Disorder in Human Oscillatory Systems*, Rensing, L., an der Heiden, U., Mackey, M.C. (eds.), Springer, Berlin

Holden, A.V. (1997): The restless heart of a spiral, Nature **387**, 655–657

Kelley, A. (1967): The stable, center-stable, center, center-unstable and unstable manifolds, J. Diff. Equ. **3**, 546–570

Koepchen, H.P. (1991): Physiology of rhythms and control systems: an integrative approach. In: Haken and Koepchen (1991), pp. 3–20

Krinskii, V.I. (1966): Spread of excitation in an inhomogeneous medium, Biofizika **11**, 676–683; (1968): Fibrillation in excitable media, Systems Theory Research (Prob. Kyb.) **20**, 46–65 (1978): Mathematical models of cardiac arrhythmias (spiral waves), Pharmac. Ther. B. **3**, 539–555

Krinskii, V.I., Kholopov, A.V. (1967a): Conduction of impulses in excitable tissue with continuously distributed refractoriness, Biofizika **12**, 669–675; (1967b): Echo in excitable tissue, Biofizika **12**, 524–528

Kuramoto, Y. (1984): *Chemical Oscillations, Waves, and Turbulence*, Springer, Berlin

Llinás, R. , Jahnsen, H. (1982): Electrophysiology of mammalian thalamic neurons in vitro, Nature **297**, 406–408

Moe, G.K., Abildskov, J.A. (1959): Atrial fibrillation as a self-sustaining arrhythmia independent of focal discharge, Am. Heart J. **58**, 59–70

Moe, G.K., Rheinboldt, W.C., Abildskov, J.A. (1964): A computer model of atrial fibrillation, Am. Heart J. **67**, 200–220

Nakagawa, N., Kuramoto, Y. (1993): Collective chaos in a population of globally coupled oscillators, Prog. Theor. Phys. **89**, 313–323

Niedermeyer, E., Lopes da Silva, F. (1987): *Electroencephalography – Basic Principles, Clinical Applications and related Fields*, 2nd ed., Urban & Schwarzenberg, Baltimore

Okuda, K. (1993): Variety and generality of clustering in globally coupled oscillators, Physica D **63**, 424–436

Panfilov, A.V., Holden, A.V. (eds.) (1997): *Computational Biology of the Heart*, Wiley, Chichester

Pare, D., Curro'Dossi, R., Steriade, M. (1990): Neuronal basis of the parkinsonian resting tremor: a hypothesis and its implications for treatment, Neuroscience **35**, 217–226

Perlitz, V., Schmid-Schönbein, H., Schulte, A., Dolgner, J., Petzold, E.R., Kruse, W. (1995): Effektivität des autogenen Trainings, Therapiewoche **26**, 1536–1544

Pertsov, A.M., Davidenko, J.M., Salomonsz, R., Baxter, W.T., Jalife, J. (1993): Spiral waves of excitation underlie reentrant activity in isolated cardiac muscle, Circ. Res. **72**, 631–650

Pliss, V. (1964): Principal reduction in the theory of stability of motion, Izv. Akad. Nauk. SSSR Math. Ser. **28**, 1297–1324 (in Russian)

Sakaguchi, H., Shinomoto, S., Kuramoto, Y. (1987): Local and global self-entrainments in oscillator lattices, Prog. Theor. Phys. **77**, 1005–1010; (1988): Mutual entrainment in oscillator lattices with nonvariational type interaction, Prog. Theor. Phys. **79**, 1069–1079

Schmid-Schönbein, H., Ziege, S. (1991): The high pressure system of the mammalian circulation as a dynamic self-organizing system. In: Haken and Koepchen (1991), pp. 77–96

Schmid-Schönbein, H., Ziege, S., Rütten, W., Heidtmann, H. (1992): Active and passive modulation of cutaneous red cell flux as measured by Laser Doppler anemometry, Vasa **32**, 38–47 (Suppl.)

Sears, T.A., Stagg, D. (1976): Short-term synchronization of intercostal motoneurone activity, J. Physiol. (London) **263**, 357–381

Singer, W., Gray, C.M. (1995): Visual feature integration and the temporal correlation hypothesis, Annu. Rev. Neurosci. **18**, 555–586

Smith, J.M., Cohen, R.J. (1984): Simple finite-element models account for wide range of cardiac dysrhythmias, Proc. Natl. Acad. Sci. USA **81**, 233–237

Steriade, H., Jones, E.G., Llinás, R. (1990): *Thalamic Oscillations and Signaling*, John Wiley & Sons, New York

Strafella, A., Ashby, P., Munz, M., Dostrovsky, J.O., Lozano, A.M., Lang, A.E. (1997): Inhibition of Voluntary Activity by Thalamic Stimulation in Humans: Relevance for the Control of Tremor, Movement Disorders **12**, 727–737

Strogatz, S.H. (1994): *Nonlinear Dynamics and Chaos*, Addison-Wesley, Reading, MA

Strogatz, S.H., Mirollo, R.E. (1988a): Phase-locking and critical phenomena in lattices of coupled nonlinear oscillators with random intrinsic frequencies, Physica D **31**, 143–168; (1988b): Collective Synchronisation in lattices of non-linear oscillators with randomness, J. Phys. A **21**, L699–L705

Tass, P. (1996a): Resetting biological oscillators – a stochastic approach, J. Biol. Phys. **22**, 27–64; (1996b): Phase resetting associated with changes of burst shape, J. Biol. Phys. **22**, 125–155; (1997): Phase and frequency shifts in a population of phase oscillators, Phys. Rev. E **56**, 2043–2060

Tass, P., Haken, H. (1996): Synchronization in networks of limit cycle oscillators, Z. Phys. B **100**, 303–320

Volkmann, J., Joliot, M., Mogilner, A., Ioannides, A.A., Lado, F., Fazzini, E., Ribary, U., Llinás, R. (1996): Central motor loop oscillations in parkinsonian resting tremor revealed by magnetoencephalography, Neurology **46**, 1359–1370

Volkmann, J., Sturm, V. (1998): Indication and results of stereotactic surgery for advanced Parkinson's disease, Crit. Rev. Neurosurg. **8**, 209–216

Volkmann, J., Sturm, V., Freund, H.-J. (1998): Die subkortikale Hochfrequenzstimulation zur Behandlung von Bewegungsstörungen, Akt. Neurologie **25**, 1–9

von Holst, E. (1935): Über den Prozess der zentralnervösen Koordination, Pflügers Archiv **236**, 149–158; (1939): Die relative Koordination als Phänomen und als Methode zentralnervöser Funktionsanalysen, Erg. Physiol. **42**, 228–306

Winfree, A.T. (1970): An integrated view of the resetting of a circadian clock, J. Theor. Biol. **28**, 327–374; (1980): *The Geometry of Biological Time*, Springer, Berlin; (1987): *When Time Breaks Down: The Three-Dimensional Dynamics of Electrochemical Waves and Cardiac Arrhythmias*, Princeton University Press, Princeton; (1994): Electrical turbulence in three-dimensional heart muscle, Science **266**, 1003–1006

Witkowski, F.X., Leon, L.J., Penkoske, P.A., Giles, W.R., Spano, M.L., Ditto, W.L., Winfree, A.T. (1998): Spatiotemporal evolution of ventricular fibrillation, Nature **392**, 78–82

Wunderlin, A., Haken, H. (1981): Generalized Ginzburg–Landau equations, slaving principle and center manifold theorem, Z. Phys. B **44**, 135–141

2. Resetting an Ensemble of Oscillators

Aizawa, Y. (1976): Synergetic approach to the phenomena of mode-locking in nonlinear systems, Prog. Theor. Phys. **56**, 703–716

Cohen, A.H., Rossignol, S., Grillner, S. (eds.) (1988): *Neural Control of Rhythmic Movements in Vertebrates*, John Wiley & Sons, New York

Elble, R.J., Koller, W.C. (1990): *Tremor*, The Johns Hopkins University Press, Baltimore

Elphik, C., Tirapegui, E., Brachet, M., Coullet, P. Iooss, G. (1987): A simple global characterization for normal forms of singular vector fields, Physica D **29**, 95–127

Gardiner, C.W. (1985): *Handbook for Stochastic Methods for Physics, Chemistry and the Natural Sciences*, 2nd ed., Springer, Berlin

Gardner, F.M. (1966): *Phaselock techniques*, John Wiley & Sons, New York

Glass, L., Mackey, M.C. (1988): *From Clocks to Chaos. The Rhythms of Life*, Princeton University Press

Haken, H. (1977): *Synergetics. An Introduction*, Springer, Berlin; (1983): *Advanced Synergetics*, Springer, Berlin; (1996) *Principles of Brain Functioning, A Synergetic Approach to Brain Activity, Behavior and Cognition*, Springer, Berlin

Haken, H., Kelso, J.A.S., Bunz, H. (1985): A theoretical model of phase transitions in human hand movements, Biol. Cybern. **51**, 347–356

Haken, H., Sauermann, H., Schmid, C., Vollmer, H.D. (1967): Theory of laser noise in the phase locking region, Z. Phys. **206**, 369–393

Iooss, G., Adelmeyer, M. (1992): *Topics in Bifurcation Theory and Applications*, Advanced Series in Nonlinear Dynamics, Vol. 3, World Scientific, Singapore

Kuramoto, Y. (1984): *Chemical Oscillations, Waves, and Turbulence*, Springer, Berlin

Matthews, P.C., Strogatz, S.H. (1990): Phase diagram for the collective behavior of limit-cycle oscillators, Phys. Rev. Lett. **65**, 1701–1704

Milnor, J. (1965): *Topology From the Differentiable Viewpoint*, Univ. of Va. Press, Charlottesville

Murray, J. D. (1989): *Mathematical Biology*, Springer, Berlin

Risken, H. (1989): *The Fokker–Planck Equation, Methods of Solution and Applications*, 2nd ed., Springer, Berlin

Shiino, M., Frankowicz, M. (1989): Synchronization of infinitely many coupled limit-cycle type oscillators, Physics Letters A **136**, 103–108

Singer, W., Gray, C.M. (1995): Visual feature integration and the temporal correlation hypothesis, Annu. Rev. Neurosci. **18**, 555–586

Steriade, H., Jones, E.G., Llinás, R. (1990): *Thalamic Oscillations and Signaling*, John Wiley & Sons, New York

Tass, P. (1996a): Resetting biological oscillators – a stochastic approach, J. Biol. Phys. **22**, 27–64; (1996b): Phase resetting associated with changes of burst shape, J. Biol. Phys. **22**, 125–155; (1997): Phase and frequency shifts in a population of phase oscillators, Phys. Rev. E **56**, 2043–2060

Tass, P., Haken, H. (1996): Synchronization in networks of limit cycle oscillators, Z. Phys. B **100**, 303–320

Winfree, A.T. (1970): An integrated view of the resetting of a circadian clock, J. Theor. Biol. **28**, 327–374; (1980): *The Geometry of Biological Time*, Springer, Berlin

Yamaguchi, Y., Shimizu, H. (1984): Theory of self-synchronization in the presence of native frequency distribution and external noises, Physica D **11**, 212–226

3. Synchronization Patterns

Abeles, M. (1982): *Local cortical circuits. An elektrophysiological study*, Springer, Berlin

Aertsen, A. (ed.) (1993): *Brain theory*, Elsevier, Amsterdam

Aizawa, Y. (1976): Synergetic approach to the phenomena of mode-locking in nonlinear systems, Prog. Theor. Phys. **56**, 703–716

Arbib, A. (ed.) (1995): *The handbook of brain theory and neural networks*, MIT Press, Cambridge

Arnold, V.I. (1983): *Geometrical methods in the theory of ordinary differential equations*, Springer, Heidelberg

Aulbach, B. (1984): *Continuous and Discrete Dynamics near Manifolds of Equilibria*, LNM 1058, Springer, Heidelberg

Beurle, R.L. (1956): Properties of a mass of cells capable of regenerating pulses, Philos. Trans. Soc. London, Ser. A **240**, 55–94

Braitenberg, V., Schüz, A. (1991): *Anatomy of the Cortex*, Springer, Berlin

Carr, J. (1981): *Applications of Centre Manifold Theory*, Appl. Math. Sciences 35, Springer

Chawanya, T., Aoyagi, T., Nishikawa, I., Okuda, K., Kuramoto, Y. (1993): A model for feature linking via collective oscillations in the primary visual cortex, Biol. Cybern. **68**, 483–490

Cowan, J.D. (1987): Brain mechansims underlying visual hallucinations. In: Paines, D. (ed.), *Emerging syntheses in science*, Addison-Wesley, New York, 123–131

Creutzfeldt, O.D. (1983): *Cortex Cerebri*, Springer, Berlin

Crick, F. (1984): Function of the thalamic reticular complex: the searchlight hypothesis, Proc. Natl. Acad. Sci. USA **81**, 4586–4590

Daido, H. (1992a): Quasientrainment and slow relaxation in a population of oscillators with random and frustrated interactions, Phys. Rev. Lett. **68**, 1073–1076; (1992b): Order function and macroscopic mutual entrainment in uniformly coupled limit-cycle oscillators, Prog. Theor. Phys. **88**, 1213–1218; (1993): A solvable model of coupled limit-cycle oscillators exhibiting partial perfect synchrony and novel frequency spectra, Physica D **69**, 394–403; (1994): Generic scaling at the onset of macroscopic mutual entrainment in limit-cycle oscillators with uniform all-to-all coupling, Phys. Rev. Lett. **73**, 760–763; (1996): Onset of cooperative entrainment in limit-cycle oscillators with uniform all-to-all interactions: bifurcation of the order function, Physica D **91**, 24–66

Eckhorn, R., Bauer, R., Jordan, W., Brosch, M., Kruse, W., Munk, M., Reitboeck, H.J. (1988): Coherent oscillations: a mechanism of feature linking in the visual cortex?, Biol. Cybern. **60**, 121–130

Eckhorn, R., Frien, A., Bauer, R., Woelbern, T., Kehr, H. (1993): High Frequency 60–90 Hz oscillations in primary visual cortex of awake monkey, Neuro Rep. **4**, 243–246

Edelman, G.M. (1992): *Bright air, brilliant fire*, Penguin Books, London

Eggermont, J.J. (1990): *The correlative brain. Theory and experiment in neural interaction*, Springer, Berlin

Elphik, C., Tirapegui, E., Brachet, M., Coullet, P. Iooss, G. (1987): A simple global characterization for normal forms of singular vector fields, Physica D **29**, 95–127

Engel, A.K., König, P., Gray, C.M., Singer, W. (1990): Synchronization of oscillatory responses: a mechanism for stimulus-dependent assembly formationin cat visual cortex. In: *Parallel Processing in Neural Systems and Computers*, Eckmiller, R., Hartmann, G., Hauske, G. (eds.), Elsevier, North Holland

Ermentrout, G.B., Cowan, J. (1979): A mathematical theory of visual hallucination patterns, Biol. Cybern. **34**, 137–150

Ermentrout, G.B., Rinzel, J. (1981): Waves in a simple, excitable or oscillatory reaction–diffusion model, J. Math. Biol. **11**, 269–294

FitzHugh, R. (1961): Impulses and physiological states in theoretical models of nerve membrane, Biophys. J. **1**, 445–466

Freeman, W.J. (1975): *Mass action in the nervous system*, Academic Press, New York

Gerstner, W., Ritz, R., Hemmen, J.L. van (1993): A biologically motivated and analytically soluble model of collective oscillations in the cortex, Biol. Cybern. **68**, 363–374

Glass, L., Mackey, M.C. (1988): *From Clocks to Chaos, The Rhythms of Life*, Princeton University Press

Golomb, D., Hansel, D., Shraiman, B., Sompolinsky, H. (1992): Clustering in globally coupled phase oscillators,

Golomb, D., Wang, X.J., Rinzel, J. (1996): Propagation of spindle waves in a thalamic slice model, J. Neurophysiol. **75**, 750–769 Phys. Rev. A **45**, 3516–3530

Gray, C.M., Singer, W. (1987): Stimulus specific neuronal oscillations in the cat visual cortex: a cortical function unit, Soc. Neurosci. **404**, 3; (1989): Stimulus-specific neuronal oscillations in orientation columns of cat visual cortex, Proc. Natl. Acad. Sci. USA **86**, 1698–1702

Gray, C.M., König, P., Engel, A.K., Singer, W. (1989): Oscillatory responses in cat visual cortex exhibit inter-columnar synchronization which reflects global stimulus properties, Nature **338**, 334–337

Griffith, J.S. (1963): A field theory of neural nets: I: Derivation of field equations, Bull. Math. Biophys. **25**, 111–120; (1965): A field theory of neural nets: II: Properties of field equations, Bull. Math. Biophys. **27**, 187–195

Grossberg, S., Somers, D. (1991): Synchronized oscillations during cooperative feature linking in a cortical model of visual perception, Neural Networks **4**, 453–466

Guckenheimer, J., Holmes, P. (1990): *Nonlinear Oscillations, Dynamical Systems, and Bifurcations of Vector Fields*, Berlin, Heidelberg

Haken, H. (1964): A nonlinear theory of laser noise and coherence I, Z. Phys. **181**, 96–124; (1970): *Laser Theory*, Springer, Berlin; (ed.) (1973): *Synergetics* (Proceedings of a Symposium on Synergetics, Elmau 1972), B.G. Teubner, Stuttgart; (1975): Generalized Ginzburg–Landau equations for phase transition-like phenomena in lasers, nonlinear optics, hydrodynamics and chemical reactions, Z. Phys. B **21**, 105–114; (1977): *Synergetics, An Introduction*, Springer, Berlin; (1979): Pattern formation and pattern recognition – an attempt at a synthesis. In: *Pattern formation by dynamic systems and pattern recognition*, H. Haken (ed.), Springer, Berlin, 2–13; (1983) *Advanced Synergetics*, Springer, Berlin; (1988): *Information and Self-Organization*, Springer, Berlin; (1991): *Synergetic computers and cognition*, Springer, Berlin; (1996a): *Principles of Brain Functioning, A Synergetic Approach to Brain Activity, Behavior and Cognition*, Springer, Berlin; (1996b): Slaving principle revisited, Physica D **97**, 95–103

Haken, H., Graham, R. (1971): Synergetik – Die Lehre vom Zusammenwirken, Umschau **6**, 191

Haken, H., Wunderlin, A. (1982): Slaving principle for stochastic differential equations with additive and multiplicative noise and for discrete noisy maps, Z. Phys. B **47**, 179–187

Haken, H., Kelso, J.A.S., Bunz, H. (1985): A theoretical model of phase transitions in human hand movements, Biol. Cybern. **51**, 347–356

Hakim, V., Rappel, W. (1992): Dynamics of the globally coupled complex Ginzburg–Landau equation, Phys. Rev. A **46**, R7347–R7350

Han, S. K., Kurrer, C., Kuramoto, Y. (1995): Dephasing and bursting in coupled neural oscillators, Phys. Rev. Lett. **75**, 3190–3193

Hebb, D.O. (1949): *Organization of Behavior*, Wiley, New York

Hansel, D., Mato, G., Meunier, C. (1993a): Clustering and slow switching in globally coupled phase oscillators, Phys. Rev. E **48**, 3470–3477; (1993b): Phase dynamics of weakly coupled Hodgkin–Huxley neurons, Europhys. Lett., **23**, 367–372

Hirsch, M.W., Smale, S. (1974): *Differential Equations, Dynamical Systems, and Linear Algebra*, Academic Press, San Diego

Hirsch, M., Pugh, C., Shub, M. (1976): *Invariant Manifolds*, Lecture Notes Math. **583**, Springer, Berlin

Hodgkin, A.L., Huxley, A. F. (1952): A quantitative description of membrane current and its application to conduction and excitation in nerve, J. Physiol. (London) **117**, 500–544

Hopfield, J.J. (1982): Neural networks and physical systems with emergent collective computational abilities, Proc. Natl. Acad. Sci. **79**, 2554–2558

Hoppensteadt, F.C., Izhikevich, E.M. (1997): *Weakly Connected Neural Networks*, Springer, Berlin

Hubel, D.H., Wiesel T.N. (1959): Receptive fields of single neurones in the cat's striate cortex, J. Physiol. **148**, 574–591; (1962): Receptive fields, binocular interaction and functional architecture in the cat's visual cortex, J. Physiol. **160**, 106–154; (1963): Shape and arrangement of columns in cat's striate cortex, J. Physiol. **165**, 559–568

Iooss, G. (1987): Global characterization of the normal form for a vector field near a closed orbit, J. Diff. Equ. **76**, 47–76

Iooss, G., Adelmeyer, M. (1992): *Topics in Bifurcation Theory and Applications*, Advanced Series in Nonlinear Dynamics, Vol. 3, World Scientific, Singapore

Jirsa, V.K., Haken, H. (1996): Field theory of electromagnetic brain activity, Phys. Rev. Lett. **77**, 960–963; (1997): A derivation of a macroscopic field theory of the brain from the quasi-microscopic neural dynamics, Physica D **99**, 503–526

Julesz, B. (1991): Early vision and focal attention, Rev. Mod. Phys. **63**, 735–772

Kelley, A. (1967): The stable, center-stable, center, center-unstable and unstable manifolds, J. Diff. Equ. **3**, 546–570

Kelso, J.A.S. (1981): On the oscillatory basis of movements, Bulletin of Psychonomic Society **18**, 63; (1984): Phase transitions and critical behavior in human bimanual coordination, American Journal of Physiology: Regulatory, Integrative and Comparative Physiology **15**, R1000–R1004

Kirchgässner, K. (1982): Wave-solutions of reversible systems and applications, J. Diff. Equations **45**, 113–127

Koch, C., Segev, I. (1989): *Methods in Neuronal Modeling, From Synapses to Networks*, MIT Press, Cambridge

König, P., Engel, A.K., Singer, W. (1996): Integrator or coincidence detector? The role of the cortical neuron revisited, TINS **19**, 130–137

Kreiter, A.K., Singer, W. (1992): Oscillatory neuronal responses in the visual cortex of awake macaque monkey, Eur. J. Neurosci. **4**, 369–375

Kuramoto, Y. (1991): Collective synchronization of pulse-coupled oscillators and excitable units, Physica D **50**, 15–30; (1984): *Chemical Oscillations, Waves, and Turbulence*, Springer, Berlin

Langenberg, U., Kessler, K., Hefter, H., Cooke, J.D., Brown, S.H., Freund, H.-J. (1992): Effects of delayed visual feedback during sinusoidal visuomotor tracking, Soc. Neurosci. Abstr. Suppl. **5**, 209

Livingstone, M.S. (1991): Visually evoked oscillations in monkey striate cortex, Soc. Neurosci. Abstr. **17**, 73

Malsburg, C. von der, Schneider, W. (1986): A neural cocktail-party processor, Biol. Cybern. **54**, 29–40

Marr, D. (1976): Early processing of visual information, Philos. Trans. R. Soc. Lond. [Biol] **275**, 483–524

Matthews, P.C., Strogatz, S.H. (1990): Phase diagram for the collective behavior of limit-cycle oscillators, Phys. Rev. Lett. **65**, 1701–1704

McCulloch, W., Pitts, W. (1943): A logical calculus of the ideas immanent in nervous activity, Bull. Math. Biophys. **5**, 115–133

Meinhardt, H. (1982): *Models of Biological Pattern Formation*, Academic Press, London

Milner, P.M. (1974): A model for visual shape recognition, Psychol. Rev. **81**, 521–535

Mirollo, R. E., Strogatz, S. H. (1990): Synchronization of pulse-coupled biological oscillators, SIAM J. Appl. Math. **50**, 1645–1662

Müller, B., Reinhardt, J. (1990): *Neural Networks, An Introduction*, Springer, Berlin

Murray, J. D. (1989): *Mathematical Biology*, Springer, Berlin

Nagumo, J.S., Arimoto, S., Yoshizawa, S. (1962): An active pulse transmission line simulating nerve axon, Proc. IRE **50**, 2061–2071

Murthy, V.N., Fetz, E.E. (1992): Coherent 25- to 35-Hz oscillations in the sensorimotor cortex of awake behaving monkeys, Proc. Natl. Acad. Sci. USA **89**, 5670–5674

Nakagawa, N., Kuramoto, Y. (1993): Collective chaos in a population of globally coupled oscillators, Prog. Theor. Phys. **89**, 313–323

Neuenschwander, S., Varela, F.J. (1993): Visually triggered neuronal oscillations in birds: an autocorrelation study of tectal activity, Eur. J. Neurosci. **5**, 870–881

Nicholis, S., Wiesenfeld, K. (1992): Ubiquitous neutral stability of splay-phase states, Phys. Rev. A **45**, 8430–8435

Nicolis, G., Prigogine, I. (1977): *Self-Organization in Nonequilibrium Systems*, Wiley, New York

Niebur, E., Schuster, H.G., Kammen, D.M. (1991): Collective frequencies and metastability in networks of limit-cycle oscillators with time delay, Phys. Rev. Lett. **67**, 2753–2756

Nunez, P.L. (1974): The brain wave equation: a model for the EEG, Math. Biosci. **21**, 279–297; (1981): *Electric fields of the brain*, Oxford University Press; (1995): Neocortical dynamics and human EEG rhythms, Oxford University Press

Okuda, K (1993): Variety and generality of clustering in globally coupled oscillators, Physica D **63**, 424–436

Omidvar, O.M. (ed.) (1995): *Progress in neural networks*, Vol. **3**, Ablex Publishing Corporation, Norwood, New Jersey

Orban, G.A. (1984): *Neuronal Operations in the Visual Cortex*, Springer, Berlin

Perkel, D.H., Bullock, T.H. (1968): Neural coding, Neurosci. Res. Prog. Sum. **3**, 405–527

Plant, R.E. (1978): The effects of calcium^{++} on bursting neurons, Biophys. J. **21**, 217–237; (1981): Bifurcation and resonance in a model for bursting nerve cells, J. Math. Biol. **11**, 15–32

Pliss, V. (1964): Principal reduction in the theory of stability of motion, Izv. Akad. Nauk. SSSR Math. Ser. **28**, 1297–1324 (in Russian)

Ramachandran, V.S. (1988): Perception of shape from shading Nature **331**, 163–166

Reichardt, W.E., Poggio, T (eds.) (1981): *Theoretical approaches in neurobiology*, MIT Press, Cambridge

Rieke, F., Warland, D., de Ruyter van Stevenick, R., Bialek, W. (1997): *Spikes: Exploring the Neural Code*, MIT Press, Cambridge

Rinzel, J. (1986): On different mechanisms for membrane potential bursting, Proc. Sympos. on Nonlinear Oscillations in Biology and Chemistry, Salt Lake City 1985, Lect. Notes in Biomath. Springer, Berlin **66**, 19–33

Roelfsema, P.R., Engel, A.K., König, P., Singer, W. (1997): Visuomotor integration is associated with zero time-lag synchronization among cortical areas, Nature **385**, 157–161

Sakaguchi, H., Shinomoto, S., Kuramoto, Y. (1987): Local and global self-entrainments in oscillator lattices, Prog. Theor. Phys. **77**, 1005–1010; (1988): Mutual entrainment in oscillator lattices with nonvariational type interaction, Prog. Theor. Phys. **79**, 1069–1079

Sandstede, B., Scheel, A., Wulff, C. (1997): Center-manifold reduction for spiral waves, C. R. Acad. Sci. Paris (Série I, Équations aux dérivées partielles/Partial Differential Equations) **324**, 153–158

Schillen, T.B., König, P. (1994): Binding by temporal structure in multiple feature domains of an oscillatory neuronal network, Biol. Cybern. **70**, 397–405

Schöner, G., Haken, H., Kelso, J.A.S. (1986): A stochastic theory of phase transitions in human hand movement, Biol. Cybern. **53**, 247–257

Schuster, H.G., Wagner, P. (1990a): A model for neuronal oscillations in the visual cortex. 1. Mean-field theory and derivation of the phase equations. Biol. Cybern. **64**, 77–82; (1990b): A model for neuronal oscillations in the visual cortex. 2. Phase description of the feature dependent synchronization. Biol. Cybern. **64**, 83–85

Shiino, M., Frankowicz, M. (1989): Synchronization of infinitely many coupled limit-cycle type oscillators, Physics Letters A **136**, 103–108

Shimizu, H., Yamaguchi, Y., Tsuda, I., Yano, M. (1985): Pattern recognition based on holonic information dynamics: towards synergetic computers. In: *Complex systems – operational approaches*, Haken, H. (ed.), Springer, Berlin

Singer, W. (1989): Search for coherence: a basic principle of cortical self-organization, Concepts Neurosci. **1**, 1–26

Singer, W., Gray, C.M. (1995): Visual feature integration and the temporal correlation hypothesis, Annu. Rev. Neurosci. **18**, 555–586

Sompolinsky, H., Golomb, D., Kleinfeld, D. (1991): Cooperative dynamics in visual processing, Phys. Rev. A **43**, 6990–7011

Stephan, K.M., Binkofski, F., Halsband, U., Dohle, C., Wunderlich, G., Schnitzler, A., Tass, P., Posse, S., Herzog, H., Sturm, V., Zilles, K., Seitz, R.J., Freund, H.-J.: The role of ventral medial wall motor areas in bimanual coordination, Brain, **122**, 351–368

Steriade, H., Jones, E.G., Llinás, R. (1988): *Thalamic Oscillations and Signaling*, Wiley, New York

Strogatz, S.H. (1994): *Nonlinear Dynamics and Chaos*, Addison-Wesley, Reading, MA

Strogatz, S.H., Mirollo, R.E. (1988a): Phase-locking and critical phenomena in lattices of coupled nonlinear oscillators with random intrinsic frequencies, Physica D **31**, 143–168; (1988b): Collective Synchronisation in lattices of non-linear oscillators with randomness, J. Phys. A **21**, L699–L705; (1993): Splay states in globally coupled Josephson arrays: analytical prediction of Floquet multipliers, Phys. Rev. E **47**, 220–227

Swift, J.W., Strogatz, S.H., Wiesenfeld, K (1992): Averaging of globally coupled oscillators, Physica D **55**, 239–250

Tass, P. (1995a): Cortical pattern formation during visual hallucinations, J. Biol. Phys. **21**, 177–210; (1995b): Phase and frequency shifts of two nonlinearly coupled oscillators, Z. Phys B **99**, 111–121; (1997a): Phase and frequency shifts in a population of phase oscillators, Phys. Rev. E **56**, 2043–2060; (1997b): Oscillatory cortical activity during visual hallucinations, J. Biol. Phys. **23**, 21–66

Tass, P., Haken, H. (1996a): Synchronization in networks of limit cycle oscillators, Z. Phys. B **100**, 303–320; (1996b): Synchronized oscillations in the visual cortex – a synergetic model, Biol. Cybern. **74**, 31–39

Tass, P., Wunderlin, A., Schanz, M. (1995): A theoretical model of sinusoidal forearm tracking with delayed visual feedback, J. Biol. Phys. **21**, 83–112

Tass, P., Kurths, J., Rosenblum, M.G., Guasti, G., Hefter, H. (1996): Delay-induced transitions in visually guided movements, Phys. Rev. E **54**, R2224–R2227

Thom, R. (1972): *Stabilité structurelle et morphogénèse – Essai d'une théorie générale des modèles*, W.A. Benjamin, Inc., Reading, Massachusetts

Treisman, A. (1980): A feature-integration theory of attention, Cogn. Psychol. **12**, 97–136; (1986): Properties, parts and objects. In: *Handbook of perception and human performances*, Boff, K., Kaufman, L., Thomas, I. (eds.), Wiley, New York

Vaadia, E., Haalman, I., Abeles, M., Bergman, H., Prut, Y., Slovin, H., Aertsen, A. (1995): Dynamics of neuronal interactions in monkey cortex in relation to behavioural events, Nature **373**, 515–518

Vanderbauwhede, A. (1989): Center Manifolds, Normal Forms and Elementary Bifurcations, Dyn. Rep. **2**, 89–169

Wiesenfeld, K., Hadley, P. (1989): Attractor crowding in oscillator arrays, Phys. Rev. Lett. **62**, 1335–1338

Wilson, H.R., Cowan, J.D. (1972): Excitatory and inhibitory interactions in localized populations of model neurons, Biophysical Journal **12**, 1–24; (1973): A mathematical theory of the functional dynamics of cortical and thalamic nervous tissue, Kybernetik **13**, 55–80

Winfree, A. T. (1967): Biological rhythms and the behavior of populations of coupled oscillators, J. Theor. Biol. **16**, 15–42; (1980): *The Geometry of Biological Time*, Springer, Berlin

Wischert, W., Wunderlin, A., Pelster, A., Olivier, M., Groslambert, J. (1994): Delay-induced instabilities in nonlinear feedback systems, Phys. Rev. E **49**, 203–219

Wunderlin, A., Haken, H. (1975): Scaling theory for nonequilibrium systems, Z. Phys. B **21**, 393–401; (1981): Generalized Ginzburg–Landau equations, slaving principle and center manifold theorem, Z. Phys. B **44**, 135–141

Yamaguchi, Y., Shimizu, H. (1984): Theory of self-synchronization in the presence of native frequency distribution and external noises, Physica D **11**, 212–226

4. Stochastic Model

Best, E.N. (1979): Null space in the Hodgkin–Huxley equations: a critical test, Biophys. J. **27**, 87–104

Gardiner, C.W. (1985): *Handbook of Stochastic Methods for Physics, Chemistry and the Natural Sciences*, 2nd. ed., Springer, Berlin

Haken, H. (1977): *Synergetics, An Introduction*, Springer, Berlin; (1983): *Advanced Synergetics*, Springer, Berlin

Kuramoto, Y. (1984): *Chemical Oscillations, Waves, and Turbulence*, Springer, Berlin

Risken, H. (1989): *The Fokker–Planck Equation, Methods of Solution and Applications*, Springer, Berlin

5. Clustering in the Presence of Noise

Crawford, J.D. (1994): Amplitude expansions for instabilities in populations of globally-coupled oscillators, J. Stat. Phys. **74**, 1047–1084; (1995): Scaling and singularities in the entrainment of globally coupled oscillators, Phys. Rev. Lett. **74**, 4341–4344

Daido, H. (1994): Generic scaling at the onset of macroscopic mutual entrainment in limit-cycle oscillators with uniform all-to-all coupling, Phys. Rev. Lett. **73**, 760–763; (1996): Onset of cooperative entrainment in limit-cycle oscillators with uniform all-to-all interactions: bifurcation of the order function, Physica D **91**, 24–66

Hämäläinen, M., Hari, R., Ilmoniemi, R.J., Knuutila, J., Lounasmaa, O.V. (1993): Magnetoencephalography – theory, instrumentation, and applications to non-invasive studies of the working brain, Reviews of Modern Physics, **65**, 413–497

Haken, H. (1983): *Advanced Synergetics*, Springer, Berlin; (1996): *Principles of Brain Functioning, A Synergetic Approach to Brain Activity, Behavior and Cognition*, Springer, Berlin S

Hirsch, M.W., Smale, S. (1974): *Differential Equations, Dynamical Systems, and Linear Algebra*, Academic Press, San Diego

Kelley, A. (1967): The stable, center-stable, center, center-unstable and unstable manifolds, J. Diff. Equ. **3**, 546–570

Kuramoto, Y. (1984): *Chemical Oscillations, Waves, and Turbulence*, Springer, Berlin

Niedermeyer, E., Lopes da Silva, F. (1987): *Electroencephalography – Basic Principles, Clinical Applications and related Fields*, 2nd ed., Urban & Schwarzenberg, Baltimore

Nunez, P. (1995): *Neocortical Dynamics and Human EEG Rhythms*, Oxford University Press, New York

Pliss, V. (1964): Principal reduction in the theory of stability of motion, Izv. Akad. Nauk. SSSR Math. Ser. **28**, 1297–1324 (in Russian)

Strogatz, S.H., Mirollo, R.E. (1991): Stability of incoherence in a population of coupled oscillators, J. Stat. Phys. **63**, 613–635

Strogatz, S.H., Mirollo, R.E., Matthews, P.C. (1992): Coupled nonlinear oscillators below the synchronization threshold: relaxation by generalized Landau damping, Phys. Rev. Lett. **68**, 2730–2733

Wiesenfeld, K., Hadley, P. (1989): Attractor crowding in oscillator arrays, Phys. Rev. Lett. **62**, 1335–1338

6. Single Pulse Stimulation

Glass, L., Mackey, M.C. (1988): *From Clocks to Chaos. The Rhythms of Life*, Princeton University Press

Haken, H. (1977): *Synergetics. An Introduction*, Springer, Berlin; (1983): *Advanced Synergetics*, Springer, Berlin

Winfree, A.T. (1980): *The Geometry of Biological Time*, Springer, Berlin

7. Periodic Stimulation

Arnold, V.I. (1983): *Geometrical methods in the theory of Ordinary Differential Equations*, Springer, Berlin

Ayers, A.L., Selverston, A.I. (1979): Monosynaptic entrainment of an endogenous pacemaker network: A cellular mechanism for von Holst's magnet effect, J. Comp. Physiol. **129**, 5–17

Baconnier, P., Benchetrit, G., Demongeot, J., Pham Dinh, T. (1983): Simulation of the entrainment of the respiratory rhythm by two conceptually different models. In: *Rhythms in Biology and Other Fields of Application*, Cosnard, M., Demongeot, J., Le Breton, A. (eds.), Springer, Berlin, 2–16

Creutzfeldt, O.D. (1983): *Cortex Cerebri*, Springer, Berlin

Eccles, J., Dimitrijevic, M.R. (1985): *Upper motor neuron functions and dysfunctions*, Karger, Basel

Elble, R.J., Koller, W.C. (1990): *Tremor*, John Hopkins University Press, Baltimore

Elble, R.J., Higgins, C., Hughes, L. (1992): Phase resetting and frequency entrainment of essential tremor, Experimental Neurology **116**, 355–361

Fallert, M., Mühlemann, R. (1971): Der Hering–Breuer Reflex bei künstlicher Beatmung des Kaninchens. I. Die Auslösung der reflektorischen Inspirationen durch den Respirator, Pfluegers Arch. **330**, 162–174

Glass, L., Mackey, M.C. (1988): *From Clocks to Chaos. The Rhythms of Life*, Princeton University Press

Glass, L., Guevara, M.R., Shrier, A., Perez, R. (1983): Bifurcation and Chaos in a periodically stimulated cardiac oscillator, Physica D **7**, 89–101

Glass, L., Guevara, M.R., Bélair, J., Shrier, A. (1984): Global bifurcations of a periodically forced biological oscillator, Phys. Rev. **29**, 1348–1357

Guckenheimer, J., Holmes, P. (1990): *Nonlinear Oscillations, Dynamical Systems, and Bifurcations of Vector Fields*, Berlin, Heidelberg

Guevara, M.R., Glass, L., Shrier, A. (1981): Phase locking, period-doubling bifurcations, and irregular dynamics in periodically stimulated cardiac cells, Science **214**, 1350–1353

Guevara, M.R., Shrier, A., Glass, L. (1988): Phase-locked rhythms in periodically stimulated heart cell aggregates, Am. J. Physiol. **254**, H1–H10

Guttman, R., Feldman, L., Jakobsson, E. (1980): Frequency entrainment of squid axon membrane, J. Memb. Biol. **56**, 104–118

Haken, H. (1977): *Synergetics. An Introduction*, Springer, Berlin; (1983): *Advanced Synergetics*, Springer, Berlin

Hoppensteadt, F.C. (1986): *An Introduction to the Mathematics of Neurons*, Cambridge University Press, New York

Jalife, J., Moe, G.K. (1976): Effect of electrotonic potential on pacemaker activity of canine Purkinje fibers in relation to parasystole, Circ. Res. **39**, 801–808; (1979) A biologic model for parasystole, Am. J. Cardiol. **43**, 761–772

Jalife, J., Michaels, D.C. (1985): Phase-dependent interactions of cardiac pacemakers as mechanisms of control and synchronization in the heart. In: *Cardiac Electrophysiology and Arrhythmias*, Zipes, D.P., jalife, J. (eds.), Grune and Stratton, Orlando, 109–119

Keener, J.P., Hoppensteadt, F.C., Rinzel, J. (1981): Integrate and fire models of nerve membranes response to oscillatory inputs, SIAM J. Appl. Math. **41**, 503–517

Levy, M.N., Iano, T., Zieske, H. (1972): Effects of repetitive bursts of vagal activity on heart rate, Circ. res. **30**, 186–195

Perkel, D.H., Schulman, J.H., Bullock, T.H., Moore, G.P., Segundo, J.P. (1964): Pacemaker neurons: Effects of regularly spaced synaptic input, Science **145**, 61–63

Petrillo, G.A., Glass, L., Trippenbach, T. (1983): Phase locking of the respiratory rhythm in cats to a mechanical ventilator. Can. J. Physiol. Pharmacol. **61**, 599–607

Petrillo, G.A., Glass, L. (1984): A theory for phase locking respiration in cats to a mechanical ventilator. Am. J. Physiol. **246**, R311–320

Pinsker, H.M. (1977): Aplysia bursting neurons as endogenous oscillators. II: Synchronization and entrainment by pulsed inhibitory synaptic input, J. Neurophysiol. **40**, 544–552

Pittendrigh, C.S. (1965): On the mechanism of entrainment of a circadian rhythm by light cycles. In: *Circadian Clocks*, Aschoff, J. (ed.), North Holland, Amsterdam, 277–297

Reid, J.V.O. (1969): The cardiac pacemaker: Effects of regularly spaced nervous input, Am. Heart J. **78**, 58–64

Siegfried, J., Hood, T. (1985): Brain Stimulation Procedures in Dystonic, Hypertonic, Dyskinetic, and Hyperkinetic Conditions. In: Eccles and Dimitrijevic (1985), 79–90

Van der Tweel, L.H., Meijler, F.L., Van Capelle, F.J.L. (1973): Synchronization of the heart, J. Appl. Physiol. **34**, 283–287

Vibert, J.-F., Caille, D., Segundo, J.P. (1981): Respiratory oscillator entrainment by periodic vagal afferents, Biol. Cybern. **41**, 119–130

Volkmann, J., Sturm, V. (1998): Indication and results of stereotactic surgery for advanced Parkinson's disease, Crit. Rev. Neurosurg. **8**, 209–216

Volkmann, J., Sturm, V., Freund, H.-J. (1998): Die subkortikale Hochfrequenzstimulation zur Behandlung von Bewegungsstörungen, Akt. Neurologie **25**, 1–9

Winfree, A.T. (1980): *The Geometry of Biological Time*, Springer, Berlin

Ypey, D.L., Van Meerwijk, W.P.M., DeHaan, R.L. (1982): Synchronization of cardiac pacemaker cells by electrical coupling. In: *Cardiac rate and Rhythm*, Bouman, L.N., Jongsma, H.J. (eds.), Martinus Nijhof, The Hague, 363–395

8. Data Analysis

Ahonen, A.I., Hämäläinen, M.S., Kajola, M.J., Knuutila, J.E.T., Lounasmaa, O.V., Simola, J.T., Tesche, C.D., Vilkman, V.A. (1991): Multichannel SQUID systems for brain research, IEEE Trans. Magn. **27**, 2786–2792

Andrä, W., Nowak, H. (eds.) (1998): *Magnetism in Medicine*, Wiley-VCH, Berlin

Basar, E. (1998a): *Brain Oscillations*, Springer, Berlin; (1998b): *Integrative Brain Function*, Springer, Berlin

Berger, H. (1929): Über das Elektroenkephalogramm des Menschen, Arch. Psychiatr. Nervenkr. **87**, 527–570

Clarke, C.J.S., Janday, B.S. (1989): The solution of the biomagnetic inverse problem by maximum statistical entropy, Inverse Problems **95**, 483–500

Clarke, C.J.S., Ioannides, A.A., Bolton, J.P.R. (1990): Localised and distributed source solutions for the biomagnetic inverse problem I. In: *Advances in Biomagnetism*, Williamson, S.J., Hoke, M., Stroink, G., Kotani, M. (eds.), Plenum Press, New York, 587–590

Cohen, D. (1972): Magnetoencephalography: Detection of the brain's electrical activity with a superconducting magnetometer, Science **175**, 664–666

Cooper, R., Osselton, J.W., Shaw, J.C. (1984): *Elektroenzephalographie: Technik und Methoden*, 3rd. rev. ed., Gustav Fischer Verlag, Stuttgart

Creutzfeldt, O.D. (1983): *Cortex Cerebri*, Springer, Berlin

Eckhorn, R., Bauer, R., Jordan, W., Brosch, M., Kruse, W., Munk, M., Reitboeck, H.J. (1988): Coherent oscillations: a mechanism of feature linking in the visual cortex?, Biol. Cybern. **60**, 121–130

Eckhorn, R., Dicke, P., Kruse, W., Reitboeck, H.-J. (1990): Stimulus-related facilitation and synchronization among visual cortical areas. In: *Nonlinear Dynamics and Neural Networks*, Schuster, H.G., Singer, W. (eds.), VCH, Weinheim

Elble, R.J., Koller, W.C. (1990): *Tremor*, John Hopkins University Press, Baltimore

Feldman, M.S. (1985): Investigation of the natural vibrations of machine elements using Hilbert transform, Sov. Machine Sci. **2**, 44–47; (1994): Nonlinear system vibration analysis using Hilbert transform. I. Free vibration analysis method "FREEVIB", Mech. Systems and Signal Processing **8**, 119–127

Feldman, M.S., Rosenblum, M.G. (1988): Computer program for determination of nonlinear elastic and damping properties of a vibrating system. In: *Proceedings of the Workshop "Software in Machine Building CAD Systems"*, 89. Izhevsk (In Russian)

Freeman, W.J. (1975): *Mass action in the nervous system: Examination of the neurophysiological basis of adaptive behavior through the EEG*, Academic Press, London

Freund, H.-J. (1983): Motor unit and muscle activity in voluntary motor control, Physiological Reviews **63**, 387–436

Friedrich, R., Fuchs, A., Haken, H. (1992): Spatio-temporal EEG-patterns. In: *Rhythms in biological systems*, Haken, H., Köpchen, H.P. (eds.), Springer, Berlin

Friedrich, R., Uhl, C. (1992): Synergetic analysis of human electroencephalograms: Petit-mal epilepsy. In: *Evolution of dynamical structures in complex systems*, Friedrich, R., Wunderlin, A. (eds.), Springer, Berlin

Fuchs, A., Friedrich, R., Haken, H., Lehmann, D. (1987): Spatio-temporal analysis of multichannel α-EEG map series. In: *Computational systems – natural and artificial*, Haken, H. (ed.), Springer, Berlin

Fuchs, M., Wagner, M., Wischmann H.A., Dössel, O. (1995): Cortical current imaging by morphologically constrained reconstructions. In: *Biomagnetism: Fundamental Research and Clinical Applications*, Baumgartner, C., Deeke, L., Stroink, G., Williamson, S.J. (eds.), Elsevier Science Publishers, Amsterdam, 299–301

Glass, L., Mackey, M.C. (1988): *From Clocks to Chaos. The Rhythms of Life*, Princeton University Press

Gray, C.M., Singer, W. (1987): Stimulus specific neuronal oscillations in the cat visual cortex: a cortical function unit, Soc. Neurosci. **404**, 3; (1989): Stimulus-specific neuronal oscillations in orientation columns of cat visual cortex, Proc. Natl. Acad. Sci. USA **86**, 1698–1702

Greenblatt, R.E. (1993): Probabilistic reconstruction of multiple sources in the bioelectromagnetic inverse problem, Inverse Problems **9**, 271-284

Haken, H. (1977): *Synergetics. An Introduction*, Springer, Berlin; (1983): *Advanced Synergetics*, Springer, Berlin

Haken, H., Koepchen, H.P. (eds.) (1991): *Rhythms in Physiological Systems*, Springer, Berlin

Hämäläinen, M., Hari, R., Ilmoniemi, R.J.,Knuutila, J., Lounasmaa, O.V. (1993): Magnetoencephalography – theory, instrumentation, and applications to non-invasive studies of the working human brain, Rev. Mod. Phys. **65**, 413–497

Hari, R., Salmelin, R. (1997): Human cortical oscillations: a neuromagnetic view through the skull, TINS **20**, 44–49

Haykin, S. (1986): *Adaptive Filter Theory*, Prentice Hall, Englewood Cliffs, New Jersey

Hefter, H., Logogian, E., Witte, O.W., Reiners, K., Freund, H.-J. (1992): Oscillatory activity in different motor subsystems in palatal myoclonus. A case report, Acta Neurol. Scand. **86**, 176–183

Helmholtz, H., von (1853): Über einige Gesetze der Verteilung elektrischer Ströme in körperlichen Leitern, mit Anwendung auf die tierisch-elektrischen Versuche, Ann. Phys. Chem. **89**, 211–233, 353–377

Hildebrandt, G. (1982): The time structure of autonomous processes, In: *Biological Adaptation*, Hildebrandt, G., Hensel, H. (eds.), Georg Thieme, Stuttgart (1987): The autonomous time structure and its reactive modifications in the human organism, In: *Temporal Disorder in Human Oscillatory Systems*, Rensing, L., an der Heiden, U., Mackey, M.C. (eds.), Springer, Berlin

Ioannides, A.A., Bolton, J.P.R., Clarke, C.J.S. (1990): Continuous probabilistic solutions to the biomagnetic inverse problem, Inverse Problems **6**, 523–542

Jansen, B.H., Brandt, M.E. (1993): *Nonlinear dynamical analysis of the EEG*, World Scientific, Singapore

Müller, M.M., Junghöfer, M., Elbert, T., Rochstroh, B. (1997): Visually induced gamma-band responses to coherent and incoherent motion: a replication study, NeuroReport **8**, 2575–2579

Niedermeyer, E., Lopes da Silva, F. (1987): *Electroencephalography – Basic Principles, Clinical Applications and related Fields*, 2nd ed., Urban & Schwarzenberg, Baltimore

Nunez, P.L. (1995): *Neocortical dynamics and human EEG rhythms*, Oxford University Press, Oxford

Ogawa, S., Lee, T.M., Nayak, A.S., Glynn, P. (1990): Oxygenation-sensitive contrast in magnetic resonance image of rodent brain at high magnetic fields, Magnetic Resonance in Medicine **14**, 68–78

Otnes, R.K., Enochson, L. (1972:) *Digital Time Series Analysis*, John Wiley & Sons, New York

Panter, P. (1965): *Modulation, Noise, and Spectral Analysis*, McGraw–Hill, New York

Parks, T.W., Burrus, C.S. (1987): *Digital Filter Design*, John Wiley & Sons, New York

Pascual-Marqui, R.D., Michel, C.M., Lehmann, D. (1994): Low resolution electromagnetic tomography: a new method for localising electrical activity in the brain, Int. J. Psychophysiol. **18**, 49–65

Perlitz, V., Schmid-Schönbein, H., Schulte, A., Dolgner, J., Petzold, E.R., Kruse, W. (1995): Effektivität des autogenen Trainings, Therapiewoche **26**, 1536–1544

Pikovsky, A.S., Rosenblum, M.G., Kurths, J. (1996): Synchronization in a population of globally coupled chaotic oscillators, Europhys. Lett. **34**, 165–170

Rosenblum, M., Kurths, J. (1998): Analysing synchronization phenomena from bivariate data by means of the Hilbert transform. In: *Nonlinear Analysis of Physiological Data*, Kantz, H., Kurths, J., Mayer–Kress, G. (eds.), Springer, Berlin, 91–99

Rosenblum, M.G., Pikovsky, A.S., Kurths, J. (1996): Phase Synchronization of Chaotic Oscillators, Phys. Rev. Lett. **76**, 1804–1807

Rosenblum, M.G., Firsov G.I., Kuuz, R.A., Pompe, B. (1998): Human Postural Control: Force Plate Experiments and Modelling. In: *Nonlinear Analysis of Physiological Data*, Kantz, H., Kurths, J., Mayer-Kress, G. (eds.), Springer, Berlin, 283–306

Schäfer, C., Rosenblum, M.G., Kurths, J., Abel, H.-H. (1998): Heartbeat Synchronized with Ventilation, Nature **392**, 239–240

Schmid-Schönbein, H., Ziege, S. (1991): The high pressure system of the mammalian circulation as a dynamic self-organizing system. In: Haken and Koepchen (1991), pp. 77–96

Schmid-Schönbein, H., Ziege, S., Rütten, W., Heidtmann, H. (1992): Active and passive modulation of cutaneous red cell flux as measured by Laser Doppler anemometry, Vasa **32**, 38–47 (Suppl.)

Selesnick, I.W., Lang, M., Burrus, C.S. (1996): Constrained Least Square Design of FIR Filters without Specified Transition Bands, IEEE Transactions on Signal Processing **44**, 1879–1892

Singer, W. (1989): Search for coherence: a basic principle of cortical self-organization, Concepts Neurosci. **1**, 1–26

Singer, W., Gray, C.M. (1995): Visual feature integration and the temporal correlation hypothesis, Annu. Rev. Neurosci. **18**, 555–586

Steriade, H., Jones, E.G., Llinás, R. (1990): *Thalamic Oscillations and Signaling*, John Wiley & Sons, New York

Stratonovich, R.L. (1963): *Topics in the Theory of Random Noise*, Gordon and Breach, New York

Tass, P., Kurths, J., Rosenblum, M.G., Guasti, G., Hefter, H. (1996): Delay-induced transitions in visually guided movements, Phys. Rev. E **54**, R2224–R2227

Tass, P., Rosenblum, M.G., Weule, J., Kurths, J., Pikovsky, A., Volkmann, J., Schnitzler, A., Freund, H.-J. (1998): Detection of $n:m$ phase locking from noisy data: Application to magnetoencephalography, Phys. Rev. Lett. **81**, 3291–3294

Toga, A.W., Mazziotta, J.C. (eds.) (1996): *Brain Mapping – The Methods*, Academic Press, San Diego

Uhl, C., Kruggel, F., Opitz, B., Yves von Cramon, D. (1998): A New Concept for EEG/MEG Signal Analysis: Detection of Interacting Spatial Modes, Human Brain Mapping **6**, 137–149

von Holst, E. (1935): Über den Prozess der zentralnervösen Koordination, Pflügers Archiv **236**, 149–158; (1939): Die relative Koordination als Phänomen und als Methode zentralnervöser Funktionsanalysen, Erg. Physiol. **42**, 228–306

Wang, J., Williamson, S.J., Kaufman, L. (1995): Spatio-temporal model of neural activity of the human brain based on the MNLS inverse. In: *Biomagnetism: Fundamental Research and Clinical Applications*, Baumgartner, C., Deeke, L.,

Stroink, G., Williamson, S.J. (eds.), Elsevier Science Publishers, Amsterdam, 299–301

Winfree, A.T. (1980): *The Geometry of Biological Time*, Springer, Berlin

9. Modelling Perspectives

Benabid, A.L., Pollak, P., Louveau, A., Henry, S., De Rougemont, J. (1987): Combined (thalamotomy and stimulation) stereotaxic surgery of the VIM thalamic nucleus for bilateral Parkinson's disease, Proc. Am. Soc. Stereotaxis Funct. Neurosurg., Montreal, Applied Neurophysiology **50**, 344–346

Best, E.N. (1979): Null space in the Hodgkin–Huxley equations: a critical test, Biophys. J. **27**, 87–104

Creutzfeldt, O.D. (1983): *Cortex Cerebri*, Springer, Berlin

Eckhorn, R., Bauer, R., Jordan, W., Brosch, M., Kruse, W., Munk, M., Reitboeck, H.J. (1988): Coherent oscillations: a mechanism of feature linking in the visual cortex?, Biol. Cybern. **60**, 121–130

Engel, J., Pedley, T. A. (eds.) (1997): *Epilepsy : A comprehensive textbook*, Lippincott-Raven, Philadelphia

Ermentrout, G.B. (1981): $n : m$ phase-locking of weakly coupled oscillators, J. Math. Biol. **12**, 327–342

Ermentrout, G.B. (1994): In: Ventriglia, F. (ed.): Neural Modeling and Neural Networks, Pergamon Press, Oxford, 79–119

Ermentrout, G.B., Rinzel, J. (1981): Waves in a simple, excitable or oscillatory reaction–diffusion model, J. Math. Biol. **11**, 269–294

Farmer, J.D. (1981): Spectral broadening of period-doubling bifurcation sequences, Phys. Rev. Lett. **47**, 179–182 (1981)

Freund, H.-J. (1983): Motor unit and muscle activity in voluntary motor control, Physiological Reviews **63**, 387–436

Gray, C.M., Singer, W. (1987): Stimulus specific neuronal oscillations in the cat visual cortex: a cortical function unit, Soc. Neurosci. **404**, 3; (1989): Stimulus-specific neuronal oscillations in orientation columns of cat visual cortex, Proc. Natl. Acad. Sci. USA **86**, 1698–1702

Guttman, R., Lewis, S., Rinzel, J. (1980): Control of repetitive firing in squid axon membrane as a model for a neurone oscillator, J. Physiol. (London) **305**, 377–395

Haken, H. (1983):*Advanced Synergetics*, Springer, Berlin

Hodgkin, A.L., Huxley, A.F. (1952): A quantitative description of membrane current and its application to conduction and excitation in nerve, J. Physiol. (London) **117**, 500–544

Hoppensteadt, F.C., Izhikevich, E.M. (1997): *Weakly Connected Neural Networks*, Springer, Berlin

Izhikevich, E.M. (1998): Phase models with explicit time delays, Phys. Rev. E **58**, 905–908

König, P., Engel, A.K., Singer, W. (1996): Integrator or coincidence detector? The role of the cortical neuron revisited, TINS **19**, 130–137

Kuramoto, Y. (1984): *Chemical Oscillations, Waves, and Turbulence*, Springer, Berlin

Lamarre, Y., DeMontigny, C., Dumont, M. , Weiss, M. (1971): Harmaline-induced rhythmic activity of cerebellar and lower brain stem neurons, Brain Res. **32**, 246–250

Lamarre, Y., Joffroy, A.J., Dumont, M., DeMontigny, C. , Grou, F., Lind, P. (1975): Central mechanisms of tremor in some feline and primate models, can. J. neurol. Sci. **2**, 227–233

Lamarre, Y., Joffroy, A.J. (1979): Experimental tremor in monkey: activity of thalamic on precentral cortical neurons in the absence of peripheral feedback, Adv. Neurol. **24**, 109–122

Lee, R.G., Stein, R.B. (1981): Resetting of tremor by mechanical perturbations: a comparison of essential tremor and parkinsonian tremor, Ann. Neurol. **10**, 523–531

Lenz, F.A., Kwan, H.C., Martin, R.L., Tasker, R.R., Dostrovsk, J.O., Lenz, Y.E. (1994): Single unit analysis of the human ventral thalamic nuclear group. Tremor-related activity in functionally identfied cells, Brain **117**, 531–543

Llinás, R. , Jahnsen, H. (1982): Electrophysiology of mammalian thalamic neurons in vitro, Nature **297**, 406–408

Murray, J. D. (1989): *Mathematical Biology*, Springer, Berlin

Nieuwenhuys, R., Voogd, J., van Huijzen, Chr. (1991): *Das Zentralnervensystem des Menschen*, 2nd ed., Springer, Berlin

Pikovsky, A.S. (1985): Phase synchronization of chaotic oscillations by a periodic external field, Sov. J. Commun. Tchnol. Electron. **30**, 85 (1985)

Pikovsky, A.S., Rosenblum, M.G., Kurths, J. (1996): Synchronization in a population of globally coupled chaotic oscillators, Europhys. Lett. **34**, 165–170

Rosenblum, M.G., Pikovsky, A.S., Kurths, J. (1996): Phase Synchronization of Chaotic Oscillators, Phys. Rev. Lett. **76**, 1804–1807

Schuster, H.G., Wagner, P. (1989): Mutual entrainment of two limit cycle oscillators with time delayed coupling, Prog. Theor. Phys. **81**, 939–945

Steriade, H., Jones, E.G., Llinás, R. (1990): *Thalamic Oscillations and Signaling*, John Wiley & Sons, New York

Stratonovich, R.L. (1963): *Topics in the Theory of Random Noise*, Gordon and Breach, New York

Tass, P., Haken, H. (1996): Synchronization in networks of limit cycle oscillators, Z. Phys. B **100**, 303–320

Winfree, A.T. (1967): Biological rhythms and the behavior of populations of coupled oscillators, J. Theor. Biol. **16**, 15–42

10. Neurological Perspectives

Alberts, W.W., Wright, E.J., Feinstein, B. (1969): Cortical potentials and parkinsonian tremor, Nature **221**, 670–672

Andy, O.J. (1983): Thalamic stimulation for control of movement disorders, Appl. Neurophysiol. **46**, 107–113

Barker, A.T., Jalinous, R., Freeston, I.L. (1986): Non-invasive magnetic stimulation of the human cortex, Lancet **I**, 1106–1107

Bathien, N., Rondot, P., Toma, S. (1980): Inhibition and synchronization of tremor induced by a muscle twitch, J. Neurol. Neurosurg. Psychiatry **43**, 713–718

Benabid, A.L., Pollak, P., Louveau, A., Henry, S., De Rougemont, J. (1987): Combined (thalamotomy and stimulation) stereotaxic surgery of the VIM thalamic nucleus for bilateral Parkinson's disease, Proc. Am. Soc. Stereotaxis Funct. Neurosurg., Montreal, Applied Neurophysiology **50**, 344–346

Benabid, A.L., Pollak, P., Gervason, C., Hoffmann, D., Gao, D.M., Hommel, M., Perret, J.E., De Rougemont, J. (1991): Long-term suppression of tremor by chronic stimulation of the ventral intermediate thalamic nucleus, The Lancet **337**, 403–406

Benabid, A.L., Pollak, P., Seigneuret, E., Hoffmann, D., Gay, E., Perret, J. (1993): Chronic VIM thalamic stimulation in Parkinson's disease, essential tremor and extra-pyramidal dyskinesia, Acta Neurochir. Suppl. (Wien) **58**, 39–44

Benabid, A.L., Pollak, P., Gao, D., Hoffmann, D., Limousin, P., Gay, E., Payen, I., Benazzouz, A. (1996): Chronic electrical stimulation of the ventralis intermedius nucleus of the thalamus as a treatment of movement disorders, J. Neurosurg. **84**, 203–214

Bergman, H., Wichmann, T., Karmon, B. DeLong, M.R. (1994): The primate subthalamic nucleus. II: Neuronal activity in the MPTP model of parkinsonism, J. Neurophysiol. **72**, 507–520

Bergman, H., Raz, A., Feingold, A., Nini, A., Nelken, I., Hansel, D., Ben-Pazi, H., Reches, A. (1998): Physiology of MPTP tremor, Movement Disorders **13**, Suppl. 3, 29–34

Blond, S., Caparros-Lefebvre, D., Parker, F., Assaker, R., Petit, H., Guieu, J.-D., Christiaens, J.-L. (1992): Control of tremor and involuntary movement disorders by chronic stereotactic stimulation of the ventral intermediate thalamic nucleus, J. Neurosurg. **77**, 62–68

Britton, T.C., Thompson, P.D., Day, B.L., Rothwell, J.C., Findley, L.J., Marsden, C.D. (1993a): Modulation of postural tremors at the wrist by supramaximal electrical median nerve shocks in essential tremor, Parkinson's disease and normal subjects mimicking tremor, J. Neurol. Neurosurg. Psychiatry **56**, 1085–1089; (1993b): Modulation of postural wrist tremors by magnetic stimulation of the motor cortex in patients with Parkinson's disease or essential tremor and in normal subjects mimicking tremor, Annals of Neurology **33**, 473–479

Caparros-Lefebvre, D., Ruchoux, M.M., Blond, S., Petit, H., Percheron, G. (1994): Long term thalamic stimulation in Parkinson's disease, Neurology **44**, 1856–1860

Classen, J., Schnitzler, A., Binkofski, F., Werhahn, K.J., Kim, Y.-S., Kessler, K., Benecke, R. (1997): The motor syndrome associated with exaggerated inhibition within the primary motor cortex of patients with hemiparetic stroke, Brain **120**, 605–619

Cohen, A.H., Rossignol, S., Grillner, S. (eds.) (1988): *Neural Control of Rhythmic Movements in Vertebrates*, John Wiley & Sons, New York

Cook, A.W., Weinstein, S.P. (1973): Chronic dorsal column stimulation in multiple sclerosis, N.Y. St. J. Med. **73**, 2868–2872

Cooper, I.S. (1953): Ligation of anterior choroidal artery for involuntary movements – parkinsonism, Psynchriatric Quart. **27**, 317–319; (1956): *The Neurosurgical Alleviation of Parkinsonism*, Charles C. Thomas, Springfield; (1973): Effect of chronic stimulation of anterior cerebellum on neurological disease, Lancet **i**, 1321

Cooper, I.S., Bravo, G.J. (1958): Anterior choroidal occlusion, chemo-pallidectomy and chemo-thalamectomy in Parkinsonism, In: Fileds, W.S. (ed.): *Pathogenesis and Treatment of Parkinsonism*, Charles C. Thomas, Springfield

Creutzfeldt, O.D. (1983): *Cortex Cerebri*, Springer, Berlin

Deuschl, G., Lucking, C.H., Schenk, E. (1987): Essential tremor: electrophysiological and pharmacological evidence for a subdivision, J. Neurol. Neurosurg. Psychiatry **50**, 1435–1441

Eccles, J., Dimitrijevic, M.R. (1985): *Upper motor neuron functions and dysfunctions*, Karger, Basel

Eckhorn, R., Bauer, R., Jordan, W., Brosch, M., Kruse, W., Munk, M., Reitboeck, H.J. (1988): Coherent oscillations: a mechanism of feature linking in the visual cortex?, Biol. Cybern. **60**, 121–130

Elble, R.J, Higgins, C., Moody, C.J. (1987): Stretch reflex oscillations and essential tremor, J. Neurol. Neurosurg. Psychiatry **50**, 691–698

Elble, R.J., Koller, W.C. (1990): *Tremor*, John Hopkins University Press, Baltimore

Elble, R.J., Higgins, C., Hughes, L. (1992): Phase resetting and frequency entrainment of essential tremor, Experimental Neurology **116**, 355–361

Foerster, O. (1936): Motorische Felder und Bahnen, In: *Handbuch der Neurologie*, Bumke, H., Foerster, O. (eds.), Springer, Berlin, vol. 6, 1–357

Freund, H.-J. (1983): Motor unit and muscle activity in voluntary motor control, Physiological Reviews **63**, 387–436; (1987a): Central Rhythmicities in Motor Control and Its Perturbances, In: *Temporal Disorder in Human Oscillatory Systems*, Rensing, L., an der Heiden, U., Mackey, M.C. (eds.), Springer, Berlin, 79–82; (1987b): Abnormalities of motor behavior after cortical lesions in man, In: *The Nervous System: Higher Functions of the Brain*, Mountcastle, V.B. (Section ed.), Plum, F. (Vol. ed.), Sect. 1, vol. 5 of Handbook of Physiology, Williams and Wilkins, Baltimore 763–810

Fritsch, G., Hitzig, E. (1870): The electrical excitability of the cerebrum, Arch. Anat. Physiol. **37**, 300–332

Gildenberg, P.L. (1985): Selective Brain Surgery Procedures for Motor Disorders, In: Eccles and Dimitrijevic (1985)

Gray, C.M., Singer, W. (1987): Stimulus specific neuronal oscillations in the cat visual cortex: a cortical function unit, Soc. Neurosci. **404**, 3; (1989): Stimulus-specific neuronal oscillations in orientation columns of cat visual cortex, Proc. Natl. Acad. Sci. USA **86**, 1698–1702

Gybels, J., Van Roost, D. (1985): Spinal Cord Stimulation for the Modification of Dystonic and Hyperkinetic Conditions: A Critical Review, In: Eccles and Dimitrijevic (1985), 56–70

Haken, H. (1996): *Principles of Brain Functioning, A Synergetic Approach to Brain Activity, Behavior and Cognition*, Springer, Berlin

Hari, R., Salmelin, R. (1997): Human cortical oscillations: a neuromagnetic view through the skull, TINS **20**, 44–49

Hassler, R., Riechert, T., Mundinger, F., Umbach, W., Ganglberger, J.A. (1960): Physiological observations in stereotaxic operations in extrapyramidal motor disturbances, Brain **83**, 337–350

Laitinen, L.V., Bergenheim, T., Hariz, M.I. (1992): Leksell's posteroventral pallidotomy in the treatment of Parkinson's disease, J. Neurosurg. **76**, 53–61

Lamarre, Y., DeMontigny, C., Dumont, M. , Weiss, M. (1971): Harmaline-induced rhythmic activity of cerebellar and lower brain stem neurons, Brain Res. **32**, 246–250

Lamarre, Y., Joffroy, A.J., Dumont, M., DeMontigny, C. , Grou, F., Lind, P. (1975): Central mechanisms of tremor in some feline and primate models, can. J. neurol. Sci. **2**, 227–233

Lamarre, Y., Joffroy, A.J. (1979): Experimental tremor in monkey: activity of thalamic on precentral cortical neurons in the absence of peripheral feedback, Adv. Neurol. **24**, 109–122

Lee, R.G., Stein, R.B. (1981): Resetting of tremor by mechanical perturbations: a comparison of essential tremor and parkinsonian tremor, Ann. Neurol. **10**, 523–531

Lenz, F.A., Kwan, H.C., Martin, R.L., Tasker, R.R., Dostrovsk, J.O., Lenz, Y.E. (1994): Single unit analysis of the human ventral thalamic nuclear group. Tremor-related activity in functionally identfied cells, Brain **117**, 531–543

Limousin, P., Pollak, P., Benazzouz, A., Hoffmann, D., Le Bas, J.F., Broussolle, E., Perret, J.E., Benabid, A.L. (1995): Effect on parkinsonian signs and symptoms of bilateral subthalamic nucleus stimulation, Lancet **345**, 91–95

Llinás, R. , Jahnsen, H. (1982): Electrophysiology of mammalian thalamic neurons in vitro, Nature **297**, 406–408

Meyer, B.-U. (1992): *Magnetstimulation des Nervensystems – Grundlagen und Ergebnisse der klinischen und experimentellen Anwendung*, Springer, Berlin

Mundinger, F. (1977a): Neue stereotaktisch-funktionelle Behandlungsmethode des Torticollis spasmodicus mit Hirnstimulatoren, Med. Klin. **72**, 19820–1986; (1977b): Die Behandlung chronischer Schmerzen mit Hirnstimulatoren, Dtsch. Med. Wochenschrift **102**, 1724–1729

Niedermeyer, E., Lopes da Silva, F. (1987): *Electroencephalography – Basic Principles, Clinical Applications and related Fields*, 2nd ed., Urban & Schwarzenberg, Baltimore

Nieuwenhuys, R., Voogd, J., van Huijzen, Chr. (1991): *Das Zentralnervensystem des Menschen*, 2nd ed., Springer, Berlin

Ohye, C., Narabayashi, H. (1979): Physiological study of presumed ventralis intermedius neurons in the human thalamus, J. Neurosurg. **50**, 290–297

Pare, D., Curro'Dossi, R., Steriade, M. (1990): Neuronal basis of the parkinsonian resting tremor: a hypothesis and its implications for treatment, Neuroscience **35**, 217–226

Pascual-Leone, A., Valls-Sole, J., Toro, C., Wassermann, E.M., Hallett, M. (1994): Resetting of essential tremor and postural tremor in Parkinson's disease with transcranial magnetic stimulation, Muscle Nerve **17**, 800–807

Penfield, W., Evans, J. (1934): Functional deficits produced by cerebral lobectomies, Res. Publ. Assoc. Res. Nerv. Ment. Dis. **13**, 352–377

Penfield, W., Boldrey, E. (1937): Somatic motor and sensory representation in the cerebral cortex of man as studied by electrical stimulation, Brain **60**, 389–443

Poewe, W.H., Lees, A.J., Stern, G.M. (1986): Low-dose L-Dopa therapy in Parkinson's disease: A 6-year follow-up study, Neurology **36**, 1528–1530

Rothwell, J. (1994): *Control of Human Voluntary Movement*, 2nd ed., Chapman & Hall, London

Schnitzler, A., Benecke, R. (1994): The silent period after transcranial magnetic stimulation is of exclusive cortical origin: evidence from isolated cortical ischemic lesions in man, Neurosci. Lett. **180**, 41–45

Siegfried, J., Lippitz, B. (1994): Bilateral chronic electrostimulation of ventroposterolateral pallidum: a new therapeutic approach for alleviating all parkinsonian symptoms, Neurosurgery **35**(6), 1129–1130

Singer, W. (1989): Search for coherence: a basic principle of cortical self-organization, Concepts Neurosci. **1**, 1–26

Singer, W., Gray, C.M. (1995): Visual feature integration and the temporal correlation hypothesis, Annu. Rev. Neurosci. **18**, 555–586

Steriade, H., Jones, E.G., Llinás, R. (1990): *Thalamic Oscillations and Signaling*, John Wiley & Sons, New York

Strafella, A., Ashby, P., Munz, M., Dostrovsky, J.O., Lozano, A.M., Lang, A.E. (1997): Inhibition of Voluntary Activity by Thalamic Stimulation in Humans: Relevance for the Control of Tremor, Movement Disorders **12**, 727–737

Tasker, R.R., Siqueira, J., Hawrylyshyn, P., Organ, L.W. (1983): What happened to VIM thalamotomy for Parkinon's disease?, Applied Neurophysiology **46**, 68–83

Tass, P., Rosenblum, M.G., Weule, J., Kurths, J., Pikovsky, A., Volkmann, J., Schnitzler, A., Freund, H.-J. (1998): Detection of $n : m$ phase locking from noisy data: Application to magnetoencephalography, Phys. Rev. Lett. **81**, 3291–3294

Vodovnik, L., Kralj, A., Bajed, T. (1985): Modification of abnormal motor control with functional electrical stimulation of peripheral nerves, In: Eccles and Dimitrijevic (1985)

Volkmann, J., Joliot, M., Mogilner, A., Ioannides, A.A., Lado, F., Fazzini, E., Ribary, U., Llinás, R. (1996): Central motor loop oscillations in parkinsonian resting tremor revealed by magnetoencephalography, Neurology **46**, 1359–1370

Volkmann, J., Sturm, V. (1998): Indication and results of stereotactic surgery for advanced Parkinson's disease, Crit. Rev. Neurosurg. **8**, 209–216

Volkmann, J., Sturm, V., Freund, H.-J. (1998): Die subkortikale Hochfrequenz-stimulation zur Behandlung von Bewegungsstörungen, Akt. Neurologie **25**, 1–9

Volkmann, J., Sturm, V., Weiss, P., Kappler, J., Voges, J., Koulousakis, A., Lehrke, R., Hefter, H. Freund, H.-J. (1998): Bilateral high-frequency stimulation of the internal globus pallidus in advanced Parkinson's disease, Ann. Neurol. **44**(6), 953–961

Zilles, K., Rehkämper, G. (1988): *Funktionelle Neuroanatomie*, 3rd. rev. ed., Springer, Berlin

11. Epilogue

Abeles, M. (1982): *Local cortical circuits. An elektrophysiological study*, Springer, Berlin

Ahonen, A.I., Hämäläinen, M.S., Kajola, M.J., Knuutila, J.E.T., Lounasmaa, O.V., Simola, J.T., Tesche, C.D., Vilkman, V.A. (1991): Multichannel SQUID systems for brain research, IEEE Trans. Magn. **27**, 2786–2792

Benabid, A.L., Pollak, P., Gervason, C., Hoffmann, D., Gao, D.M., Hommel, M., Perret, J.E., De Rougemont, J. (1991): Long-term suppression of tremor by chronic stimulation of the ventral intermediate thalamic nucleus, The Lancet **337**, 403–406

Benabid, A.L., Pollak, P., Seigneuret, E., Hoffmann, D., Gay, E., Perret, J. (1993): Chronic VIM thalamic stimulation in Parkinson's disease, essential tremor and extra-pyramidal dyskinesia, Acta Neurochir. Suppl. (Wien) **58**, 39–44

Benabid, A.L., Pollak, P., Gao, D., Hoffmann, D., Limousin, P., Gay, E., Payen, I., Benazzouz, A. (1996): Chronic electrical stimulation of the ventralis intermedius nucleus of the thalamus as a treatment of movement disorders, J. Neurosurg. **84**, 203–214

Blond, S., Caparros-Lefebvre, D., Parker, F., Assaker, R., Petit, H., Guieu, J.-D., Christiaens, J.-L. (1992): Control of tremor and involuntary movement disorders by chronic stereotactic stimulation of the ventral intermediate thalamic nucleus, J. Neurosurg. **77**, 62–68

Caparros-Lefebvre, D., Ruchoux, M.M., Blond, S., Petit, H., Percheron, G. (1994): Long term thalamic stimulation in Parkinson's disease, Neurology **44**, 1856–1860

Clarke, C.J.S., Janday, B.S. (1989): The solution of the biomagnetic inverse problem by maximum statistical entropy, Inverse Problems **95**, 483–500

Clarke, C.J.S., Ioannides, A.A., Bolton, J.P.R. (1990): Localised and distributed source solutions for the biomagnetic inverse problem I. In: *Advances in Biomagnetism*, Williamson, S.J., Hoke, M., Stroink, G., Kotani, M. (eds.), Plenum Press, New York, 587–590

Classen, J., Schnitzler, A., Binkofski, F., Werhahn, K.J., Kim, Y.-S., Kessler, K., Benecke, R. (1997): The motor syndrome associated with exaggerated inhibition within the primary motor cortex of patients with hemiparetic stroke, Brain **120**, 605–619

Cohen, D. (1972): Magnetoencephalography: Detection of the brain's electrical activity with a superconducting magnetometer, Science **175**, 664–666

Creutzfeldt, O.D. (1983): *Cortex Cerebri*, Springer, Berlin

Crick, F. (1984): Function of the thalamic reticular complex: the searchlight hypothesis, Proc. Natl. Acad. Sci. USA **81**, 4586–4590

Elble, R.J., Koller, W.C. (1990): *Tremor*, John Hopkins University Press, Baltimore

Engel, J., Pedley, T. A. (eds.) (1997): *Epilepsy : A comprehensive textbook*, Lippincott-Raven, Philadelphia

Freund, H.-J. (1983): Motor unit and muscle activity in voluntary motor control, Physiological Reviews **63**, 387–436; (1987): Abnormalities of motor behavior after cortical lesions in man, In: *The Nervous System: Higher Functions of the Brain*, Mountcastle, V.B. (Section ed.), Plum, F. (Vol. ed.), Sect. 1, vol. 5 of Handbook of Physiology, Williams and Wilkins, Baltimore 763–810

Fritsch, G., Hitzig, E. (1870): The electrical excitability of the cerebrum, Arch. Anat. Physiol. **37**, 300–332

Fuchs, M., Wagner, M., Wischmann H.A., Dössel, O. (1995): Cortical current imaging by morphologically constrained reconstructions. In: *Biomagnetism: Fundamental Research and Clinical Applications*, Baumgartner, C., Deeke, L., Stroink, G., Williamson, S.J. (eds.), Elsevier Science Publishers, Amsterdam, 299–301

Glass, L., Mackey, M.C. (1988): *From Clocks to Chaos. The Rhythms of Life*, Princeton University Press

Greenblatt, R.E. (1993): Probabilistic reconstruction of multiple sources in the bioelectromagnetic inverse problem, Inverse Problems **9**, 271-284

Haken, H. (1977): *Synergetics, An Introduction*, Springer, Berlin; (1983) *Advanced Synergetics*, Springer, Berlin; (1996): *Principles of Brain Functioning, A Synergetic Approach to Brain Activity, Behavior and Cognition*, Springer, Berlin

Hämäläinen, M., Hari, R., Ilmoniemi, R.J.,Knuutila, J., Lounasmaa, O.V. (1993): Magnetoencephalography – theory, instrumentation, and applications to non-invasive studies of the working human brain, Rev. Mod. Phys. **65**, 413–497

Hebb, D.O. (1949): *Organization of Behavior*, Wiley, New York

Hubel, D.H., Wiesel T.N. (1959): Receptive fields of single neurones in the cat's striate cortex, J. Physiol. **148**, 574–591

Ioannides, A.A., Bolton, J.P.R., Clarke, C.J.S. (1990): Continuous probabilistic solutions to the biomagnetic inverse problem, Inverse Problems **6**, 523–542

Julesz, B. (1991): Early vision and focal attention, Rev. Mod. Phys. **63**, 735–772

Llinás, R. , Jahnsen, H. (1982): Electrophysiology of mammalian thalamic neurons in vitro, Nature **297**, 406–408

MacKay, W.A. (1997): Synchronized neuronal oscillations and their role in motor processes, Trends in Cognitive Sciences **1**, 176–183

Marr, D. (1976): Early processing of visual information, Philos. Trans. R. Soc. Lond. [Biol] **275**, 483–524

Malsburg, C. von der, Schneider, W. (1986): A neural cocktail-party processor, Biol. Cybern. **54**, 29–40

Milner, P.M. (1974): A model for visual shape recognition, Psychol. Rev. **81**, 521–535

Pascual-Marqui, R.D., Michel, C.M., Lehmann, D. (1994): Low resolution electromagnetic tomography: a new method for localising electrical activity in the brain, Int. J. Psychophysiol. **18**, 49–65

Roelfsema, P.R., Engel, A.K., König, P., Singer, W. (1997): Visuomotor integration is associated with zero time-lag synchronization among cortical areas, Nature **385**, 157–161

Schnitzler, A., Benecke, R. (1994): The silent period after transcranial magnetic stimulation is of exclusive cortical origin: evidence from isolated cortical ischemic lesions in man, Neurosci. Lett. **180**, 41–45

Shimizu, H., Yamaguchi, Y., Tsuda, I., Yano, M. (1985): Pattern recognition based on holonic information dynamics: towards synergetic computers. In: *Complex systems – operational approaches*, Haken, H. (ed.), Springer, Berlin

Singer, W. (1989): Search for coherence: a basic principle of cortical self-organization, Concepts Neurosci. **1**, 1–26

Steriade, H., Jones, E.G., Llinás, R. (1990): *Thalamic Oscillations and Signaling*, John Wiley & Sons, New York

Strafella, A., Ashby, P., Munz, M., Dostrovsky, J.O., Lozano, A.M., Lang, A.E. (1997): Inhibition of Voluntary Activity by Thalamic Stimulation in Humans: Relevance for the Control of Tremor, Movement Disorders **12**, 727–737

Tass, P., Rosenblum, M.G., Weule, J., Kurths, J., Pikovsky, A., Volkmann, J., Schnitzler, A., Freund, H.-J. (1998): Detection of $n : m$ phase locking from noisy data: Application to magnetoencephalography, Phys. Rev. Lett. **81**, 3291–3294

Treisman, A. (1980): A feature-integration theory of attention, Cogn. Psychol. **12**, 97–136; (1986): Properties, parts and objects. In: *Handbook of perception and human performances*, Boff, K., Kaufman, L., Thomas, I. (eds.), Wiley, New York

Volkmann, J., Joliot, M., Mogilner, A., Ioannides, A.A., Lado, F., Fazzini, E., Ribary, U., Llinás, R. (1996): Central motor loop oscillations in parkinsonian resting tremor revealed by magnetoencephalography, Neurology **46**, 1359–1370

Volkmann, J., Sturm, V. (1998): Indication and results of stereotactic surgery for advanced Parkinson's disease, Crit. Rev. Neurosurg. **8**, 209–216

Wang, J., Williamson, S.J., Kaufman, L. (1995): Spatio-temporal model of neural activity of the human brain based on the MNLS inverse. In: *Biomagnetism: Fundamental Research and Clinical Applications*, Baumgartner, C., Deeke, L., Stroink, G., Williamson, S.J. (eds.), Elsevier Science Publishers, Amsterdam, 299–301

Winfree, A.T. (1980): *The Geometry of Biological Time*, Springer, Berlin

Author Index

Subject Index

Computer to plate: Mercedes Druck, Berlin
Binding: Buchbinderei Lüderitz & Bauer, Berlin